176 Topics in Current Chemistry

Springer-Verlag Berlin Heidelberg GmbH

Technetium and Rhenium

Their Chemistry and Its Applications

Editors: K. Yoshihara and T. Omori

With contributions by
R. Alberto, J. Baldas, K. Hashimoto,
B. Johannsen, S. Jurisson, S. Kryutchkov,
T. Omori, H. Spiess, W. A. Volkert,
K. Yoshihara

With 99 Figures and 28 Tables

 Springer

This series presents critical reviews of the present position and future trends in modern chemical research. It is addressed to all research and industrial chemists who wish to keep abreast of advances in their subject.

As a rule, contributions are specially commissioned. The editors and publishers will, however, always be pleased to receive suggestions and supplementary information. Papers are accepted for "Topics in Current Chemistry" in English.

ISBN 978-3-662-14849-5   ISBN 978-3-540-49273-3 (eBook)
DOI 10.1007/978-3-540-49273-3

Library of Congress Catalog Card Number 74-644622

© Springer-Verlag Berlin Heidelberg 1996
Originally published by Springer-Verlag Berlin Heidelberg New York in 1996.
Softcover reprint of the hardcover 1st edition 1996

The use of general descriptive names, registered names, trademarks, etc. in this publication does not imply, even in the absence of a specific statement, that such names are exempt from the relevant protective laws and regulations and therefore free for general use.

Typesetting: Macmillan India Ltd., Bangalore-25
SPIN: 10495095   51/3020 - 5 4 3 2 1 0 - Printed on acid-free paper

Guest Editors

Prof. Dr. *Kenji Yoshihara*
Tohoku University, Department of Chemistry,
Sendai 980-77, Japan

Prof. Dr. *Takashi Omori*
Shizuoka University, Radiochemistry Research
Laboratory, Shizuoka 422, Japan

Editorial Board

Attention
all "Topics in Current Chemistry" readers:

A file with the complete volume indexes Vols.22 (1972) through 175 (1995) in delimited ASCII format is available for downloading at no charge from the Springer EARN mailbox. Delimited ASCII format can be imported into most databanks.

The file has been compressed using the popular shareware program "PKZIP" (Trademark of PKware Inc., PKZIP is available from most BBS and shareware distributors).

This file is distributed without any expressed or implied warranty.

To receive this file send an e-mail message to:
SVSERV@VAX.NTP. SPRINGER.DE
The message must be:"GET/CHEMISTRY/TCC_CONT.ZIP".

SVSERV is an automatic data distribution system. It responds to your message. The following commands are available:

HELP	returns a detailed instruction set for the use of SVSERV
DIR (name)	returns a list of files available in the directory "name",
INDEX (name)	same as "DIR",
CD <name>	changes to directory "name",
SEND <filename>	invokes a message with the file "filename",
GET <filename>	same as "SEND".

For more information send a message to:
INTERNET:STUMPE@SPINT. COMPUSERVE.COM

Preface

The discovery of the elements 43 and 75 was reported by Noddack et al. in 1925, just seventy years ago. Although the presence of the element 75, rhenium, was confirmed later, the element 43, masurium, as they named it, could not be extracted from naturally occurring minerals. However, in the cyclotron-irradiated molybdenum deflector, Perrier and Segré found radioactivity ascribed to the element 43. This discovery in 1937 was established firmly on the basis of its chemical properties which were expected from the position between manganese and rhenium in the periodic table. However, ten years later in 1937, the new element was named technetium as the first artificially made element.

Technetium then became available in a weighable quantity because of uranium nuclear fission leading to the production of ^{99}Tc in nuclear reactors. The total amount of ^{99}Tc in the world at the end of 1993 is estimated to be 78 tons, more abundant than rhenium on the earth.

Of the Group Seven Elements, technetium and rhenium have closer similarities to each other compared to manganese. Despite a variety of their valence states and chemical bond natures the chemistry of both elements has been developing in recent years. The background of the development is obviously related to the use of technetium radiopharmaceuticals for diagnosis and of rhenium for therapy in nuclear medicine. Impact from other research fields often animates activity in a research field and sometimes creates an interdisciplinary field. We can find a good relationship among chemistry, pharmacy and medicine on technetium and rhenium aimed at the final goal of overcoming one of the great enemies of human kind - cancer.

Bearing this situation in mind we wished to describe the present status of studies on chemistry and its applications of technetium and rhenium. A part of this book was planned before the "Topical Symposium on the Behavior and Utilization of Technetium '93" was held in Sendai, Japan in March 1993, but the planning of the book in this style was accelerated after the symposium by suggestions from our friends. The editors are grateful for the cooperation from the contributers and the publisher.

Japan, March 16, 1995

Kenji Yoshihara
Takashi Omori

Table of Contents

Recent Studies on the Nuclear Chemistry of Technetium

Kenji Yoshihara

Department of Chemistry, Tohoku University Sendai 980–77, Japan

Table of Contents

The present status of the nuclear chemistry of technetium and its applications are reviewed. Recent re-evaluations have given longer half-lives for ^{97}Tc and ^{98}Tc than those previously found. Nuclear reactions useful in the radioisotope production of technetium are described, emphasizing, in particular, positron-emmitting ^{94m}Tc and ^{96}Tc/^{183}Re among the multitracers produced by heavy ion bombardment of gold. Understanding ^{99}Tc isomer excitation and de-excitation processes is essential to deriving a nuclear model which includes the nucleus–outer electron interaction. The nuclear synthesis of technetium in stars and the shortening of the half-life of ^{99}Tc with increasing temperature are presented. The 'Molecular rocket' and shock wave phenomena induced by nuclear reactions are further up-to-date subjects to which attention is directed. The results concerned with technetium are summarized and discussed.

Topics in Current Chemistry, Vol. 176
© Springer-Verlag Berlin Heidelberg 1996

1 Introduction

Nuclear reactions provide energy for heavenly bodies burning in the universe. Since Einstein introduced his theory of relativity at the beginning of this century, it has been accepted that energy and matter are not independent of each other, but rather interchangeable; the mass lost in nuclear reactions is converted into energy.

Chemical elements including technetium are being produced in nuclear reactions occurring in the stars today. This has been proved by observing of the presence of technetium in some stars [1]. Technetium has no stable isotopes and none of the technetium isotopes has a half-life long enough to survive the age of the universe. So the technetium observed must have been synthesized by nuclear processes in the stars.

Nuclear reactions involving technetium have been actively studied until today. Our interest in the nuclear chemistry of technetium is based on various reasons. Technetium was the first artificially produced element in the periodic table, a weighable amount of technetium (^{99}Tc) is now available, and ^{99m}Tc is one of the most important radionuclides in nuclear medicine. In addition, technetium is an element of importance from a nuclear safety point of view.

Moreover, novel applications of nuclear reactions involving technetium are being developed. Among them, multitracer utilization, molecular rocket reactions and shock wave enhancement phenomena are up-to-date examples. All these aspects are presented in this review paper.

2 Long-Lived Technetium Nuclides: Redetermination of Their Half-Lives

There had been some confusion about the discovery of element number 43 until in 1937 Perrier and Segré succeeded in producing it by deuteron irradiation of molybdenum placed in a cyclotron. A Japanese chemist by the name of Ogawa believed that he had succeeded in discovering this element in 1908, but in vain. Afterwards, in 1925 the Noddack group claimed to have discovered this element, but their claim turned out to be false.

The suggestion made by Masuda and Qi [2] that natural technetium was present in tungsten and rhenium sources caused a stir in nuclear chemistry circles. According to data given in the "Table of Radioactive Isotopes" [3] no technetium isotopes could have survived the period of 4.5 billion years which is commonly believed to be the age of the earth. Even if the (n, γ) reaction of natural ^{96}Ru produces ^{97}Ru, a precursor of ^{97}Tc, accumulation of a weighable amount of the latter nuclide would require a fairly high flux of neutrons. If the previously determined half-lives of ^{97}Tc [4] and ^{98}Tc [5, 6] were not correct, there might be a small possibility for existence of 'natural technetium'. In order to examine this possibility, Kobayashi et al. [7] redetermined the half-lives of ^{97}Tc and ^{98}Tc, two nuclides of geochemical and astrophysical significance.

The half-life of ^{97}Tc was first reported by Katcoff [4] $(2.6 \times 10^6$ yr). He measured the characteristic X-rays of its EC decay daughter ^{97}Mo using a proportional counter with critical X-ray absorbers to discriminate against X-rays of ^{97m}Tc (half-life: 90d). The number of ^{97}Tc atoms was estimated from the parent ^{97}Ru activity and the branching ratio of the ^{97}Ru EC decay (0.041%). This determination involved an uncertainty in proportional counting of the sample and it depended on the accuracy of the branching ratio. Moreover, the sample was contaminated with ^{98}Tc, whose half-life was described in a private communication from O'Kelly [5] to Nuclear Data Tables. The value was based on the Master's Thesis work of Goeking [6] in which deuteron irradiation of natural Mo was used. The activity of ^{98}Tc was determined from the photopeak areas of two γ-rays (652 keV and 745 keV) in a γ-ray spectrum obtained using an NaI(Tl) detector.

In their new report, Kobayashi et al. [7] produced ^{97}Tc by ^{95}Mo $(\alpha, 2n)$ ^{97}Ru(EC) ^{97}Tc processes by irradiating an enriched ^{95}Mo (94.9%) target with 22 MeV α particles in the SF-cyclotron at the University of Tokyo. After cooling, the irradiated target was dissolved in warm hydrogen peroxide and nitric acid was added to the solution. This was first passed through an SnO$_2$ column and then an anion exchange column. Metallic magnesium was added to the final eluate to reduce the molybdate and the solution was filtered. The filter paper was dried and counted with an HP-Ge detector. Fig. 1 shows the half-life determined to be $(4.0 \pm 0.3) \times 10^6$ yr. In this determination the number of ^{97}Tc atoms in the source was estimated to be $(1.65 \pm 0.08) \times 10^{14}$, based on the ^{97}Ru 216 keV γ-ray intensity and the ^{97}Ru–^{97}Tc parent–daughter relationship. The observed photopeak counting ratio A(MoK$_\alpha$)/A(TcK$_\alpha$) was corrected by the photopeak counting efficiency, the fluorescence yield and the internal conversion coefficient, thus leading to the absolute disintegration rate ratio

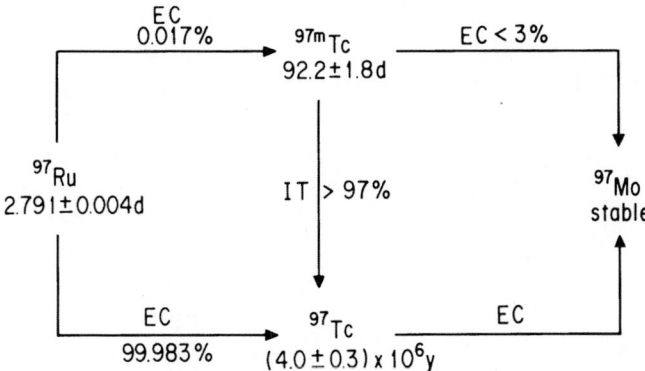

Fig. 1. Decay of ^{97}Ru leading to ^{97m,g}Tc. The data obtained in Ref. 7 are shown

A^0 (^{97}Tc)/A^0 (^{97m}Tc). This new half-life is considerably longer than the previous one determined by Katcoff, but it is still not long enough to survive, considering the earth's age.

Kobayashi et al. [7] also measured the half-life of ^{98}Tc produced by irradiation of enriched ^{97}Mo (92.9%) with 13 MeV deutrons. The ^{98}Tc thus obtained by the (d, n) reaction was purified by anion exchange and measured with an HP-Ge detector. The estimated value of the half-life was 9×10^6 yr, longer than the previous value, but also still not long enough to survive to the present.

3 Nuclear Reactions of Technetium

^{99m}Tc is a very important radionuclide in nuclear medicine. Production of ^{99}Mo–^{99m}Tc generators is carried out commercially (1) by nuclear fission of ^{235}U or (2) by neutron irradiation of ^{98}Mo. The disadvantage of the first method, however, is that a large quantity of fission-product waste has to be separated from the ^{99}Mo–^{99m}Tc and discarded; the disadvantage of the second method is that the ^{99m}Tc produced has a low specific activity. Both of these weaknesses have to be overcome. The (n, γ) production method, coupled with a specifically designed concentration and purification system, has been proposed recently [8].

^{99m}Tc can be produced by other nuclear reactions of ^{99}Tc, which is accumulated by reactor operation in kilogram quantities. Cyclic milking using a ^{99}Tc(n, p) ^{99}Mo(β^-) ^{99m}Tc(IT) ^{99}Tc cycle has been proposed [9]. This approach utilizes fast neutrons in the Omega Project at the Japan Atomic Energy Research Institute. Another proposal for producing ^{99m}Tc was based on the ^{99}Tc (n, n') ^{99m}Tc reaction coupled with the Szilard–Chalmers (SzC) method. The (n, n') reaction, with an activation cross section of 2.42×10^{-29} m^2 (Yamagishi et al. [10]) for reactor neutrons, can produce 3×10^{12} Bq/gTc (or 80 Ci/gTc) in a 6-hour irradiation with a neutron flux of 10^{17} m^2 s^{-1} and SzC enrichment factor of 1000.

Recently Roesch et al. [11] at Juelich studied the high purity production of positron-emitting technetium isotope ^{94m}Tc (half-life: 52 m, β^+ 70%, E_{max} 2.47 MeV). This is a promising radionuclide for PET, since it could substitute the SPECT nuclide ^{99m}Tc. A highly enriched ^{94}Mo isotope (93.9%) was irradiated with protons of energies between 5 and 20 MeV. The excitation functions for ^{94}Mo(p, n) $^{94m, g}$Tc are shown in Fig. 2. This occurs simultaneously with the ^{94}Mo(p, 2n) $^{93m, g}$Tc reaction in the energy range 14–19 MeV. The maximum cross section for the ^{94}Mo(p, n) ^{94m}Tc reaction is 4.8×10^{-29} m^2 at 12 MeV. The expected thick-target yield from the cross-section data is 2 GBq(54 mCi)/µAh which is much better than the 7.2 mCi/µ Ah value obtained for the natural Mo(p, n) reaction [12]. The m/g isomer ratio is 9 at low energy, but it decreases with increasing energy and is about 1 at 19 MeV.

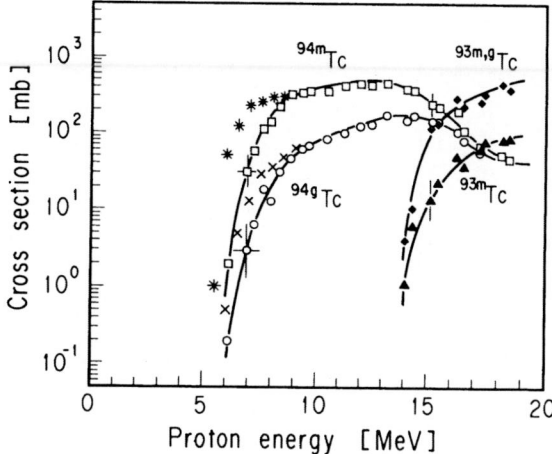

Fig. 2. Excitation functions of ^{95}Mo(p, xn)-processes leading to the formation of ^{94m}Tc, ^{94g}Tc, ^{93m}Tc and ^{93m,g}Tc [11]. Some data of other authors are included (✳, ×). The unit in the ordinate mb = 10^{-31} m^{-2}

Fassbender et al. [12a] studied the excitation functions of ^{93}Nb(^{3}He, 2n) ^{94m,g}Tc processes up to 35 MeV using the Juelich compact cyclotron, together with those of $^{93m,g/95m,g}$Tc. The maximum cross section (90 mb) for ^{94m}Tc production was obtained at about 16 MeV. The thick-target yield of ^{94m}Tc was 33 MBq/μAh. ^{94m}Tc was recovered from the target niobium by a thermochromatographic procedure at 1200 °C in the separation yield greater than 80%.

Stimulated by the increasing demand for ^{95m}Tc (half-life: 61 d), redetermination of an excitation function of the ^{95m}Tc production reaction was carried out [13]. A thick plate of enriched ^{95}Mo (96.5%), electrodeposited on an aluminum foil, was used as the target, and this was bombarded with protons with energies up to 28 MeV using a tandem accelerator at the Japan Atomic Energy Research Institute. The cross-section curve of the ^{95}Mo(p, n) ^{95m}Tc reaction showed a maximum of 2×10^{-29} m^2 at 11 MeV (Fig. 3). This (p, n) production method showed less contamination from the ground component ^{95g}Tc than the ^{96}Mo$(p, 2n)$ ^{95m}Tc method used by Hogan [14, 14a], but its production yield was smaller. Nagame et al. [15] analyzed the isomeric ratio in ^{95}Mo(p, n) ^{95m,g}Tc using a statistical model. They found a discrepancy between the experimental and calculated values due to the contribution of a pre-equilibrium process in the high energy region.

An excitation function of ^{99}Tc$(p, 3n)$ ^{97}Ru up to 100 MeV was determined by a Russian group [16]. The product nuclide ^{97}Ru can be used in nuclear medicine.

Novel applications of radiotracers have been developed by the RIKEN group [17, 18] using heavy-ion irradiation in a ring cyclotron. High energy heavy-ion reactions of Au, Ag and Cu lead to the production of many radionuclides with widely distributed mass numbers (A). The reaction cross section is plotted against A in Fig. 4 in which a gold target was bombarded by ^{14}N with an

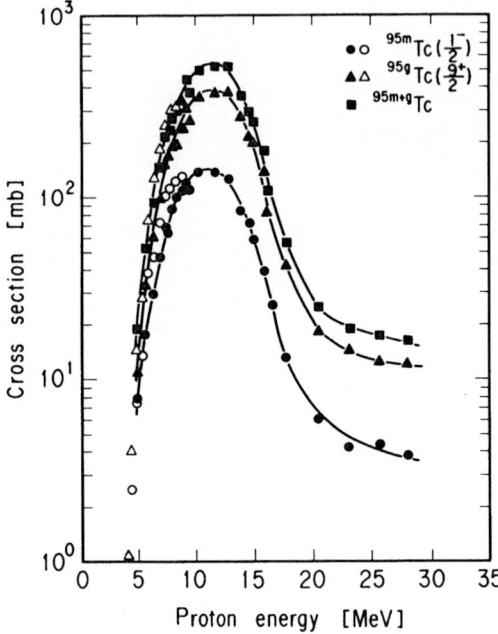

Fig. 3. Excitation functions of ^{95}Mo(p, n) ^{95m,g}Tc reactions. Open circles and triangles are taken from the data of Skakun et al. The unit in the ordinate mb = 10^{-31} m^{-2}. (Reprinted with permission from Ref. 13. Copyright (1991) Elsevier Science Ltd)

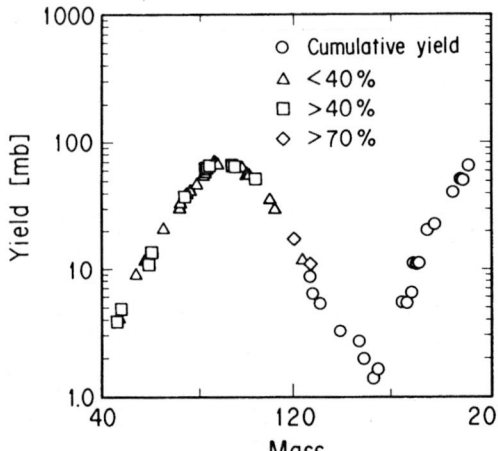

Fig. 4. Mass distribution of the product of the ^{14}N(35 MeV/u) + ^{197}Au reaction [19]. Open circles indicate cumulative yields. The percentage is for the measured nuclides in the reaction products. The unit in the ordinate mb = 10^{-31} m^{-2}

energy of 35 MeV/necleon [19]. The cross section curve reveals two components: fission and fragmentation in the lower and higher energy region, respectively. In Fig. 4, Tc is located in the vicinity of a maximum of the former component, and Re is found in the latter. Both Tc and Re have cross sections of the same order of magnitude.

These radionuclides can be used for investigating the behavior of various radionuclides simultaneously in material and life sciences (the so-called 'multitracer' method). Among the applicable radionuclides, [96]Tc(half-life: 4.35 d) and [183]Re(half-life : 70 d) are particularly interesting because they are easily separated by distillation of a solution of target gold in aqua regia and can be used for chemical, biological and medical experiments. These radionuclides were used for studying transport behavior through a supported liquid membrane [20]. A rapid increase in the biological and radiopharmaceutical applications of these radionuclides may be expected in the future.

4 [99]Tc Isomer Excitation and De-excitation

Photo-excitation and de-excitation are basic processes in nuclear reactions. A Japanese–Hungarian cooperation investigating these processes has yielded good results during the past few years [21–26]. These studies used weighable amounts of ^{99}Tc to look at the (γ, γ') reaction that leads to the production of the nuclear isomer ^{99m}Tc by electron linear accelerator irradiation.

Sekine et al. [22] determined integrated cross sections of the nuclear isomer excitation [99]Tc(γ, γ') ^{99m}Tc at various energies. The integrated cross sections were saturated above 15 MeV, as shown in Fig. 5 [25]. This observed tendency is not compatible with the earlier results obtained by Russian researchers [27,

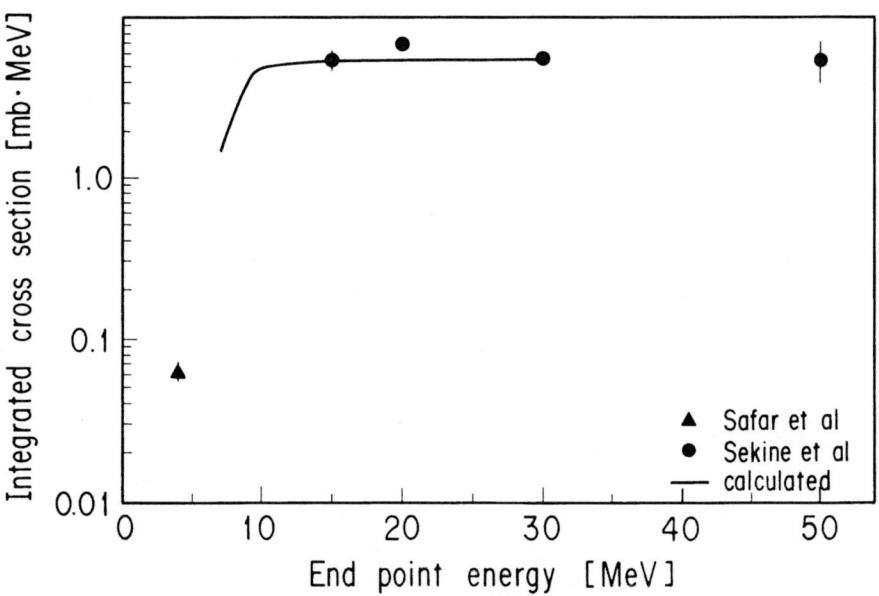

Fig. 5. Integrated cross sections for the [99]Tc(γ, γ') ^{99m}Tc reaction [25]. The calculated one is shown by a solid line. The unit in the ordinate mb $= 10^{-31}$ m^{-2}

28] who depicted the well-known double-humped excitation function for ^{115}In and ^{103}Rh. The experiments performed by Sekine et al. [22] on the ^{115}In(γ, γ') ^{115m}In reaction could not reproduce the Russian results, and both experimental and theoretical studies carried out be Sáfár et al. [23] also confirmed that the double-humped excitation function was not correct for the ^{103}Rh(γ, γ') ^{103m}Rh reaction. Thus, the double-humped excitation function, a belief held for more than 35 years, has been disproved. The second peak reported previously for ^{115}In, ^{103}Rh, ^{197}Au, ^{89}Y, or ^{107}Ag [29] was most probably the result of an uncontrollable (n, n') contribution which often accompanies in the (γ, xn) reaction. From these studies, a new picture has emerged about the general behavior of the (γ, γ') reactions in the region of giant resonance.

De-excitation of ^{99m}Tc has specific features. This nuclide decays with a half-life of 6 hours, but its half-life varies slightly according to environmental conditions [30] or chemical states [31, 32]. Moreover, the emission probabilities of characteristic X-rays just after the isomeric transition ^{99m}Tc → ^{99}Tc are influenced by environmental factors [33] which result in a change of the $K\beta_1/K\alpha$ X-ray intensity ratio [34].

One aspect of the decay rate of ^{99m}Tc has been suggested by Hinneburg et al. [35]. Its low-lying level structure is shown in Fig. 6. The energy difference of 2.17 keV between the 142.7 and 140.5 keV nuclear levels is similar to that between the M5 and L3 atomic levels (2.42 keV), but this degree of similarity is not enough to cause an inverse NEET (nuclear transition by electronic transition) [36]. However, de-excitation is possible by an electron-bridge (EB) mechanism [35] which involves a third order correction to the internal conversion (IC). Emission of a photon and the simultaneous ejection of an orbital

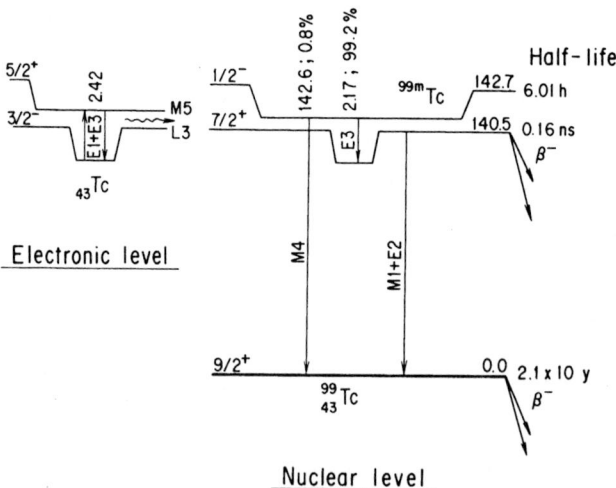

Fig. 6. The low-lying level structure of ^{99}Tc and the relevant atomic transition close to the isomeric transition in energy (keV) [25]

electron, as a result of the exchange of a virtual photon between the electron and the nucleus, are considered. The external radiation field stimulates an electronic rather than a nuclear transition, because the width of the electronic transition is much larger than that of the nuclear transition. A shortening of the half-life of ^{99m}Tc in a high radiation field has been studied experimentally and a 2% decrease in the half-life was detected [37].

5 Shortening of the Half-Life of Technetium in Stars

The origin of chemical elements has been explained by various nuclear synthesis routes, such as hydrogen or helium burning, and α-, e-, s-, r-, p- and x-processes. ^{99}Tc is believed to be synthesized by the s (slow)-process in stars. This process involves successive neutron capture and β^- decay at relatively low neutron densities; neutron capture rates in this process are slow as compared to β-decay rates. The nuclides near the β-stability line are formed from the iron group to bismuth.

It has been calculated that the decay rate of ^{99}Tc inside a star is dramatically enhanced because of β-decay channels (Fig. 7) from thermally populated photo-excited states at high temperature [38]. The calculation was based on a simple shell model, taking into account both continum and bound state β-decays. The

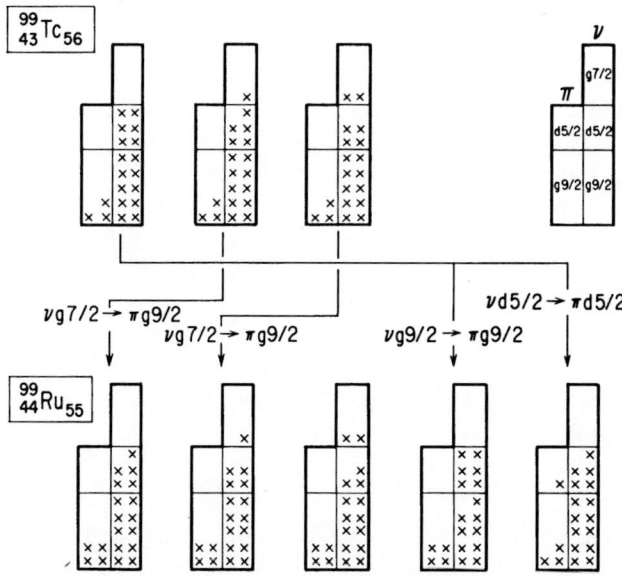

Fig. 7. Thermally populated β-decay channels from ^{99}Tc to ^{99}Ru [38]. Relevant proton(π) and neutron(ν) shells are shown as compartments filled with nucleons ($\times$)

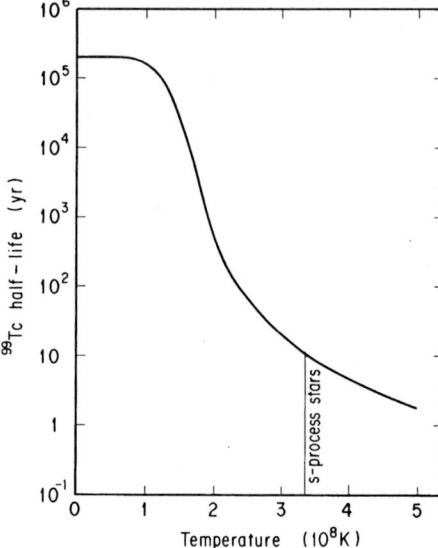

Fig. 8. Predicted β-decay half-lives of ^{99}Tc in stellar interiors as a function of temperature [38]

dependence of the predicted half-life of ^{99}Tc on temperature is shown in Fig. 8. Its decrease begins at about 1×10^8 K, and falls to only about 10 years at the average s-process temperature of $(3.3 \pm 0.5) \times 10^8$ K [39]. The relevance of bound-state β-decay in nucleosynthesis has been mentioned by several researchers, and it was observed recently [40].

Although the half-life of ^{99}Tc in steller interiors is remarkably decreased, a substantial amount of the isotope can survive the s-process. Observations have revealed that more than 50 stars contain technetium in their outer envelope. According to other calculations, the production of neutrons in the competitive processes of neutron capture and β-decay is even more enhanced at such high temperatures, and this fact almost compensates for the depletion of ^{99}Tc [41].

A comparison between the observed ^{99}Tc abundance and that of the analogous isotonic nuclei ^{97}Nb–^{97}Mo could be used as an indicator of the stellar mixing lifetime [41].

6 Molecular Rockets and Shock Waves Induced by Nuclear Reactions

In the past few years, the new concept of a molecular rocket reaction has been developed [42–47]. The system designed for this purpose consists of a metallocene rocket and a cyclodextrin cannon which includes the metallocene molecule. The central metal atom of the enclosed metallocene is irradiated with neutrons or photons so that the nuclear recoil after the nuclear reaction pushes the metal atom which, in turn, pushes the molecule as a whole. The energetic molecule is ejected from the cyclodextrin cavity just like a rocket. This energetic rocket molecule undergoes unique reactions with surrounding molecules or inclusion compound systems.

Matsue et al. [43] attempted to study the molecular rocket reaction in a ruthenocene-β-cyclodextrin inclusion compound using the $^{100}Ru(\gamma, p)$ ^{99m}Tc reaction. They noticed a parallel relationship between chemical processes and nuclear-recoil-induced processes in the non-included ruthenocene compound, as shown in Fig. 9. In the nuclear-recoil-induced processes no dimerization can be observed because of the extremely low concentration of the product, whereas in the chemical processes dimerization is possible, as demonstrated by Apostolidis et al. [48]. When ruthenocene included in β-cyclodextrin is irradiated with γ-rays, a part of the ruthenocene molecule is converted to $[TcCp_2 \cdot]$ which escapes from the β-cyclodextrin cavity. The $[TcCp_2 \cdot]$ rocket thus produced can attack neighboring inclusion compounds so as to extract the enclosed ruthenocene molecules and abstract H or Cp (Cp: cyclopentadienyl radical). This process is shown schematically in Fig. 10.

Very recently Matsue and his coworkers [49] studied molecular rocket phenomena in modified cyclodextrin inclusion compounds. In this case, a ruthenocene molecule was connected with the β-cyclodextrin by an ester bridge. When Ru is irradiated with γ-rays, the $^{104}Ru(\gamma, n)$ ^{103}Ru reaction provides recoil energy to the ^{103}Ru. As the weak bonding of the ester bridge is easily broken by the motion of the Ru atom, the ruthenocene molecule labeled with ^{103}Ru is found outside of the cyclodextrin cavity after the recoil event. The behavior of the system of a molecular rocket with a bridge connection can be compared with

(a) Chemical Processes

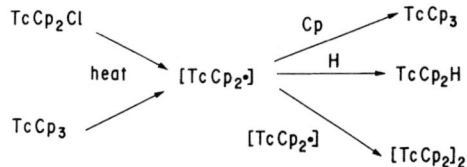

(b) Nuclear Processes

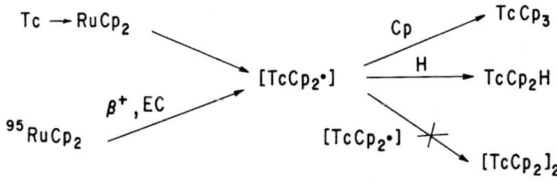

Fig. 9. Formation of the $TcCp_2 \cdot$ radical from **a** chemical and **b** nuclear processes, and abstraction of Cp or H [43]

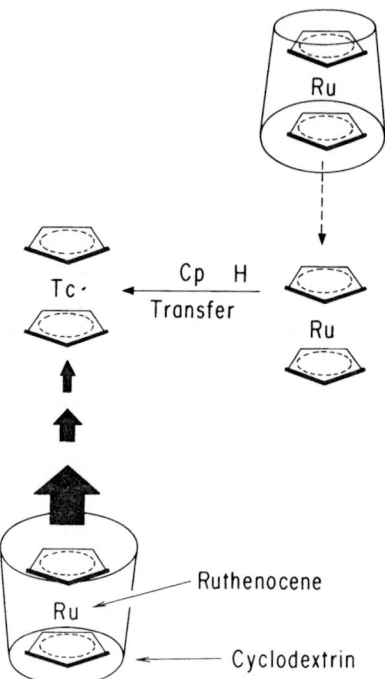

Ruthenocene

Cyclodextrin

Fig. 10. A metallocene molecular rocket extracting other metallocenes in neighboring inclusion complexes. As an example ^{100}Ru(γ, p) ^{99m}Tc reaction, which produces the ^{99m}TcCp$\cdot$ radical, is shown [43]

that of a system without such a bridge connection. They examined the energy dependence of the yield of ruthenocene labeled with ^{103}Ru and found that there were two components: direct and indirect molecular rockets. The latter involves reflection of the recoil atom by the cyclodextrin wall and recombination with its original partners.

In recoil implantation induced by a nuclear reaction, a specific and interesting phenomenon – ascribable to a shock wave effect due to high pressure and high temperature occurring simultaneously – has been reported. The high energy chemical reactions accompanied by implantation, i.e. either ion implantation or recoil implantation, were thought to have the same aspects as those associated with hot atom chemical reactions [50]. However, this traditional view has been completely disproved in the case of the high energy recoil implantation reactions carried out at Tohoku University [51–56].

A recoil-implanted metal atom (radioactive *M) undergoes the following chemical reaction with a metal coordination compound in a film of catcher material:

$$*M + ML_n \rightarrow *ML_n + M$$

where L is a ligand and n is the number of ligands coordinated to the metal. In a series of experiments using M(acac)$_3$ (M = Al, Cr, Fe, Co, Tc and Ru; acac = acetylacetone), the recoil-implanted atom *M follows adaptation and

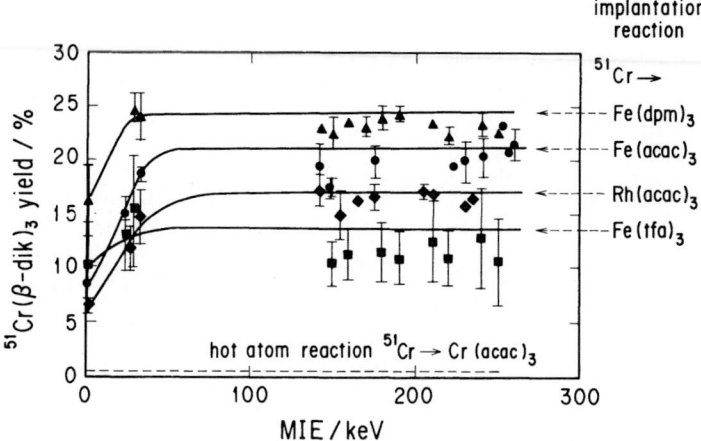

Fig. 11. Displacement yield of ^{51}Cr-β-diketonate by recoil implantation as a function of mean implantation energy (MIE) [54]

competition rules [57, 58]. The adaptation rule deals with the metal (*M) – ligand correlation, whereas the competition rule describes the metal (*M)–metal(M) competition in a displacement reaction induced by the recoil-implanted atom. Implantation work using recoil Tc and other metals has revealed these characteristic features. The competition rule could be clearly visualized by the linear relation between the yield and the force constant ratio $K(\text{*M–O})/K(\text{M–O})$, because the metal–oxygen bonding force constant is a critical factor in the formation of the complex.

In studying the energy dependence of the yield of *ML_n, application of a thin film technique to the target system of implantation led to an exciting finding. Fig. 11 shows the results for ^{51}Cr implantation in tris(β-diketonato) iron(III) or Rh(III) [54]. The yield steeply increases with increasing energy (1–20 keV) and is saturated at 100 keV. The saturation value is much higher than the value obtained in the hot atom reaction inside the target at the same energy. The enhancement factor, that is the ratio of the implantation reaction yield to the hot atom reaction yield at the same energy, is 10–20 in the plateau region. This enhancement was further confirmed in the case of the ^{105}Rh(acac)$_3$ formation yield (7%) which was obtained from ^{106}Pd(γ, p) ^{105}Rh implantation in Rh(acac)$_3$. The yield of ^{102}Rh(acac)$_3$ obtained from the ^{103}Rh(γ, n) ^{102}Rh hot atm reaction in the same target system was only 0.7%, one order of magnitude smaller than the implantation yield under identical irradiation conditions [52]. This apparently strange phenomenon was explained by a collision cascade enhancement model. As shown in Fig. 12, implantation of a high energy recoil atom produces a collision cascade which is followed by the generation of high pressure and high temperature pulses at the position of the implanted atom [53, 54]. Chemical reactions leading to formation of the coordination compound should be affected drastically by an attack of the pulses. A mathematical

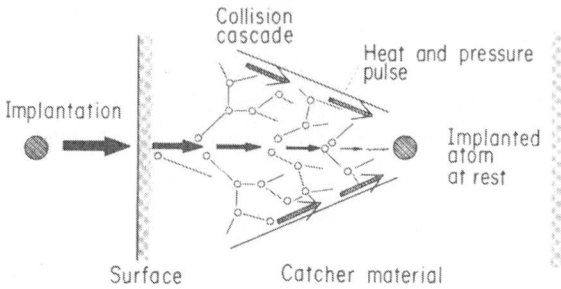

Fig. 12. Collision cascades induced by a recoil-implanted atom with the generation of heat and pressure pulses [56]

formula based on the model, involving an attenuation term and a damage concentration factor, can reproduce the shape of the energy dependence curve of the implantation reaction yield [54]. This enhancement phenomenon must be closely correlated to a shock wave-induced event because the enhancement of the $*ML_n$ yield is caused by high pressure and high temperature pulses at the top end of the conical disturbed zone produced by the collision cascade (Fig. 12). Thus the effect indicates the presence of a shock wave (in the forward direction) which is generated secondarily by the action of the high energy recoil-implanted atom. This is an interesting discovery, i.e. that the nuclear reaction can participate in the generation of a shock wave and affect the yield of the chemical reactions. It means that nuclear reactions are related to solid-state chemical reactions in a unique manner through shock waves at the atomic level. New techniques and new applications must now be developed in this field of research.

7 Conclusions

The present status of research into the nuclear chemistry of technetium has been described. Because of the importance of technetium in nuclear medicine, there are many groups whose interests are concentrated on clarifying the conditions for the production of technetium isotopes. However, attention should also be given to the significance of technetium in geochemistry and astrophysics. More data dealing with these aspects need to be collected. New applications of multitracers in material and biological sciences have emerged recently. In these, the role played by technetium and rhenium radionuclides is particularly interesting because of their unique chemical behavior in the field of nuclear medicine. In addition, rocket and shock wave phenomena need to be studied more actively.

Acknowledgements. The author wishes to express his heartfelt thanks to the many coworkers who have helped him in his research on technetium. In cooperative work between Japan and Hungary, he was supported by the

Japan Society for the Promotion of Science and the Hungarian Academy of Sciences. Prof. Á. Veres, Dr. L. Lakosi and Dr. J. Sáfár of the Institute of Isotopes kindly gave their time to discuss the physical significance of the processes involved in the (γ, γ') reaction.

8 References

1. Merrill PW (1952) Science 115: 484
2. Masuda A, Qi Lu (1989) Proc Japan Acad 65B: 87
3. Browne E, Firestone RB (1986) Table of Radioactive Isotopes. Shirley VS (ed) Wiley, New York
4. Katcoff S (1958) Phys Rev 111: 575
5. O'Kelly GD, Goeking CF Jr., Collins LL Jr (1973) private communication to Nuclear Data Group
6. Goeking CF Jr (1966) MSc Thesis, University of Tennessee
7. Kobayashi T, Sueki K, Ebihara M, Nakahara H, Imamura M, Masuda A (1993) Radiochim Acta 63: 29
8. Seifert S, Wagner G, Eckardt A (1994) Appl Radiat Isot 45: 577
9. Yoshihara K (1991) Workshop on Current Topics in the Behavior of Technetium (KURRI-TR-362) Research Reactor Institute Kyoto University, Kumatori, Osaka, Japan, p 42
10. Yamagishi I, Kubota M, Sekine T, Yoshihara K (1993) Radiochim Acta 63: 33
11. Roesch F, Nevgorodov AF, Qaim SM, Stoecklin G (1994) J Labelled Compd Radiopharmaceuticals 35: 267
12. Nickles RJ, Christian BT, Nunn AD, Stone CK (1993) J Labelled Compd Radiopharmaceuticals 32: 447
12a. Fassbender M, Novgorodov AF, Roesch F, Qaim SM (1994) Radiochim Acta 65: 215
13. Izumo M, Matsuoka H, Soria T, Nagame Y, Sekine T, Hata K, Baba S (1991) Appl Radiat Isot 42: 297
14. Hogan JJ (1973) J Inorg Nucl Chem 35: 705
14a. Hogan JJ (1973) J Inorg Nucl Chem 35: 1429
15. Nagame Y, Baba S, Saito T (1994) Appl Radiat Isot 45: 281
16. Zaitseva NG, Rurarz E, Bobecky M, Kim HH, Nowak K, Tethal T, Khalkin VA, Popinenkova LM (1992) Radiochim Acta 56: 59
17. Ambe F (1994) RIKEN Review 4: 31
18. Ambe S, Chen SY, Okubo Y, Kobayashi Y, Iwamoto M, Yanokura M, Ambe F (1991) Chem Lett p 149
19. Kuratomo A (1994) MSc Thesis, Nagoya University
20. Ambe S, Okubo Y, Kobayashi Y, Iwamoto M, Yanokura M, Maeda H, Ambe F (1993) Radiochim Acta 63: 49
21. Sekine T, Yoshihara K, Németh Zs, Lakosi L, Veres Á (1989) J Radioanal Nucl Chem 130: 269
22. Sekine T, Yoshihara K, Sáfár J, Lakosi L, Veres Á (1991) Appl Radiat Isot 42: 149
23. Sáfár J, Kaji H, Yoshihara K, Lakosi L, Veres Á (1991) Phys Rev C44: 1086
24. Lakosi L, Sáfár J, Veres Á, Sekine T, Kaji H, Yoshihara K (1993) J Phys G19: 1037
25. Lakosi L, Sáfár J, Veres Á, Sekine T, Yoshihara K (1993) Radiochim Acta 63: 23
26. Sáfár J, Lakosi L, Pavlicsek I, Veres Á, Sekine T, Yoshihara K (1993) J Radioanal Nucl Chem Lett 176: 285
27. Bogdankevich OV, Lazareva LE, Nikolaev FA (1956) J Exp Theor Phys (USSR) 31: 405
28. Bogdankevich OV, Lazareva LE, Moiseev AM (1960) J Exp Theor Phys (USSR) 39: 1224
29. Kruger P, Crawford TM, Goldemberg J, Barber WC (1965) Nucl Phys 62: 584
30. Mazaki H (1978) J Phys E11: 739
31. Johanssen B, Muenze R, Dostal KP, Nagel M (1981) Radiochem Radioanal Lett 47: 57
32. Bainbridge KT, Goldhaber M, Wilson E (1951) Phys Rev 84: 1260
33. Yoshihara K (1990) in: Yoshihara K (ed) Chemical Applications of Nuclear Probes. Springer Verlag, Heidelberg, p 1

34. Yamoto I, Kaji H, Yoshihara K (1986) J Chem Phys 84: 522
35. Hinneburg D, Nagel M, Brunner G (1979) Z Physik A291: 113
36. Morita M (1973) Progr Theor Phys 49: 1574
37. Bikit I, Aničin IV, Krmar M, Slivja J, Vesković M, Čonkić LJ (1993) J Phys G: Nucl Part Phys 19: 1359
38. Takahashi K, Mathews GJ, Bloom SD (1986) Phys Rev C33: 296
39. Kaeppeler F, Gallino R, Busso M, Picchio G, Raiteri CM (1990) Astrophys J 354: 630
40. Jung M, Bosch F, Beckert K, Eickhoff H, Folger H, Franzke B, Gruber A, Kienle P, Klepper O, Koenig W, Kozhuharov C, Mann R, Moshammer R, Nolden F, Schaaf U, Soff G, Spaedtke P, Steck M, Stoehlker Th, Suemmerer K (1993) Nucl Phys A553: 309
41. Mathews GJ, Takahashi K, Ward RA, Howard WM (1986) Astrophys J 302: 410
42. Matsue H, Sekine T, Yoshihara K (1990) J Radioanal Nucl Chem Lett 145: 271
43. Matsue H, Sekine T, Yoshihara K (1993) Radiochim Acta 63: 179
44. Yoshihara K, Matsue H, Sekine T (1994) Nucl Instr Meth Phys Res B91: 103
45. Matsue H, Sekine T, Yoshihara K (1994) Nucl Instr Meth Phys Res B91: 97
46. Matsue H, Yamaguchi I, Sekine T, Yoshihara K (1995) Shock Waves 4: 267
47. Yoshihara K (1995) Isotope News 488: 2
48. Apostolidis C, Kanellakopulos B, Maier R, Rebizant J, Ziegler ML (1990) J Organomet Chem 396: 315.
49. Matsue H, Yamaguchi I, Sekine T, Yoshihara K (1995) Radiochim Acta (in press)
50. Andersen T (1979) in: Harbottle G, Maddock AG (eds) Chemical Effects of Nuclear Transformations in Inorganic Systems, North-Holland, Amsterdam. p 403
51. Miyakawa A, Sekine T, Yoshihara K (1989) Radiochim Acta 48: 11
52. Miyakawa A, Homma K, Sekine T, Yoshihara K (1992) Nucl Instr Meth Phys Res B65: 452
53. Yoshihara K, Miyakawa A, Homma K, Sekine T (1992) in: Takayama K (ed) Shock Waves. Springer Verlag, Heidelberg, p 753
54. Miyakawa A, Sekine T, Yoshihara K (1993) Radiochim Acta 60: 87
55. Yoshihara K, Sekine T (1992) in: Adloff JP, Gasper PP, Imamura M, Maddock AG, Matsuura T, Sano H, Yoshihara K (1992) Kodansha-Verlag Chemie, Tokyo-Weinheim, p 344
56. Miyakawa A, Sekine T, Yoshihara K (1995) J Radioanal Nucl Chem (in press)
57. Sekine T, Sano M, Yoshihara K (1986) J Radioanal Nucl Chem Lett 107: 207
58. Yoshihara K (1989) Sci Repts Tohoku Univ First Series, LXXII: 81

Technetium in the Environment

Kenji Yoshihara

Department of Chemistry, Tohoku University, Sendai 980-77, Japan

Table of Contents

Environmental technetium is an important radioelement to which consideration must be given in order to guarantee safety in nuclear waste disposal. Various analysis methods for micro and ultramicro quantities of technetium are being developed and their merits and demerits are being evaluated. The behavior of technetium in the atmosphere, sea water, soils, plants and animals, as reported by many researchers, are summarized. In particular the interaction between technetium and organic matter (such as humic acid) is noted. Finally, the geochemical migration behavior of ruthenium and technetium in the Oklo natural reactors is described as a natural analogue to radioactive waste disposal.

Topics in Current Chemistry, Vol. 176
© Springer-Verlag Berlin Heidelberg 1996

1 Introduction

Among the known technetium isotopes with mass numbers 90–111, ^{99}Tc is a particularly important radionuclide from the viewpoint of environmental radioactivity and nuclear safety. ^{99}Tc is produced in nature in very small quantities by the spontaneous fission of ^{238}U [1], but is artificially produced in large quantities by the nuclear fission of ^{235}U during both nuclear weapon tests and reactor operation. Artificial technetium is now more abundant than naturally occurring rhenium of the same element group in the periodic table. The calculated inventory [2] of ^{99}Tc in typical PWR fuels has been estimated to be about 9.3 kg per GW(e) year. It was estimated that up to 1983 ^{99}Tc in the order of 15,000 TBq (24t) was generated by nuclear power stations worldwide [2]. As electricity generated during 1984–1993 by nuclear means was more than doubled, according to the data published in 1994, the total amount of ^{99}Tc at the end of 1993 was about 49,000 TBq (78t). In addition, the quantity of ^{99}Tc produced by nuclear weapons' explosions was estimated to be 160 TBq (0.25t).

A large neutron cross section of ^{235}U for fission (5.8×10^{-26} m^2), a high fission yield (6%) for ^{99}Tc, and a long half-life of the resulting ^{99}Tc (2.1×10^5 yr) make this radionuclide one of the principal nuclear wastes. Fig. 1 shows radioactivity of nuclear wastes plotted against cooling time in years. Tc activity is very important in the time interval 10^4–10^6 years.

In Japan two atomic bombs were dropped on Hiroshima and Nagasaki in August 1945. At that time I was a middle school boy in the city of Niigata, which had been considered as a next potential target for these new weapons; but it was fortunate for me that the third bomb was sunk together with the American warship carrying it in the Pacific Ocean following an attack by a Japanese submarine. Then the war ended, and when I became a nuclear scientist after graduating from Tohoku University, I decided to devote myself to promoting the peaceful uses of nuclear energy instead of its uses for weapons. I studied under Prof. Kenjiro Kimura at the University of Tokyo. He was a collaborator of Prof. Hevesy, the Nobel Prize winner for chemistry in 1943 and a pioneer in work on radioactive indicators (tracers) and radioactivation analysis, all peaceful uses of nuclear science. In 1954, Prof. Kimura and his coworkers analyzed the fallout nuclides on the deck of a Japanese fishing boat sailing near the Bikini Atoll at the time of US H-bomb testing. As a result of excellent radiochemical analysis, he contributed to the radiation safety of people in the Pacific countries. One of his students, Prof. Nagao Ikeda at Tsukuba University found ^{99}Tc in the soil of Nagasaki nearly forty years after the bombarding. The historical stream initiated by Profs. Hevesy and Kimura encouraged me and many others to move in the direction of the peaceful utilization of nuclear methods.

It is necessary to pay attention to environmental radioactivity when peaceful applications of nuclear methods are developed for various fields of science and technology. The importance of technetium in environmental safety was socially recognized not so long ago. The Commission of the European Communities

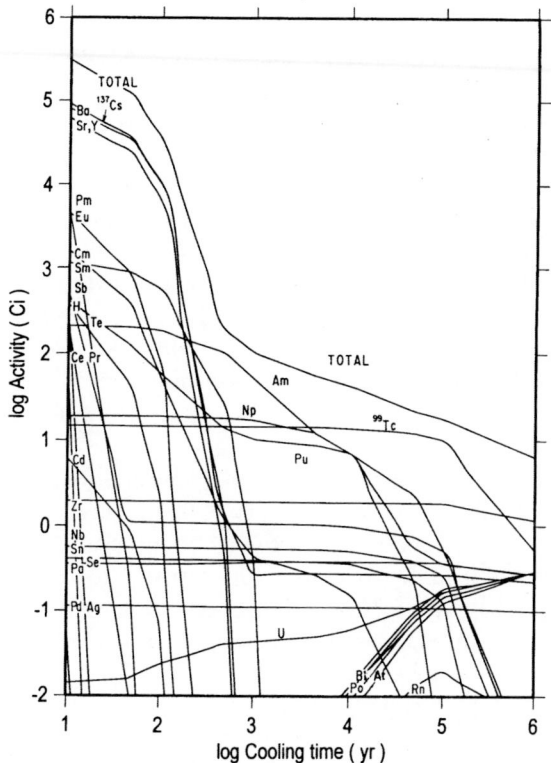

Fig. 1. Decay of high level nuclear wastes from spent fuel as a function of storage time. Radioactivity in curies per ton of spent fuel (PWR, 3.3% enriched ^{235}U, burnup 33,000 MWD/MTU at 30 MW/MTU, 5 year cooling, 99.5% U, Pu recovered)

organized a seminar on "The Behaviour of Technetium in the Environment" in 1984 [3]. Afterwards, aspects of environmental technetium were included in migration research from the viewpoints of geochemistry and radiation safety. In the series of international conferences on "Chemistry and Migration Behaviour of Actinides and Fission Products in the Geosphere" starting in Munich in 1987, technetium appeared as one of the important topics.

This review does not aim to give a complete survey of the literature; rather its emphasis is placed on currently interesting topics in environmental technetium studies.

2 Methods of Analysis

Analysis of technetium is usually made after chemical separation of the element and it requires a chemical yield tracer (^{99m}Tc or ^{95m}Tc). These sometimes make technetium analysis troublesome and inaccurate. The fact that ^{99}Tc is a pure β-emitter (if 0.0005% γ-emission is neglected) with an energy of 0.29 MeV makes it less detectable compared to other γ-emitting nuclides.

Highly sensitive determination of ^{99}Tc is possible using today's advanced mass spectroscopic methods. However, orthodox determination methods of ^{99}Tc involving radiometric techniques or even activation analysis are still used because they are simple and can be done without expensive machines. The detection limits of typical analytical methods are listed in Table 1.

Liquid scintillation counting has been used frequently for the measurement of environmental technetium. The specimens to be analyzed are treated by chemical procedures to obtain a technetium-bearing sample solution, which is mixed with a cocktail for scintillation counting. A low background scintillation counter with an anticoincidence system can be used for high precision measurements at a detection limit of 1–25 mBq.

Activation analysis for ^{99}Tc was developed by Foti et al. [4] using the (n, γ) reaction. But due to the short half-life (15 s) of the ^{100}Tc formed, there were considerable difficulties and the results generally lacked good precision and reliability. Consequently the detection limit listed in Table 1 seems to be too optimistic. Mincher and Baker [5] separated ^{99}Tc from mixed fission products and determined it by neutron activation analysis, but the sensitivity was not so good (0.3 μg) because of high background counting due to impurities. Compared to the (n, γ) reaction, (n, n') and (γ, γ') reaction analyses developed by Ikeda et al. [6] and Sekine et al. [7], respectively, look less sensitive, but are more reliable when the amounts of technetium to be measured are more than 1–10 Bq equivalent. In particular, the (γ, γ') reaction is very specific because of its unique character.

Radiometric analysis using a plastic scintillator connected to coincidence and anticoincidence circuits (e.g. a Picobeta counter) can be used to detect ^{99}Tc. The detection limit is estimated to be about 10 mBq. A slightly better detection limit (1–5 mBq) is reported for a gas flow counter with an anticoincidence shield [8].

Table 1. Detection limits of technetium in various analytical methods

Analytical method	Detection limit (mBq)
Liquid scintillation counting	30
Low background liquid scintillation counting	1–5
Low background gas flow counting	1–5
Activation analysis (n, γ)	2.5
$\qquad\qquad\qquad (n, n')$	10,000
$\qquad\qquad\qquad (\gamma, \gamma')$	630
ICP-MS	2–4
HR-ICP-MS with a ultrasonic nebulizer	0.002
RIMS	0.0005

ICP-MS (inductively coupled plasma mass spectrometry) is frequently used for determining ultratrace amounts of technetium [9]. In spite of the high cost of the equipment, this detection method is far superior to other radiometric methods as regards sensitivity. When a double focussing high-resolution system is used (HR-ICP-MS) and an ultrasonic nebulizer is introduced [10], the detection limit is in the order 0.002 mBq. The ICP-MS method has been successfully applied to the determination of environmental ^{99}Tc as well as to other long-lived radionuclides of neptunium and plutonium in the environment.

More sensitive detection is possible by laser resonance ionization mass spectrometry (RIMS) [11, 12, 12a]. This sophisticated method uses ionization in a three-step resonant excitation as shown in Fig. 2. There are three different modes of excitation (a), (b) and (c) in the figure. In (a), for example, the first step is to excite technetium from the ground state $(4d^5 5s^2\ {}^6S_{5/2})$ to the $4d^6 5p\ {}^4P_{5/2}$ level by absorption of laser light, at a wavelength of 313.12 nm, the second step is excitation from the $4d^6 5p\ {}^4P_{5/2}$ level to the $4d^6 6s\ {}^4D_{7/2}$ level by 821.13 nm light absorption, and the last step is to an autoionization state (above the ionization limit) by absorption of 670.74 nm light. Thus ionized technetium is led to an analyzer tube through an excitation electrode. As one of the aims of this spectrometric study has been the detection of technetium (^{98}Tc, ^{97}Tc) which is produced by a solar neutrino reaction in molybdenum, caution is especially needed to reduce the background due to molybdenum contamination. Chemically separated technetium still contains a large amount of molybdenum, but signals of the latter in RIMS can be suppressed by laser ionization and by time-gated detection. An example of laser ionization mass spectrometric measurements of ^{99}Tc (about 10^{14} atoms) which contains molybdenum is shown in Fig. 3. Without laser beams (a) many Mo peaks appear, but they are much reduced in number by the laser beams (b) and they almost completely disappear

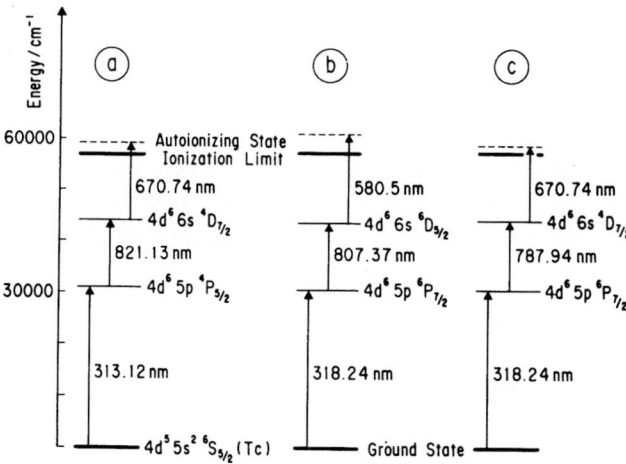

Fig. 2. Principle of laser resonance ionization of ^{99}Tc based on three different modes, (a), (b) and (c) [12]

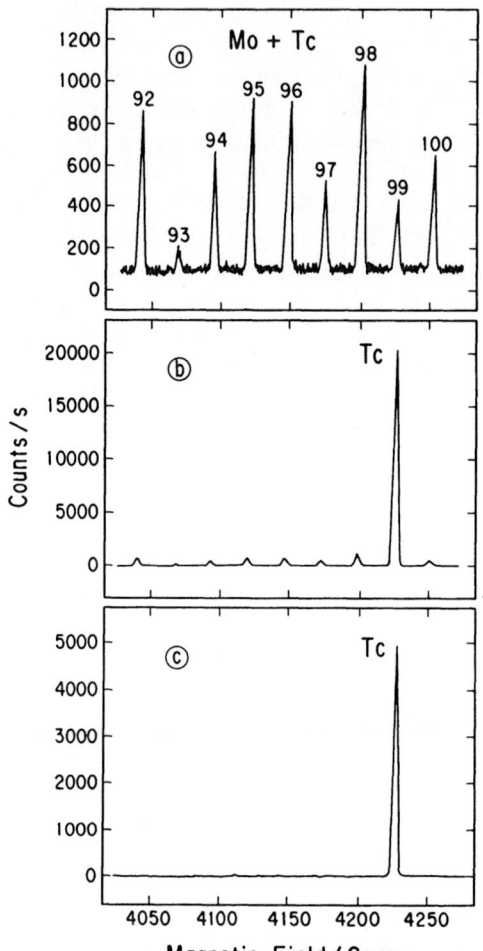

Fig. 3. Mass spectra of 10^{14} atoms of ^{99}Tc and of the stable isotopes of Mo. (a) Laser off, (b) laser on for ^{99}Tc and (c) with laser beams and gated detection [12a]

with laser beams and gated detection (c). The detection limit is 10^7 atoms (1.6 fg) of technetium in this case. With a reflection TOF mass spectrometer, the sensitivity will be in the order of 10^6 atoms. However, routine use of this method has not been made in measurements of environmental radioactivity samples because of its high cost.

3 Technetium in the Atmosphere

Technetium in rainwater was first studied by Attrep and coworkers [13, 14], but the data on ^{99}Tc were not sufficient.

The atmospheric ^{99}Tc/^{137}Cs activity ratio was much higher than expected from the nuclear fission of ^{235}U [15], and the value has increased with time as shown in Fig. 4. Interestingly in 1986 just after the Chernobyl accident the ratio (x) suddenly dropped but then rose again to the line calculated by the model in a paper by García-León et al. [16]. The ratio depends on the differing behavior of the two nuclides, which have different residence times in the atmosphere. The atmospheric residence time is 1.62 years for ^{99}Tc, a little longer than that (1.33 years) for ^{137}Cs [17].

Tagami and Uchida [18] recently determined the concentration of ^{99}Tc in rain and dry fallout by a sensitive analytical method using ICP-MS. Values ranged between 0.23 and 0.36 mBq/m^2, just slightly above the detection limit (0.16 mBq/m^2).

4 Technetium in Sea Water

Because the concentration of ^{99}Tc in sea water is usually very low, there have been only limited reliable data, except for those cases in which the ^{99}Tc concentration is relatively high due to the release of nuclear waste from spent fuel reprocessing plants. However, the advent of highly sensitive detection by ICP-MS has changed this situation.

Momoshima et al. [19, 20] developed an analytical procedure to determine ^{99}Tc in sea water by ICP-MS. ^{99}Tc was concentrated from the original sea water by the steps of filtration, reduction of pertechnetate with K$_2$S$_2$O$_5$, co-precipta-

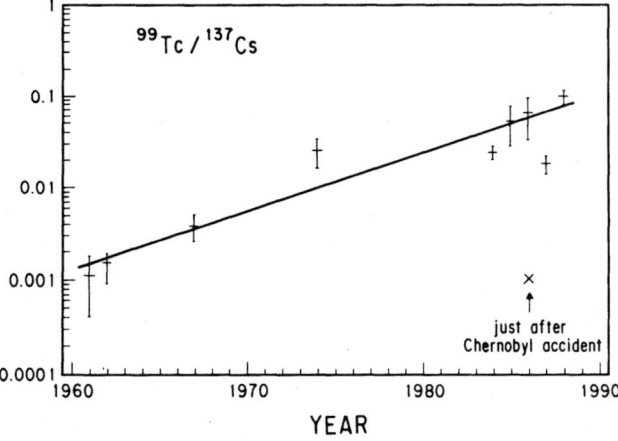

Fig. 4. Change in the ^{99}Tc/^{137}Cs activity ratio with time. Two values in 1986 represent those before and after the Chernobyl accident. (Reprinted with permission from Ref. 16. Copyright (1993) Elsevier Science Ltd)

tion with ferric hydroxide under alkaline conditions and dissolution in hydrochloric acid with H_2O_2. These were followed by the separation of the ^{99}Tc in an
aqueous solution from a precipitate of ferric hydroxide, that had been formed by
the addition of NaOH, and then purification by solvent extraction using
methylethylketone (MEK), and anion and cation exchange processes. The
samples were measured using an ICP-MS machine. Figure 5 summarizes the
analytical procedure adopted in their work. They collected coastal sea water
samples in Fukuoka, Japan and obtained ^{99}Tc values between 1 and 7mBq m^{-3},
most of them being near 1mBq m^{-3}. The mean value was one order of magnitude lower than reported earlier [21, 22]. The ^{99}Tc/^{137}Cs ratio was 2.7×10^{-4}
which was in good agreement with that calculated theoretically for ^{235}U fission.

Recently researchers of environmental radioactivity have been greatly interested in the contamination of the Arctic Sea due to radioactive waste discharges
by the former Soviet Union. The radioactivity levels of the North Sea, the
Norwegian Sea and the Greenland Sea have been affected by waste disposal
from the European reprocessing plants. Dahlgaard [23] has reported the contamination of ^{99}Tc in East Greenland polar water to be about 70 mBq m^{-3},
using the analytical procedure based on the work of Chen et al. [24, 25]. This
activity level was fairly high compared to that in the less contaminated area

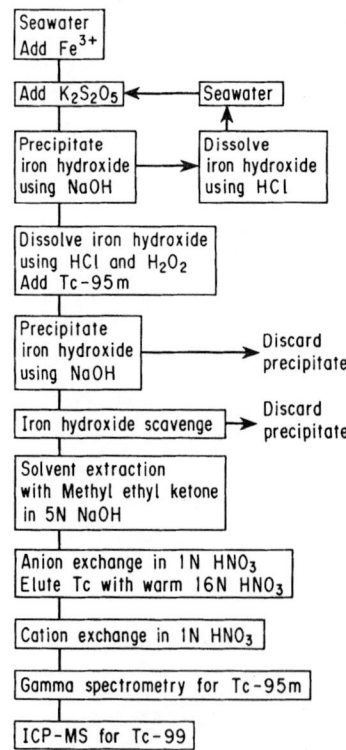

Fig. 5. Analytical procedure for very low-level ^{99}Tc in
seawater [19]

(Fukuoka, Japan [20]). A ^{99}Tc level of 63–197 mBq m^{-3} has been reported by Strand et al. [26] for the contamination of the Barents and Kara Seas.

5 Technetium in Soils

The behavior of technetium in soils is important from an ecological viewpoint, but is quite complicated and depends on many factors. Stalmans et al. [27] stated that it "is ruled by a combination of chemical, physicochemical and biological factors" and understanding of it "is needed on two main accounts: bioavailability and geochemical mobility". They also noticed that organic matter in soils and sediments play a significant role as a geochemical sink for technetium.

Chemical separation of technetium in soils is not easy, but it is fairly well-known that under aerobic conditions pertechnetate Tc(VII) is readily transferred to plants while under anaerobic conditions insoluble TcO_2 (or its hydrate) is not transferred to them. Even under aerobic conditions, however, the transfer rate decreases with time [28], indicating that soluble pertechnetate changes to insoluble forms by the action of microorganisms which produce a local anaerobic condition around themselves [29, 30]. Insoluble technetium species may be TcO_2, sulfide or complexes of organic material such as humic acid.

The formation of a humic acid complex of technetium was experimentally confirmed by a Belgian group [29]. Technetium in the chemical form of pertechnetate was reduced either by Sn^{2+} or by microbial action in the presence of humic acid and the complex formed was analyzed by gel chromatography. The radioactivity peak coincided well with the humic acid bulk peak in both cases. This fact strongly supported formation of a humic acid complex of technetium in a reduced state, but no mention was made of the valence state of the technetium. The same Belgian group [31] determined the cation exchange capacity of humic acid using a silver-thiourea system. Organic matter containing humic acid was extracted from a podzol soil using a dilute alkaline solution. This showed a lower cation exchange capacity than commercial humic acid (Fluka). As humic acid is a general name for a group of compounds which are derived from dead plants, a standard material for comparison is necessary. Purified and well characterized humic acid from the Gorleben area in Germany was used by Kim et al. [32, 33] for a migration study of actinide elements. It is clear that any study of the technetium–humic acid interaction must use well characterized humic acid.

In a recent study of the complexation of technetium with humic acid (HA) Sekine et al. [34, 35] obtained interesting results which show competition between Tc^{IV}-O(OH)$_2$ precipitate formation and Tc^{III}-HA precipitate formation during a reduction process of pertechnetate with Sn^{2+}. A weighable amount of

$^{99}TcO_4^-$ was reduced by addition of Sn^{2+} in the absence and presence respectively of humic acid at pH values of 1–11; the precipitate in the former case consisted of $Tc^{IV}O(OH)_2$ while the precipitate in the latter case was a mixture of $Tc^{IV}O(OH)_2$ and a Tc-HA complex. The Tc-HA complex was found to decompose to pertechnetate at high pH values whereas $TcO(OH)_2$ remained unchanged. From the stoichiometric relationship the valence state of technetium was suggested to be Tc(III). The amount of Tc-HA increases with increasing amounts of Sn^{2+} as shown in Fig. 6. The competition can be demonstrated formally as in Table 2. This work used purified humic acid (Gohy 573 HA) obtained from the Gorleben area in Germany.

The tracer level $Tc^{III}HA$ complex can be suspended as a colloid in an aqueous solution. Speciation of a similar complex, Am^{III} HA, in the Gorleben groundwater was performed by laser photoacoustic spectroscopy (LPAS) [33]. Considering its molecular size, however, migration of the complex is expected to be very slow when it is present in a solid phase. Similar LPAS studies of a technetium complex towards chemical speciation have been tried by our group in Sendai [36, 37].

Using ICP-MS, Morita and coworkers [38] compared the depth profile of ^{99}Tc in surface soil in Japan with those of other long-lived radionuclides, i.e. ^{137}Cs, ^{237}Np and $^{239, 240}Pu$ (Fig. 7). Surface soil is known to be rich in organic matter. The depth–concentration relation showed high retention of ^{99}Tc in the surface layer. This is contrary to the hypothesis that ^{99}Tc would have high mobility in soil because ^{99}Tc is in the chemical form of pertechnetate under highly oxidizable surface conditions. As mentioned earlier, the reducing action of microorganisms in soils can convert Tc(VII) to low valency technetium.

The results for the ICP-MS determination of ^{99}Tc in soil were compared with those for ^{137}Cs [39]. The $^{99}Tc/^{137}Cs$ ratio in soil was 3.9×10^{-3} to 3.8×10^{-2} which is considerably larger than that expected theoretically.

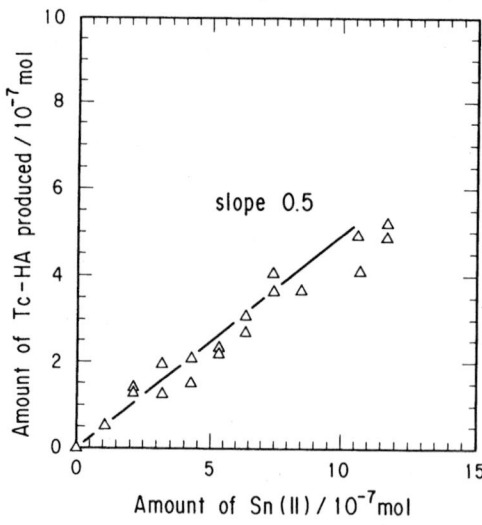

Fig. 6. Amount of Tc complexed with humic acid (HA) as a function of Sn^{2+} concentration [34]

Table 2. Competition between TcO(OH)$_2$ and Tc-HA formation routes

(A) TcO(OH)$_2$ formation route

$$3Sn(II) + 2Tc(VII) \rightarrow 3Sn(IV) + 2Tc(IV)$$

$$Tc^{IV}O^{2+} + 2H_2O \rightarrow TcO(OH)_2 + 2H^+$$

$$\downarrow$$

ppt in acid

(B) Tc-HA complex formation route

$$2Sn(II) + Tc(VII) \rightarrow 2Sn(IV) + Tc(III)$$

$$Tc(III) + HA \rightarrow Tc^{III}\text{-HA} \rightarrow \text{decomposed to } TcO_4^-$$

$$\downarrow \qquad \text{in basic conditions}$$

ppt in acid

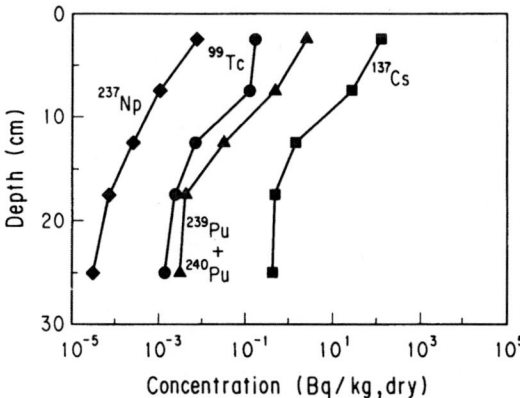

Fig. 7. Depth profiles of ^{99}Tc and other nuclides in soils [38]

Sheppard et al. [40] developed a novel method to examine the variation in Tc sorption among 34 kinds of soils. Using a 'packette and filter disc' method, the study compared the sorption of Tc under aerobic and anaerobic conditions with better control and monitoring of the samples. The geometric mean aerobic K_d values were 0 and 50 1/kg respectively for mineral and organic soils, whereas they were 18 and 68 1/kg under anaerobic conditions. The aerobic K_d increased slightly with organic matter content, whereas the anaerobic K_d increased according to clay content and pH without any definite dependence on the organic matter content.

6 Technetium in Plants

Some plants are able to concentrate technetium, a well known example being brown algae living in seawater [41,42] for which concentration factors of

250–2500 have been reported in laboratory experiments. A very high concentration factor of 10^5 was given for *Fucus vesiculosus*, which is suitable as an bioindicator for the behavior of technetium [8]. It was shown that radioactive wastes from the Sellafield (UK) reprocessing facility could be monitored using this sea plant. Calmet et al. [43] measured spatio-temporal variations of ^{99}Tc content in *Fucus serrantus* (a kind of brown algae) in the English Channel during 1982–1984 in order to determine the bioaccumulation of ^{99}Tc following the discharge of radioactive wastes from the La Hague (France) plant for reprocessing spent fuel. Fig. 8 shows the relationship between the ^{99}Tc concentration in the samples of *Fucus serrantus* and the distance from the discharge outlet. The curve in the figure assumed an exponential decrease of concentration with distance and was adjusted to give the best fit to the data.

In ^{99}Tc samples of two kinds of brown algae (one is IAEA seaweed Ag-1 *Fucus serrantus* taken from the Irish Sea near the Sellafield nuclear facility and the other is *Fucus vesiculosus* taken from the same area) measured by Koide and

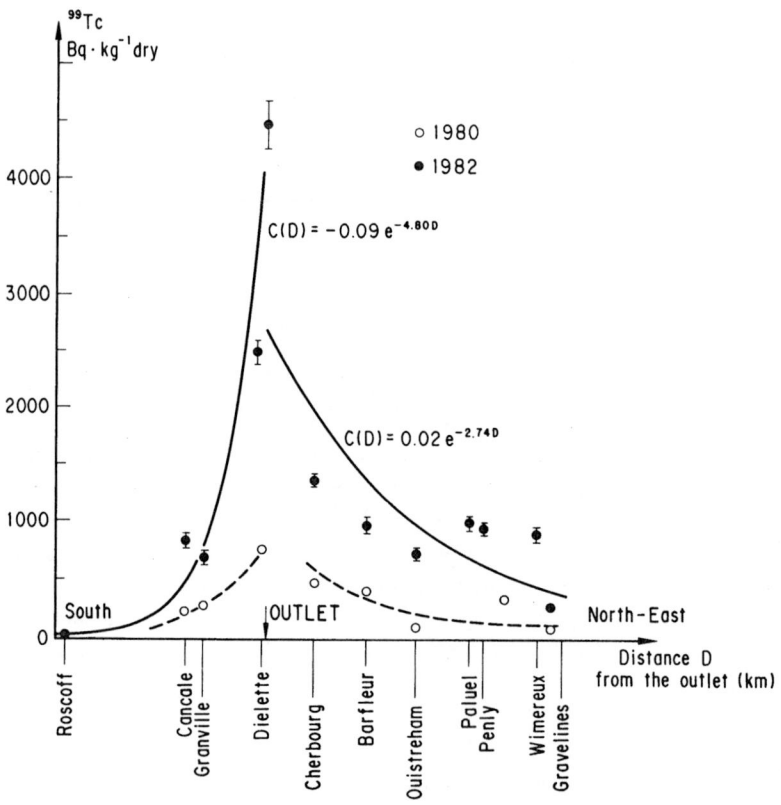

Fig. 8. Variation with location of the mean levels of ^{99}Tc in samples of *Fucus serrantus* bioindicators along the French coast of the English Channel. The discharge outlet at the La Hague reprocessing plant is shown by an arrow. (Reprinted with permission from Ref. 43. Copyright (1987) Elsevier Science Ltd)

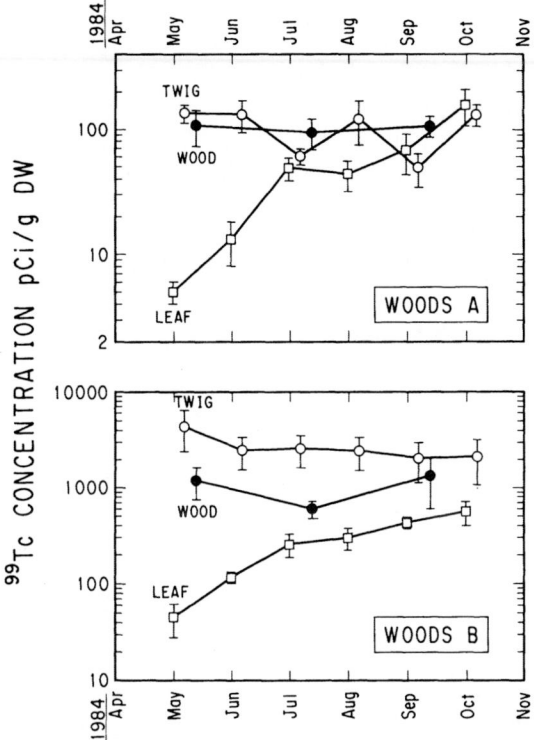

Fig. 9. Month-to-month concentrations of ^{99}Tc in tree leaves, twigs and wood from woods A and woods B at the technetium study site during 1984. Means are based on data from 4–5 quadrats. (Reprinted with permission from Ref. 48. Copyright (1986) Elsevier Science Ltd)

Goldberg [44], surprisingly high concentrations were detected: for the former 16, 000 Bq/kg, for the latter 43,000–46,000 Bq/kg. All these data indicate that technetium should be carefully recovered in reprocessing plants. The systematic recovery of technetium together with elements of the platinum group from high level wastes have been discussed by Kubota [45].

Holm and Rioseco [46] noted ^{99}Tc activity in lichen of the food chain lichen–reindeer–man. A ^{99}Tc activity concentration of $3-8 \times 10^{-5}$ Bq/kg was obtained in dry lichen during 1956–1981. Nearly the same concentration level of activity (only occasionally one order of magnitude higher) was found in reindeer.

In a kind of brown algae *Hijikia fusiforme* which is a commonly eaten sea plant in Japan, Hirano and Matsuba [47] detected ^{99}Tc activity. They estimated the concentration factor to be 5×10^3. They also studied the ^{99}Tc concentration in *Sargassum thumbergii* in relation to the discharge of radioactive wastes from a nuclear facility near the sampling sites. This alga has proved to be a good indicator of the concentration of ^{99}Tc since concentration of the element in the alga changes in direct proportion to the concentration in seawater. The concentration factor was found to be about 10,000.

Garten and Tucker [48] showed that there was a seasonal variation in the concentration of ^{99}Tc in tree leaves growing in the area contaminated with radioactive wastes from Oak Ridge National Laboratory (ORNL), although twigs and wood from the same area did not show any statistically significant change in concentration. Fig. 9 shows results obtained in 1984 for two woods, A and B, which are situated near the chemical waste pits of ORNL. In soft parts, such as leaves, the ^{99}Tc concentration rapidly increased from May to July. This is followed by a slow increase from August to September. This tendency clearly reflects plant growth. They concluded that tree wood was the major above-ground pool for ^{99}Tc.

Garten and Tucker [49] also studied the soil-to-plant transfer of ^{99}Tc. They compared plant uptake of ^{99}Tc freshly added to soil with that of ^{99}Tc that had aged in the soil for more than a decade. The combination of alkaline soil and freshly added ^{99}Tc increased the uptake into radish foliage. The plant/soil concentration ratio was found to be about 40.

7 Technetium in Animals

Transfer of technetium from seawater to animals has been studied by laboratory experiments using ^{95m}Tc. The advantages of ^{95m}Tc over ^{99}Tc are: ^{95m}Tc emits γ-rays, thus whole-body counting techniques can be used, and it has a sufficiently high specific activity because of its relatively short half life (61 d).

The behavior of technetium in animals is more complex than in plants, because uptake and excretion of ^{95m}Tc occur almost simultaneously in animals. Therefore, the concentration factor of ^{95m}Tc increases with increasing exposure time and reaches a steady-state value. Moreover, accumulation of ^{95m}Tc is very much dependent on the organs involved. As we are more concerned with chemistry than with biology, we have to limit ourselves when describing biological factors.

Pentreath [50] reported ^{95m}Tc accumulation in various species of crusacea. The final concentration factor of 785 was obtained for *Homarus gammarus L.* Using a linear differential model [51], he estimated a steady-state concentration factor (C_{ss}) of 1123.

The uptake and retention of technetium from seawater by young lobsters have been demonstrated by Swift [52]. The accumulation of ^{95m}Tc was rapid and whole body concentration factors of over 2,000 were observed in some lobsters. The kinetics of the uptake and loss of ^{95m}Tc were studied in the crab *Pachygrapus marmoratus* [53]. The steady-state concentration factor was estimated to be 18.

An interesting study on accumulation and retention of ^{95m}Tc by the edible winkle (*Littorina littorea L.*) was carried out by Swift [54] who determined a concentration factor of 45 in the whole body and a biological half-time of 115 days. The half-time depended on feeding regimes – starvation reduced the elimination of technetium in winkles. About 98% of the ^{95m}Tc in the whole body was in the flesh. In particular, the digestive gland and the kidney accumulated more than 60% of ^{95m}Tc. He compared the ^{95m}Tc concentration factor of 45 with those for ^{99}Tc (5,900–29,000), obtained after the discharge of radioactive wastes at the Sellafield reprocessing plant (R.L. Pentreath, unpublished work) and noted that the latter high value must have originated from magnification in the food pathway.

8 Technetium in Studies on Radioactive Waste Migration

There has been considerable interest recently in the migration of long-lived nuclides involving technetium. The behavior of technetium in groundwater, sorption and permeation under subterranean conditions needs to be studied for the purpose of assessing environmental safety in connection with the disposal of spent nuclear fuel. Chemical and physicochemical data on technetium under such conditions are necessary.

The solubility of $TcO_2 \cdot nH_2O$ in an aqueous solution was investigated by Meyer et al. [55–58]. Paquette and Laurence [59] reported that carbonato- and phosphato-complex formation was likely to occur. The precipitation of $TcO_2 \cdot H_2O$ is prevented in solution if the ratio

$$[CO_3^{2-}]_{total} / [Tc(IV)]_{total} > 30 \,.$$

Lemire and Garisto [60] concluded that the only important complex formed in a nearly neutral solution in $Tc(OH)(CO_3)_2^-$. Erikson et al. [61] determined the logarithm of the equilibrium constants for the equilibria involving $TcO_2 \cdot nH_2O$ species as shown in Table 3.

It was suggested that TcO_2 could be oxidized to TcO_4^- in high salinity oxygen-free water due to radiolytic effects [62]. Lieser and Rauscher [63] showed that TcO_4^- could be reduced in natural ground water under anaerobic conditions.

Table 3. Equilibria involving $TcO_2 \cdot nH_2O$ species and the equilibrium constants

Process	log K
$TcO_2 \cdot nH_2O(s) = TcO(OH)_2(aq) + (n-1)H_2O$	-8.17 ± 0.05
$TcO_2 \cdot nH_2O(s) + H_2O = TcO(OH)_3^- + H^+ + (n-2)H_2O$	-19.06 ± 0.24
$TcO(OH)_2(aq) + H_2O = TcO(OH)_3^- + H^+$	-10.89 ± 0.24
$TcO_2 \cdot nH_2O(s) + CO_2(g) = Tc(OH)_2CO_3(aq) + (n-1)H_2O$	-7.09 ± 0.08
$TcO(OH)_2(aq) + CO_2(g) = Tc(OH)_2CO_3(aq)$	-1.08 ± 0.08
$TcO_2 \cdot nH_2O(s) + CO_2(g) + H_2O = Tc(OH)_3CO_3^- + H^+ + (n-2)H_2O$	-15.35 ± 0.07
$TcO(OH)_2(aq) + CO_2(g) + H_2O = Tc(OH)_3CO_3^- + H^+$	-7.18 ± 0.07

A field experiment in a migration study of technetium in the form of TcO_4^- was carried out by Landström et al. [64] who injected Tc and radioactive [82]Br (non-sorbing tracer) in a highly permeable zone at a borehole. After 10 hours, breakthrough of the tracer was measured, but no retardation, i.e. no reduction of Tc, was observed in comparison with [82]Br.

Contrary to the above observation, however, a Swedish group [65] confirmed that in their field experiment no breakthrough of TcO_4^-, injected at a depth of 150 m, was observed, indicating that reduction of Tc(VII) to a lower valency state had occurred. This result is consistent with a laboratory experiment in which Tc sorption was fast at reducing conditions using nitrogen gas and simulated groundwater [65].

9 Technetium in the Oklo Natural Reactors

The astonishing discovery of the Oklo natural reactor in the Republic of Gabon (Africa) was reported in 1972 [66, 67]. This self-sustaining fission reactor had been in operation for a few hundred thousand years. Its operation was $(17.2 \pm 0.2) \times 10^8$ yr years ago, when the [235]U (half-life: 7.04×10^8 yr, abundance: 0.72%) content in natural uranium was at the 5% level. This natural reactor was predicted by Kuroda in 1956 [68]. To date, 16 reactor zones have been found in the same area. The reactors stopped functioning long, long ago, but isotopic abundance anomalies have remained because of the nuclear reactions that had taken place in them. The CEA (Commissariat a l'Énergie Atomique) and the CEC (Commission of the European Community) are cooperating in a research project on waste disposal in which the Oklo reactor zones are considered as a natural analogue. These relics of natural reactors can provide ideas on how to manage nuclear wastes with safety. We can or, more properly, we have to learn from Nature. Therefore, the Oklo reactors have been highlighted as objectives of radioactive waste migration and containment studies [69].

Table 4 summarizes the geographical behavior of fission products at the Oklo reactors compared with the behavior of fission products from neutron-irradiated uranium dioxide. The reactor zones RZ 1–9 are exposed on the ground, while RZ 10–16 are situated below the ground, all quite deep except RZ 15.

[99]Tc, a major fissiogenic radionuclide in the reactors, has decayed out to [99]Ru because of its relatively short half-life. If [99]Ru is investigated in detail by mass spectrometry and the contribution of fissiogenic and background ruthenium is carefully subtracted, the behavior of ^{99}Tc in the reactors can be elucidated.

Loss et al. [70] found that [99]Tc (detected as [99]Ru) in RZ 9 (exposed on the ground) was lost but was present in the peripheral rocks surrounding the reactor zone. The ^{99}Tc had been fractionated from the Ru within 10^6 years of the end of

Table 4 Comparison of the characteristic behavior of fission products in the Oklo ores with those in irradiated uranium dioxide

Fission product	Oklo RZ 1–9	Oklo RZ 10–16	Irradiated UO$_2$
Rare earths	Partial release of light RE	Well retained	Homogeneously distributed in the matrix
Ru, Tc, Rh Pd, Te	Well retained[a]; metallic inclusion unidentified	Well retained; metallic inclusion formed in the matrix	Metallic inclusion formed in the matrix
Ba, Sr, Rb	Almost disappeared	1–10% found around the boundaries between RZ and sand-stones	Oxide inclusion within the matrix
Cs	Almost disappeared	The same as above	Solid phase in cooler region of the matrix

[a] Tc: fractionated from Ru and partially retained.

the operation of the reactor. The behavior of ^{99}Tc in RZ 10 (deep underground) was somewhat different from that in RZ 9, as shown by Hidaka et al. [71] who found metallic inclusions of about 100 μm in the reactor core samples. Moreover, ^{100}Ru abundance correlated well with the deviation of ^{99}Ru which is an index of the ^{99}Tc content in the sample. This fact suggested that neutron capture by ^{99}Ru and/or ^{99}Tc occurred in the reactor zone. The behavior of the fission products produced in the Oklo reactors was recognized to be affected by the presence of organic matter such as humic acid and bitumen (with aromatic rings). Bitumen in particular is effective in preventing the widespread dispersion of fission products from the reactor core [72, 73].

10 Conclusions

The behavior of technetium in the environment has not yet been well understood in my opinion; from chemical and biological viewpoints our understanding of it is at an early stage. Considering the serious role which this element plays in the environment both at the present time and in the future, the study of environmental technetium should be promoted. There are many things to do in this regard. Both difficulties and successes are to be expected, the final goal being a contribution to human welfare.

Acknowledgements. This work was supported by many scientists who have kindly given me valuable guidance and suggestions. I dedicate this article to the late Prof. Kenjiro Kimura who opened for me the door leading to the scientific world.

Kenji Yoshihara

11 References

1. Kenna BT, Kuroda PK (1961) J Inorg Nucl Chem 23: 142
2. Luykx F (1986) in: Desmet G, Myttenaere C (eds) Technetium in the Environment. Elsevier Applied Science, New York, p 21
3. Desmet G, Myttenaere C (1986) Technetium in the Environment. Elsevier Applied Science, New York
4. Foti S, Delucchi E, Akamian V (1972) Anal Chim Acta 60: 261
5. Mincher BJ, Baker JD (1990) J Radioanal Nucl Chem 139: 273
6. Ikeda N, Seki R, Kamemoto M, Otsuji M (1989) J Radioanal Nucl Chem 131: 65
7. Sekine T, Yoshihara K, Németh Zs, Lakosi L, Veres Á (1989) J Radioanal Nucl Chem 130: 269
8. Holm E (1993) Radiochim Acta 63: 57
9. Igarashi Y, Kim CK, Takaku Y, Shiraishi K, Yamamoto M, Ikeda N (1990) Anal Sciences 6: 157
10. Morita S, Kim CK, Takaku Y, Seki R, Ikeda N (1991) Appl Radiat Isot 42: 531
11. Ruster W, Ames F, Kluge HJ, Otten EW, Rehklau D, Scheerer F, Herrmann G, Muehleck C, Riegel J, Rimke H, Sattelberger P, Trautmann N (1989) Nucl Instr Meth A281: 547
12. Sattelberger P, Mang M, Herrmann G, Riegel J, Rimke H, Trautmann N (1989) Radiochim Acta 48: 165
12a. Trautmann N (1993) Radiochim Acta 63: 37
13. Attrep M, Enochs JA, Broz LD (1971) Environ Sci Technol 5: 344
14. Ehrhardt KC, Attrep M (1978) Environ Sci Technol 12: 55
15. García-León M, Piazza C, Madurga G (1984) Int J Appl Radiat Isot 35: 961
16. García-León M, Manjon G, Sanchez-Angulo CI (1993) J Environ Radioactivity 20: 49
17. Pourchet M, Pinglot F (1979) Geophys Res Lett 6: 365
18. Tagami K, Uchida S (1995) J Radioanal Nucl Chem (in press)
19. Momoshima N, Sayad M, Takashima Y (1993) Radiochim Acta 63: 73
20. Momoshima N, Sayad M, Takashima Y (1995) J Radioanal Nucl Chem (in press)
21. Holm E, Rioseco J, Ballestra S, Walton A (1988) J Radioanal Nucl Chem 123: 167
22. Matsuoka N, Umata T, Okamura M, Shiraishi N, Momoshima N, Takashima Y (1990) J Radioanal Nucl Chem 140: 57
23. Dahlgaard H (1994) J Environ Radioactivity 25: 37
24. Chen QJ, Dahlgaard H, Hansen HJM, Aarkrog A (1990) Anal Chim Acta 228: 163
25. Chen QJ, Dahlgaard H, Nielsen SP (1994) Anal Chim Acta 285: 177
26. Strand P, Nikitin A, Rudjord AL, Salbu B, Christensen G, Foyn L, Kryshev II, Chumichev VB, Dahlgaard H, Holm E (1994) J Environ Radioactivity 25: 99
27. Stalmans M, Maes A, Cremers A (1986) in: Desmet G, Myttenaere C (eds) Technetium in the Environment. Elsevier Applied Science, London, p 91
28. Hoffman FO, Garten CT Jr, Lucas DM, Huckabee JW (1982) Environ Sci Tech 16: 214
29. Van Loon L, Stalmans M, Maes A, Cremers A, Cogneau M (1986) in: Desmet G, Myttenaere C (eds) Technetium in the Environment. Elsevier Applied Science, London, p 143
30. Landa ER, Thorvig LJ, Gast RG (1977) J Environ Quality 6: 181
31. Stalmans M, De Keijzer S, Maes A, Cremers A (1986) in: Desmet G, Myttenaere C (eds) Technetium in the Environment. Elsevier Applied Science, London, p 155
32. Kim JI, Li GH, Duschner H, Psarros N, Buckau G (1990) Z Anal Chem 338: 245
33. Kim JI, Stumpe R, Klenze R (1990) in: Yoshihara K (ed) Chemical Applications of Nuclear Probes. Springer Verlag, Heidelberg, p 129
34. Sekine T, Watanabe A, Yoshihara K, Kim JI (1993) Radiochim Acta 63: 87
35. Sekine T, Asai N (1995) J Radioanal Nucl Chem (to be published)
36. Sekine T, Hiraga M, Fujita T, Mutalilb A, Yoshihara K (1993) J Nucl Sci Tech 30: 1131
37. Fujita T, Sekine T, Hiraga M, Yoshihara K, Mutalib A, Alberto R, Kim JI (1993) Radiochim Acta 63: 45
38. Morita S, Tobita K, Kurabayashi M (1993) Radiochim Acta 63: 63
39. Tagami K, Uchida S (1993) Radiochim Acta 63: 69
40. Sheppard SC, Sheppard MI, Evenden WG (1990) J Environ Radioactivity 11: 215
41. Beasley TM, Lorz HV (1993) in: Desmet G, Myttenaere C (eds) Technetium in the Environment. Elsevier Applied Science, London, p 197
42. Jeanmarie L, Masson M, Patti F, Germain P, Cappellini L (1981) Mar Poll Bull 12: 29

43. Calmet P, Patti F, Charmasson S (1987) J Environ Radioactivity 5: 57
44. Koide M, Goldberg ED (1985) J Environ Radioactivity 2: 261
45. Kubota M (1993) Radiochim Acta 63: 91
46. Holm E, Rioseco J (1987) J Environ Radioactivity 5: 343
47. Hirano S, Matsuba M (1993) Radiochim Acta 63: 79
48. Garten CT Jr, Tucker CS, Walton BT (1986) J Environ Radioactivity 3: 163
49. Garten CT Jr, Tucker CS, Scott TG (1986) J Environ Radioactivity 4: 91
50. Pentreath RJ (1981) in: Impacts of Radionuclide Releases into the Marine Environment. IEAE, Vienna, p 241 (STI/PUB/565)
51. Pentreath RJ (1975) in: Design of Radiotracer Experiments in Marine Biological Systems. IEAE, Vienna, p 137 (Tech. Rep. Series No. 167)
52. Swift DJ (1985) J Environ Radioactivity 2: 229
53. Conversi A (1985) J Environ Radioactivity 2: 161
54. Swift DJ (1989) J Environ Radioactivity 9: 31
55. Meyer RE, Arnold WD, Case FI (1986) Valence effects on solubililty and sorption. The solubility of Tc(IV) oxides NUREG/CR-4309 ORNL-6199
56. Meyer RE, Arnold WD, Case FI (1987) The solubility of electrodeposited Tc-(IV) oxides NUREG/CR-4865 ORNL-6734
57. Meyer RE, Arnold WD, Case FI, O'Kelly GD (1988) Thermodynamic properties of Tc(IV) oxides: solubilities and the electrode potential of the Tc(VII)/Tc(IV) oxides couple NUREG/CR-5108 ORNL-6480
58. Meyer RE, Arnold WD, Case FI, O'Kelly GD (1989) Solubilities of Tc(IV) oxides and the electrode potential of the Tc(VII)/Tc(IV)-oxide couple NUREG/CR-5235 ORNL-6503
59. Paquett J, Lawrence WE (1989) Can J Chem 63: 2369
60. Lemire RJ, Garisto F (1989) The solubility of U, Np, Pu, Th and Tc in a geological disposal vault for used nuclear fuel AECL-10009
61. Eriksen TE, Ndalamba P, Bruno J, Caceci M (1992) Radiochim Acta 58/59: 67
62. Lieser KH, Bauscher C, Nakashima T (1987) Radiochim Acta 42: 191
63. Lieser KH, Bauscher C (1987) Radiochim Acta 42: 205
64. Landström O, Klockars CE, Holmberg KE, Westerberg S (1978) Swedish Nuclear Fuel and Waste Management Co. SKB Technical Report TR-110, Stockholm
65. Byegård J, Albinsson Y, Skarnemark G, Skålberg M (1992) Radiochim Acta 58/59: 239
66. Bodu R, Bouzigues H, Morin N, Pfiffelmann JP (1972) CR Acad Sci Paris 275D: 1731
67. Ruffenach JC, Menes J, Lucas M, Hagemann R, Nief G (1975) The Oklo Phenomenon. IAEA, Vienna, p 371
68. Kuroda PK (1956) J Chem Phys 25: 781; 1295
69. Curtis D, Bejamin T, Gancarz A, Loss R, Rosman K, DeLaeter J, Delmore JE, Maeck WJ (1989) Appl Geochem 4: 49
70. Loss RD, Rosman KJR, DeLaeter JR, Curtis DB, Benjamin TM, Gancarz AJ, Maeck WJ, Delmore JE (1989) Chem Geol 76: 71
71. Hidaka H, Shinotsuka K, Holliger P (1993) Radiochim Acta 63: 19
72. Nagy B, Gauthier-Lafaye F, Holliger P, Davis DW, Mossman DJ, Leventhal JS, Rigali MJ, Parnell J (1991) Nature 354: 472
73. Nagy B, Gauthier-Lafaye F, Holliger P, Mossman DJ, Leventhal JS, Rigali MJ (1993) Geology 21: 655

The Chemistry of Technetium Nitrido Complexes

John Baldas

Australian Radiation Laboratory, Lower Plenty Road, Yallambie, Victoria 3085, Australia

Table of Contents

Topics in Current Chemistry, Vol. 176
© Springer-Verlag Berlin Heidelberg 1996

John Baldas

4 Technetium(VII) Nitrido Complexes

A characteristic feature of the chemistry of technetium in the $+5-+7$ oxidation states is the formation of a wide variety of highly stable complexes containing the terminal $Tc\equiv N$ bond. The $Tc\equiv N$ bond is short (1.585–1.65 Å) and formally triple with one σ component and two π components formed by overlap of p_x and p_y orbitals on N with Tc d_{xz}, d_{yz} orbitals. Coordination numbers of 5, 6 and 7 and square pyramidal, trigonal bipyramidal, distorted octahedral and pentagonal pyramidal and bipyramidal geometries have been established by X-ray crystallography. The nitrido ligand exerts a strong *trans* influence, with the *trans* ligand either absent or only weakly bound. In octahedral complexes, the NTc–L$_{trans}$ bond is 0.1–0.3 Å longer than NTc–L$_{cis}$. In the IR spectrum $v(TcN)$ is observed as a strong sharp band at 1100–1000 cm^{-1}. [Tc^{V}N]$^{2+}$ complexes have been prepared with diverse coordination spheres and are monomeric except for one example with the cyclic Tc$_4$N$_4$ core. Monomeric [Tc^{VI}N]$^{3+}$ (d^1) complexes are readily detected by EPR spectroscopy and this technique has been extensively used to identify new species and study exchange reactions. Substitution of [Tc^{VI}NX$_4$]$^-$ (X = Cl, Br) by thiols, phosphines, amines and nitrogen heterocycles in organic solvents generally yields the reduced [Tc^{V}N]$^{2+}$ complexes. In aqueous solution dimerization of [Tc^{VI}N]$^{3+}$ occurs to yield EPR silent [NTc^{VI}–O–Tc^{VI}N]$^{4+}$ and [NTc^{VI}(μ-O)$_2$Tc^{VI}N]$^{2+}$ complexes, including the structurally characterized (AsPh$_4$)$_4$[Tc$_4$N$_4$(O)$_2$(ox)$_6$] and (AsPh$_4$)$_2$[(TcNX$_2$)$_2$(μ-O)] (X - Cl, Br). Novel [Tc^{VII}N]$^{4+}$ complexes are Cs[TcN(O$_2$)$_2$Cl] and (AsPh$_4$)$_2$[{TcN(O$_2$)$_2$}$_2$(ox)]. ^{99m}TcN complexes have been investigated as potential radiopharmaceuticals.

1 Introduction

In 1972, Griffith noted that, in view of the stability of rhenium nitrido complexes, it should also be possible to prepare TcN complexes [1], but it was only in 1981 that the preparation of the first $Tc\equiv N$ complexes, [Tc^{V}N(S$_2$CNEt$_2$)$_2$], [Tc^{V}NCl$_2$(PPh$_3$)$_2$] and [Tc^{V}NCl$_2$(PR$_2$Ph)$_3$] (R = Me, Et) was reported [2, 3]. Since that time, the chemistry of technetium nitrido complexes has been actively investigated and, to the author's knowledge, some 60 crystal structures have been reported or completed. Novel features are the remarkable stability of the $Tc\equiv N$ bond to protonation and cleavage by strong aqueous acids and the existence of an extensive aqueous solution chemistry. In marked contrast to Tc^{VI}O complexes, Tc^{VI}N complexes show no tendency to disproportionate in aqueous solution. Analogues of the dimeric Tc^{VI}N μ-oxo and bis(μ-oxo) and of the Tc^{VII}N nitridoperoxo complexes are not known for any other transition metal. Also, a variety of ^{99m}TcN complexes (where the Tc concentration is $10^{-8}-10^{-7}$ M) have been prepared and investigated as potential radiopharmaceuticals [4]. There is also a considerable, and growing, chemistry of technetium hydrazido (TcNNR$_2$), imido (TcNR), and related complexes which contain TcN multiple bonds. These complexes have been reviewed elsewhere [4] and are not discussed here.

The nitrido ligand (N^{3-}) is isoelectronic with the oxo ligand (O^{2-}) and the substituted imido (RN^{2-}), alkylidene (R_2C^{2-}) and alkylidyne (RC^{3-}) ligands. Due to the high negative charge, N^{3-} is a powerful π-electron donor which effectively stabilizes technetium in the +5 to +7 oxidation states. Technetium nitrido complexes are not known in lower oxidation states {a compound formulated as "TcNF" has been shown to be $(NH_4)_2[TcF_6]$ [5]}. The Tc≡N bond is short, in the range 1.59–1.7 Å, and is formally triple with one σ component and two π components formed by the overlap of the p_x and p_y orbitals on N with Tc d_{xz} and d_{yz} orbitals [6]. Although the EPR spectroscopy of $Tc^{VI}N$ (d^1) complexes has been intensively studied [7, 8], other spectroscopic studies of TcN complexes have received less attention. Kinetic studies have been few but a number of systems have been studied electrochemically.

The results obtained to date are considerable and show that the chemistry of the TcN group may well be the most varied and interesting of the transition metal nitrido complexes [1, 9, 10]. The aim of this chapter is to provide a fairly comprehensive review of the literature up to the latter part of 1994. Additional data may be found in two conference volumes [11, 12] and a recent review of Tc coordination chemistry [4]. For macroscopic studies with the long-lived ^{99}Tc ($t_{1/2} = 2.11 \times 10^5$ years) the ^{99}Tc radionuclide is denoted simply as Tc. "No carrier added" studies and radiopharmaceutical applications utilizing the short-lived ^{99m}Tc radionuclide ($t_{1/2} = 6.01$ hours) are denoted as ^{99m}Tc.

2 Technetium(V) Nitrido Complexes

The chemistry of technetium in the +5 oxidation state is dominated by complexes containing the $[TcO]^{3+}$ and $[TcN]^{2+}$ cores [4]. The lower charge on $[TcN]^{2+}$ in comparison with the isoelectronic $[TcO]^{3+}$ results in little tendency to deprotonation of coordinated amine and aqua ligands. Most $[TcN]^{2+}$ complexes are either five-coordinate square pyramidal or six-coordinate distorted octahedral. Trigonal bipyramidal complexes are also known. In the case of square pyramidal and octahedral complexes, the strong axial ligand field induced by the nitrido ligand results in a low energy, essentially nonbonding, d_{xy} orbital (b_2 in C_{4v} symmetry), which accommodates the d^2 electrons and results in diamagnetic complexes with a $^1A_1$ ground state. The two π components of the Tc≡N bond result from overlap of the p_x, p_y (e) orbitals on N with the d_{xz}, d_{yz} (e) orbitals on Tc. In general, $[TcN]^{2+}$ complexes are not readily reduced and require agents such as chlorine for oxidation [13, 14]. A number of Tc^VN complexes have been found to be only poor catalysts in the radical induced ring opening of epoxides, with cis-epoxides reacting faster than the trans isomers [15].

$[Tc^VN]^{2+}$ complexes may be prepared by the reaction of TcO_4^-/ligand with $NH_2NH_2 \cdot HCl$ as the reducing agent and source of the nitrido ligand [2, 3]. Two general synthetic methods are by ligand exchange of $[Tc^V NCl_2(PPh_3)_2]$ [3], and by reduction/exchange of $[Tc^{VI}NX_4]^-$ (X = Cl, Br) [4, 16]. The Tc≡N bond is readily detected in the IR spectrum by the presence of a generally sharp $v(TcN)$ absorption at 1100–1000 cm^{-1} [4], which is shifted by ~30 cm^{-1} on ^{15}N labelling [17] [for $v(Tc^{14}N)$ at 1050 cm^{-1} the diatomic approximation gives $\Delta^{15}N = -31$ cm^{-1}].

2.1 Halo, Cyano and Thiocyanato Complexes

Spectroelectrochemical studies at −60°C show a reversible one-electron reduction of the intensely coloured $[Tc^{VI}NX_4]^-$ (X = Cl, Br) to $[Tc^V NX_4]^{2-}$, but the colourless reduced species have not been isolated [18].

The cyano complex $(AsPh_4)_2 trans\text{-}[TcN(OH_2)(CN)_4] \cdot 5H_2O$ **(1)** is prepared by the addition of KCN to a solution of $AsPh_4[Tc^{VI}NCl_4]$ in MeCN and may be recrystallized from hot water [19]. The original report of the formation of $[TcNCl_{4-n}(CN)_n]^-$ ($n = 0$–2) on addition of conc. HCl to the $Cs_2[TcNCl_5]/CN^-$ reaction mixture has been shown not to be due to oxidation of $[TcN(OH_2)(CN)_4]^{2-}$ but to the cleavage of $[Tc_2^{VI}N_2O_2(CN)_4]^{2-}$ present in the mixture [20]. The anion in **1** has distorted octahedral geometry with a Tc≡N bond length of 1.60(1) Å, a very long NTc–OH$_2$ length of 2.559(9) Å, a near-linear N≡Tc–OH$_2$ angle of 177.9(5)° and the Tc atom 0.35 Å above the equatorial plane [19]. These values are similar to those of the isostructural $(AsPh_4)_2 trans\text{-}$ $[ReN(OH_2)(CN)_4] \cdot 5H_2O$ [21] (Table 1). The pK$_{a1}$ value of the coordinated water in **1** is not known, but the long bond length indicates very low acidity and high kinetic lability. In the case of $[Re^V O(OH_2)(CN)_4]^-$ and $[Re^V N(OH_2)(CN)_4]^{2-}$ [ORe–OH$_2$ 2.142(7), NRe–OH$_2$ 2.496(7) Å] the strong *trans* effect of the nitrido ligand is apparent in the pK$_{a1}$ values of 1.4 and 11.7 for the oxo and nitrido complexes, respectively, and a reaction rate constant for the substitution of the *trans* water by NCS$^-$ some 9×10^5 times greater for the nitrido complex [22]. The high value of $v(TcN)$ (1100 cm^{-1}) in **1** is due to the ability of CN$^-$ to act as an effective π-acceptor as well as a good σ-donor.

1   **2**

The structurally characterized $Cs_2Na trans\text{-}[TcN(CN)_4(N_3)]$ [$v(TcN)$ 1066 cm^{-1}] is obtained by ligand exchange of $Cs_2Na[TcN(CN)_4Cl]$ [$v(TcN)$ 1037 cm^{-1}] with a large excess of NaN$_3$, and $Cs_2K[TcN(CN)_5]$ [$v(TcN)$ 1067 cm^{-1}] may be isolated from the $[TcN(tu)_4Cl]Cl/KCN/Cs^+$ reaction [23]. The IR spectrum of $Cs_2K[TcN(CN)_5]$ shows three well-defined $v(CN)$ absorptions at 2150, 2128 and 2117 cm^{-1} ($A_1^{axial} + A_1^{radial} + E$) consistent with C_{4v} symmetry. Application of the Cotton-Kraihanzel (CK) force field method [24] yields stretching parameters of 17.32 and 17.21 mdyn Å^{-1} for the *trans* and the equatorial cyano groups, respectively, and a CN CN interaction parameter of 0.08 mdyn Å^{-1} [23]. The CK C≡N stretching parameters may be compared with the C≡N stretching force constant of 14.57 mdyn Å^{-1} for $K_5[Tc^I(CN)_6]$ [25], where the low oxidation state of Tc results in greater π-acceptance by cyanide.

The preparation of $R_2[TcN(NCS)_4]$ (R = AsPh$_4$, NBu$_4$) has been reported [26, 27], but a crystal structure determination has shown that the product crystallized from MeCN/EtOH is the *trans*-aqua complex $(AsPh_4)_2[TcN(OH_2)(NCS)_4]\cdot$EtOH [28]. Similarly, $(NEt_4)_2[TcN(NCS)_4(MeCN)]$ is prepared by reaction of $[TcNCl_4]^-$ with NCS$^-$ and crystallization from MeCN [16]. In the $[TcN(NCS)_4(MeCN)]^{2-}$ anion in the crystal, the Tc atom is equally disordered over two sites with the N atom of the MeCN ligand sharing equally the *trans*-axial coordination sites with the nitrido N atom [29]. Reaction of NCS$^-$ with $[TcNCl_2(PPh_3)_2]$ yields yellow $[TcN(NCS)_2(PPh_3)_2]$, which on reflux in MeCN is converted to orange-red crystals of [2]·1/2MeCN [30]. The Tc≡N bond length is 1.629(4) Å and the weak binding of the MeCN ligand is apparent in the long Tc–NCMe bond length of 2.491(4) Å and the formation of the five-coordinate complex on dissolution in CHCl$_3$. The MeCN ligand is bonded in a bent fashion with the Tc–C–N angle 168.6(4)°. The solid state IR spectrum of **2** shows only two very weak peaks at 2295 and 2265 cm^{-1} {ν(CN) and a combination mode [31]} attributable to the coordinated and lattice MeCN. In all three cases the thiocyanato ligands are N-bonded, as has been found for all other structurally characterized Tc complexes [4].

2.2 TcN{O$_4$}, TcN{S$_4$}, TcN{Se$_4$} and Related Complexes

A series of $[TcN(\beta\text{-diketonate})_2]$ complexes has been prepared in good yield by ligand exchange of $[TcNCl_2(PPh_3)_2]$ with acetylacetone (and the benzoyl, dibenzoyl and dipivaloyl analogues) in the presence of KHCO$_3$ [32]. The diketones are poor reducing agents and no reaction was observed with $[TcNCl_4]^-$. The complexes show a strong band at 400–200 nm in the UV-vis spectrum and ν(TcN) at 1045–1043 cm^{-1} (IR).

A considerable number of $[Tc^VN(L\text{-}L)_2]^{0/2-}$ complexes, where L-L is a bidentate 1, 1- or 1, 2-dithiolate or selenolate ligand, have been reported. Synthesis is by the reaction of $TcO_4^-/N_2H_5^+/L\text{-}L$ [2, 3], ligand exchange of $[TcNCl_2(PPh_3)_2]$ [3] or reduction/substitution of $[TcNX_4]^-$ (X = Cl, Br) [4, 16]. Examples include $[TcN(S_2CNR_2)_2]$ (**3**) [R = Et, Pr^i, Bu, Bu^i; R$_2$ = (CH$_2$)$_5$, (CH$_2$)$_2$-O-(CH$_2$)$_2$] [2, 3, 33], $[TcN(SeYCNEt_2)_2]$ (Y = S, Se), $[TcN(S_2PR_2)_2]$ (R = Ph, OPr^i] [34], $R_2[TcN(mnt)_2]$ (R = NEt$_4$, AsPh$_4$) [34, 35], $(AsPh_4)_2[TcN(SCOCOS)_2]$ [36, 37] and $(NBu_4)_2[TcN\{Se_2C=C(CN)_2\}_2]$ [38]. Of particular interest is $[TcN\{S_2CNEt(OEt)\}_2]$, as the ^{99m}Tc complex is a promising myocardial imaging agent [39] (see Sect. 2.7). Reaction of $[TcNCl_2(PPh_3)_2]$ with K(S$_2$COEt) and treatment

Table 1. Structural and v(TcN) IR data for square pyramidal and octahedral $Tc^V NL_4$ and $Tc^V NL_4 L'_{trans}$ complexes and structural and v(TcO) data for selected $Tc^V OL_4$ complexes with the same coordination environment[a]

Complex	$Tc{\equiv}N/{=}O$ Å	Tc–L Å	Tc–L_{trans} Å	$NTcL°/OTcL°$	δ-sbp[b] Å	v(TcN)/v(TcO) (cm^{-1})	Ref.
Cyano complexes							
(AsPh₄)₂[TcN(OH₂)(CN)₄]·5H₂O (**1**)	1.596(10)	2.11 av.	2.559(9) (OH₂)	99.4 av.	0.35	1100	19
(AsPh₄)₂[ReN(OH₂)(CN)₄]·5H₂O	1.639(8)	2.11 av.	2.496(7) (OH₂)	99.5 av.	0.35	1060	21
TcN/O{S₄}complexes							
[TcN(S₂CNEt₂)₂] (**3**)	1.604(6)	2.401 av.		108.1 av.	0.745(1)	1070	2
K₂[TcN(S₂CO)₂]·2H₂O (**4**)	1.621(6)	2.390 av.		107.3 av.	0.71	1060	40
AsPh₄[TcN(S₂CNEt₂)(SCOCOS)]	1.54(2)	2.393 av.		106.0 av.	0.66	1071	20
(AsPh₄)₂[TcN(SCOCOS)₂] (triclinic form)	1.613(4)	2.378(2)-2.391(2)		105.4(2)-106.1(3)	0.647	1071	36
(AsPh₄)₂[TcN(SCOCOS)₂] (monoclinic form)	1.606(7)	2.390(1)-2.398(2)		104.58(5)-104.94(5)	0.611	1045	37
AsPh₄[TcO(SCOCOS)₂]	1.646(4)	2.327(1)-2.330(1)		108.6(2)-109.9(2)	0.759	–	36
(AsPh₄)₂[TcN(mnt)₂]	1.59(1)	2.367(4)-2.419(4)		101.8(8)-106.8(8)	0.59	1060	35
AsPh₄[TcO(mnt)₂]	1.655(6)	2.310(2)-2.320(2)		107.4(2)-109.8(2)	0.742(3)	950	46
(NBu₄)₂[TcN{Se₂CC(CN)₂}₂]	1.61(1)	2.508(2)-2.528(2)		106.1(4)-109.5(4)	0.768(1)	1072	38

NEt4[TcO{Se2CC(CN)2}2]	1.67(2)	2.463(4)-2.476(4)		108.1(6)-112.4(6)	0.88	965	47
[TcN(14S4)Cl](TcNCl4)^c (5)	1.615(3)	2.405(1)-2.414(1)	2.718(1) (Cl)	95.0(1)-96.7(1)	0.236		49
[TcN(18S6)Cl](TcNCl4)^c (6)	1.707(3)	2.461(1)-2.499(1)	2.567(1) (Cl)	90.6(1)-100.9(1)	0.209		49
[TcN{14S4-(OH)2}Cl]Cl	1.95(1)	2.430(4)-2.447(4)	2.469(4) (Cl)	91.5(3)-94.1(3)	0.114		49
TcN{N4} *and other complexes with nitrogen ligands*							
[TcN(en)2Cl]BPh4	1.603(3)	2.148(2)-2.166(2)	2.7320(8) (Cl)	97.3(1)-99.4(1)	0.3231(3)	1085	51
[TcN(tad)Cl]BPh4	1.626(6)	2.150(6)-2.166(7)	2.663(2) (Cl)	94.2(3)-99.1(3)	0.2163(6)	1085	51
[TcN(en)2(OOCNHCH2CH2NH3)](BPh4)2 (7)	1.607(9)	–	2.330(6) (O)	–	0.3268(5)	–	53
[TcN(OH2)(L^1)]·2H2O^d (8, X = O, R absent)	1.617 av.	2.051(N) av. 2.127(NH) av.	2.688(4) 2.947(4) (OH2)		0.52	1080	52
[TcN(OH2)(L^2)]Cl·2H2O (8, X = H2, R = H)	1.614(2)	2.046(1)(N) 2.124 (NH) av.	2.560(2) (OH2)	98.97(7)-105.45(7)	0.3981(4)	1090	55

[a] The square basal or equatorial coordinated atoms are the same. For six-coordinate complexes the *trans* ligand is denoted as L'_{trans}

[b] Displacement of Tc above the square basal plane (*i.e.* towards the nitrido or oxo ligand). For six-coordinate complexes, the displacement of Tc above the equatorial plane

[c] Data for the cation

[d] Two independent molecules in the asymmetric unit

with aqueous ethanol result in hydrolysis of the intermediate xanthate to yield the dithiocarbonato complex $K_2[4] \cdot 2H_2O$ [40].

3   **4**

The mixed-ligand complex $AsPh_4[TcN(S_2CNEt_2)(SCOCOS)]$ is prepared in a controlled manner by substitution/reduction of $[Tc^{VI}NCl_2(S_2CNEt_2)]$ with K_2 (SCOCOS) [20]. The $v(TcN)$ IR absorption of the above complexes occurs at $1083–1045$ cm^{-1}. The ^1H NMR spectra for dithiocarbamato (and other) complexes show narrow lines consistent with diamagnetism [33, 34]. Electron impact mass spectra and mass analyzed ion kinetic energy spectra have been reported for $[TcN(SYCNEt_2)_2]$ $(Y = S, Se)$ and $[TcN\{S_2P(OPr^i)_2\}_2]$ [41]. In the mass spectrometer $[TcN(SSeCNEt_2)_2]$ undergoes thermal scrambling to give the S_4, S_3Se and SSe_3 species [42]. The negative mode FAB spectrum of $(NBu_4)_2[TcN(L-L)_2]$ $\{L-L = Se_2CC(CN)_2\}$ shows the presence of the $\{[TcN(L-L)_2](NBu_4)\}^-$ ion pair, $[TcN(L-L)_2]^-$, and of fragmentation products [38].

Treatment of $[TcN(S_2CNEt_2)_2]$, $[TcN(SCOCOS)_2]^{2-}$ and related complexes with Cl_2 or Br_2 results in oxidation to $[Tc^{VI}NX_4]^-$ [13, 14, 37]. The reactions may be followed by EPR spectroscopy (see Sect. 3.1.2), and in some cases mixed ligand $Tc^{VI}N$ species have been identified [14, 37]. Ligand exchange of $[^{99/99m}TcN(SCOCOS)_2]^{2-}$ with mnt^{2-} has been shown by HPLC to occur with the build-up of an intermediate, presumably $[TcN(SCOCOS)(mnt)]^{2-}$ [43]. Reaction of $[TcN(S_2CNEt_2)_2]$ with S_2Cl_2 (or $SOCl_2$) and $SOBr_2$ yields the pentagonal bipyramidal $[Tc(NS)X_2(S_2CNEt_2)_2]$ $(X = Cl, Br, respectively)$, in which the thionitrosyl ligand occupies one axial position and one X atom the other [44, 45].

Two or more anionic S or Se ligands provide sufficient negative charge to the $[TcN]^{2+}$ core, and all complexes have either been shown or are thought to be square pyramidal with the nitrido ligand in the apical position and the *trans* position vacant. Structural data are given in Table 1. A structural comparison of the square pyramidal $[Tc^VN(L-L)_2]^{2-}/[Tc^VO(L-L)_2]^-$ pairs with the same coordination environment $[L-L = SCOCOS, mnt, Se_2CC(CN)_2]$, and data for six-coordinate complexes in Tables 1 and 2 (see Sect. 3.1.1), shows that, although the nitrido ligand exerts the greater *trans* influence in terms of the bond length to the *trans* ligand, the oxo ligand exerts the greater steric effect [35]. For the five-coordinate complexes the $O=Tc–L$ angles $(107.4–112.4°)$ are greater than the $N\equiv Tc–L$ angles $(105.4–109.5°)$ of the basal ligands and consequently the displacement of the Tc atom above the square basal plane is greater for the oxo complexes. The Tc–L bond lengths are greater for the nitrido complexes, but this may be due to the lower core charge on $[Tc^VN]^{2+}$ compared to $[Tc^VO]^{3+}$. In the anion of $(AsPh_4)_2[TcN(mnt)_2]$ the $Tc\equiv N$ core is disordered about a pseudo-inversion centre, with unequal occupancies of the disordered sites [35].

The water-soluble thiourea complex [TcN(tu)₄Cl]Cl is formed in high yield by the reaction of [TcNCl₄]⁻/tu/MeCN/H₂O. The aqueous solution is acidic due to hydrolysis, but substitution by ligands such as ⁻S₂CNEt₂ occurs in high yield. The IR spectrum indicates that the thiourea ligands are *S*-bonded [48]. A novel series of six-coordinate cationic thiacrown ether complexes has been prepared by reduction/ligand exchange with NBu₄[TcNCl₄] [49]. Crystal structures have been determined for [Tc^V NCl(14S4)][Tc^VI NCl₄] (cation **5**), [TcNCl{16S4-(OH)₂}]Cl and [TcNCl(18S6)][TcNCl₄] (cation **6**) (Table 1).

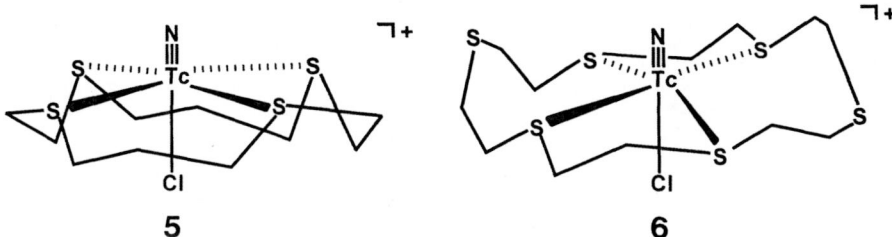

| **5** | **6** |

Interestingly, for **5** and **6**, [Tc^VI NCl₄]⁻ serves as the counterion. Technetium is coordinated by four thioether S atoms in the equatorial plane, in the 18S6 complex **6** the two 1, 10 S atoms are not coordinated. The strong *trans* influence of the nitrido ligand is apparent in the very long NTc–Cl$_{trans}$ bond lengths of 2.718(1), 2.567(1) and 2.469(4) Å for **5, 6** and the 16S4-(OH)₂ complex, respectively. The Tc≡N bond lengths refine as 1.615(3), 1.707(3) and 1.95(1) Å for the 14S4 (**5**), 18S6 (**6**), and the 16S4-(OH)₂ coordinated cations, respectively. The apparent long bond length of 1.95 Å (and to a lesser extent that of 1.707 Å) is most likely due to a crystallographic artifact arising from partial disorder of the nitrido and chloro ligands [49, 50].

2.3 Complexes with Nitrogen Ligands

A variety of Tc^V N{N₄} complexes have been prepared by ligand exchange of [TcNCl₂(PPh₃)₂] or from [TcNCl₄]⁻ (usually in the presence of an auxiliary reducing agent such as KBH₄ or PPh₃) [51, 52]. Structural data are given in Table 1. The cationic ethylenediamine complex *trans*-[TcN(en)₂Cl]BPh₄ is typical with ν(TcN) at 1085 cm⁻¹, distorted octahedral geometry, a Tc≡N bond length of 1.603(3) Å and the very long NTc–Cl$_{trans}$ bond length of 2.7320(8) Å [51]. In the linear tetraamine complex *trans*-[TcN(tad)Cl]BPh₄ (tad=1, 5, 8, 12-tetraazadodecane) the Tc–Cl distance is 2.663(2) Å. Magnetic susceptibility measurements have shown these, and the 1,3-propanediamine complex, to be diamagnetic [51]. Crystallography has shown that [TcNCl₂(PPh₃)₂] reacts with excess diethylenetriamine in benzene/ethanol under aerobic conditions to give the novel dicationic [**7**](BPh₄)₂ [53]. The mechanism of formation of the zwitterionic NH₃⁺CH₂CH₂NHCOO⁻ ligand and the cleavage of the triamine to ethylenediamine is not clear. The crystal structure shows that the zwitterionic ligand lies in a peculiar "transient state" and is stabilized by strong intramolecular hydrogen bonding [53]. Notably, [ReNCl$_2$(PPh$_3$)$_2$] results in more extensive C–N bond cleavage of the triamine to give a quantitative yield of β-alanine (H$_2$NCH$_2$CH$_2$CO$_2$H) [54].

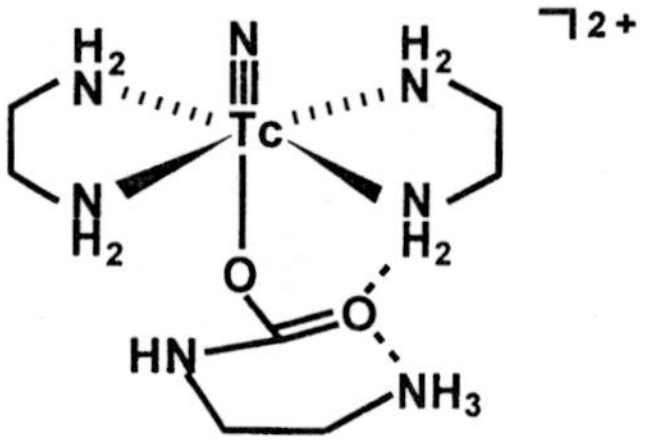

7 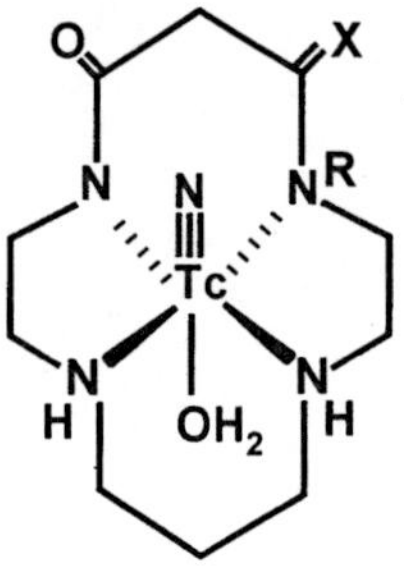   **8**

A complex with a cyclic polyamine is [TcN(cyclam)Cl]Cl (cyclam=1,4,8, 11-tetraazacyclotetradecane) [52]. In this and the above complexes, the coordinated amine groups do not undergo deprotonation. The greater acidity of amides, however, results in deprotonation. For **8** (X = O, R absent) neutrality is achieved by deprotonation of both amide groups, with the result that the Tc–N$_{amido}$ bond lengths of 2.051 Å are significantly shorter than the Tc–NH lengths of 2.126 Å[52]. The monoamido ligand results in .the cationic [**8**]Cl (X = H$_2$, R = H) for which the *trans*-NTc–OH$_2$ bond is very long at 2.560(2) Å [55].

Crystallography has shown that the product of the reaction of [TcNBr$_4$]$^-$ with 2,2'-bipyridine in ethanol is {*cis*-[Tc^{V}NBr(bpy)$_2$]}$_2$[Tc^{II}Br$_4$]. This remarkable and unprecedented reaction involves loss of a nitrido ligand and reduction of Tc^{VI} to Tc^{II} under mild conditions to yield the previously unknown [TcBr$_4$]$^{2-}$. With methanol as the solvent the product is [**9**]BPh$_4$ [56, 57].

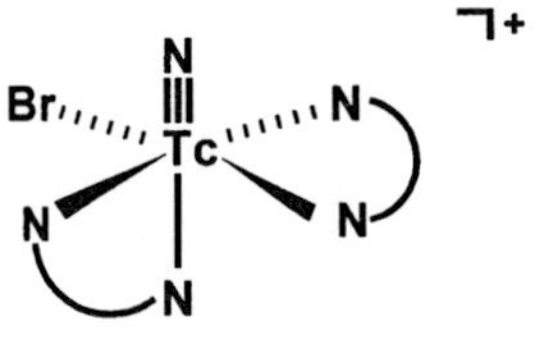

9 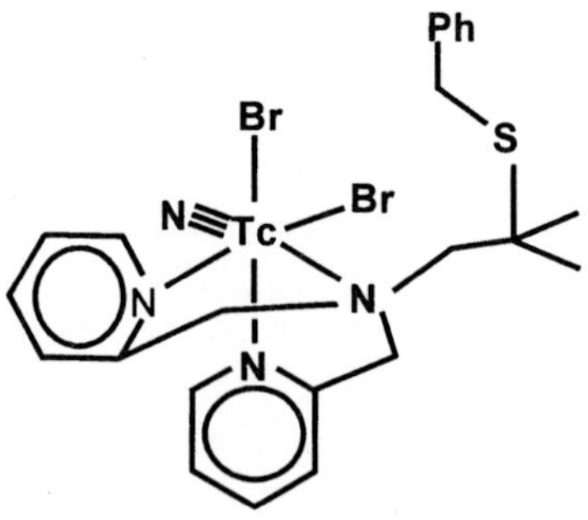   **10**

Other structurally characterized complexes are *cis*-[TcNCl(phen)$_2$]PF$_6$, in which the complex cation exhibits a pseudo two-fold symmetry axis that gives rise to reproducible enantiomeric disorder, and cis-[TcNCl(phen)$_2$]Cl·H$_2$O. Notably, reduction of [TcNCl$_4$]$^-$ occurs simply on reaction with 1,10-phenanthroline in MeOH or water at room temperature [58]. The structure of cis-[TcN(CF$_3$CO$_2$)(phen)$_2$](CF$_3$CO$_2$)·H$_2$O has been briefly mentioned [58] and the preparations of cis-[TcNX(phen)$_2$]X (X = Cl, Br) [57] and [TcNL$_4$]Cl$_2$ (L = py, imidazole) [27] have been reported. In MeOH/pyridine, AsPh$_4$[TcNCl$_4$] is reduced to give the hydroxo complex [TcN(OH)(py)$_4$]BPh$_4$, with the hydroxide coordinated trans to the nitrido ligand. The pyridine ligands in [TcN(OH)(py)$_4$]$^+$ undergo exchange with free pyridine in solution [59], as has also been found to be the case for trans-[TcO$_2$(py)$_4$]$^+$ [60]. Interestingly, the [N≡Tc–OH]$^+$ core shows similar ^{99}Tc NMR shifts to [O=Tc=O]$^+$ [59]. Mass spectrometry confirms that an NH$_4$TcO$_4$/phthalodinitrile melt yields the dark [TcN{phthalocyanine(2-)}]. Use of ^{15}NH$_4$TcO$_4$ results in the shift of v(TcN) from 1078 cm^{-1} in the unlabelled complex to 1049 cm^{-1}, and identifies NH$_4^+$ as the source of the nitrido ligand [61].

Complex 10 is prepared by substitution of [TcNBr$_2$(PPh$_3$)$_2$]. The Tc–N bond length of the tertiary amine N atom coordinated trans to the nitrido ligand is 2.47(1) Å and of the pyridine N atoms coordinated cis is 2.141 av. Å. The thioether sulfur atom is not coordinated in the solid, but the ^{1}H NMR spectrum shows that in solution there is an equilibrium between the dibromo form and one in which Br$^-$ is expelled and the thioether sulfur is coordinated [62].

2.4 Complexes with Phosphine Ligands

The important sparingly-soluble synthetic intermediate [TcNCl$_2$(PPh$_3$)$_2$] (11, E = P) may be prepared by a variety of routes including TcO$_4^-$/N$_2$H$_4$·HCl/PPh$_3$ [3], reduction/substitution of [TcNCl$_4$]$^-$ [16] or substitution of [TcN(tu)$_4$Cl]Cl [48]. Substituted hydrazines may also act as sources of N^{3-}. The formation of [TcNCl$_2$(PPh$_3$)$_2$] in high yield by the reaction of TcO$_4^-$, PPh$_3$ and PhCONHNH$_2$ followed by addition of HCl may occur via loss of HCl from the intermediate imido complex [TcCl$_3$(NH)(PPh$_3$)$_2$] [63]. The reaction of TcO$_4^-$ with H$_2$NNRC(=S)SMe (R = H, Me) in the presence of HCl and PPh$_3$ gives [TcNCl$_2$(PPh$_3$)$_2$] in high yield [64]. Also, reaction of PPh$_3$ or AsPh$_3$ with [TcNX$_4$]$^-$ gives the five-coordinate [TcNX$_2$(EPh$_3$)$_2$] (X = Cl, Br; E = P, As) in high yield. The sterically less demanding PMe$_2$Ph gives the six-coordinate [TcNX$_2$(PMe$_2$Ph)$_3$], with the cis-mer structure established by the NMR spectra [65]. The oxidation/substitution of [Tc^{III}Cl$_3$(PMe$_2$Ph)$_3$] by Na^{15}N$_3$ has been used to prepare cis-mer-[Tc^{15}NCl$_2$(PMe$_2$Ph)$_3$] (12) [17]. The presence of the trans halide ligand in the six-coordinate complexes results in v(TcN) at 1048 (X = Cl), 1028 cm^{-1} (X = Br) compared to 1095–1090 cm^{-1} for the five-coordinate complexes [65].

All these complexes readily undergo ligand exchange reactions [65], including substitution of [TcNCl$_2$(PPh$_3$)$_2$] by more basic phosphines [3]. Tri(cyanoethyl) phosphine (L) yields the anionic NBu$_4$[TcNX$_3$L] (X = Cl, Br) on reaction with

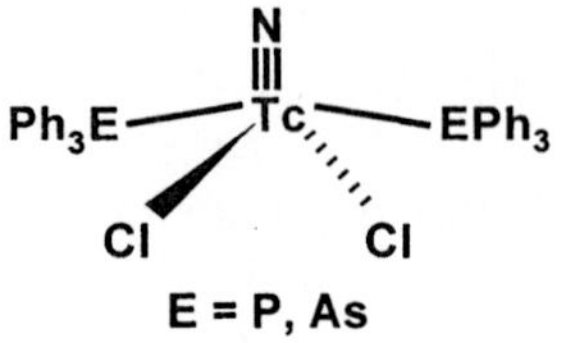

11

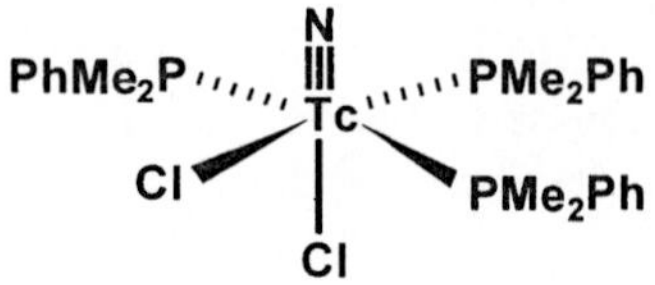

12

NBu$_4$[TcNX$_4$] [66]. The Tc≡N bond lengths in [TcNCl$_2$(EPh$_3$)$_2$] (**11**) (E = As [67], P [68]) are 1.601(5) and 1.602(8) Å, respectively, and the geometry may be regarded as intermediate between square pyramidal and trigonal bipyramidal with E–Tc–E angles of 162.0(1) (P) and 161.56° (As). The N≡Tc–As and N≡Tc–Cl angles for the arsine complex are 99.22(1) and 110.51(3)°, respectively [67]. The six-coordinate **12** has NTc–Cl bond lengths of 2.441(1) (*cis*) and 2.665(1) Å (*trans*) and a Tc≡N bond length of 1.624(4) Å [69]. Monodentate phosphine complexes readily react with S$_2$Cl$_2$ to form thionitrosyl complexes [70]. Crystal structures have been determined for *cis-mer*-[Tc^I(NS)Cl$_2$(PMe$_2$Ph)$_3$] and for [Tc^{II}(NS)Cl$_3$(PMe$_2$Ph)(OPMe$_2$Ph)] [71, 72]. A 66% yield of AsPh$_4$[TcNCl$_4$] has been isolated from a solution of [TcNCl$_2$(AsPh$_3$)$_2$] in SOCl$_2$, but the phosphine complex is largely reduced to [TcCl$_6$]$^{2-}$ [67]. Complexes with one monodentate phosphine ligand in a mixed coordination sphere are described in Sect. 2.5.

Reaction of NBu$_4$[TcNCl$_4$] with the bidentate phosphines (P-P) dppe, dmpe and Me$_2$P(CH$_2$)$_2$NMe(CH$_2$)$_3$NMe(CH$_2$)$_2$PMe$_2$ gives the cationic [TcNCl(P-P)$_2$]A (A = BPh$_4$, PF$_6$) in good yield [57, 73]. For the distorted octahedral [**13**]BPh$_4$ a Tc≡N bond length of 1.853(6) Å has been reported, but problems in the refinement noted [57]. This abnormally long length is again likely to be an artifact due to disorder between the nitrido and *trans*-chloro ligands. Cation **13** undergoes reversible electrochemical reduction with unexpected ease at −0.02 V vs SCE [73].

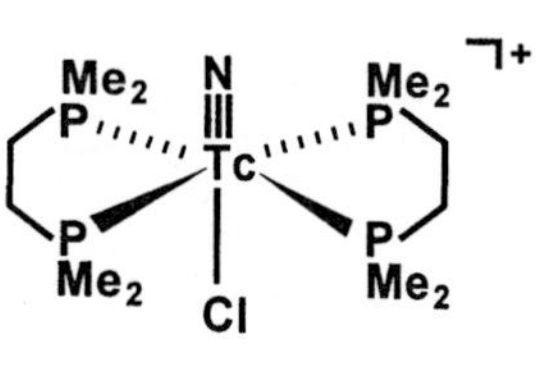

13

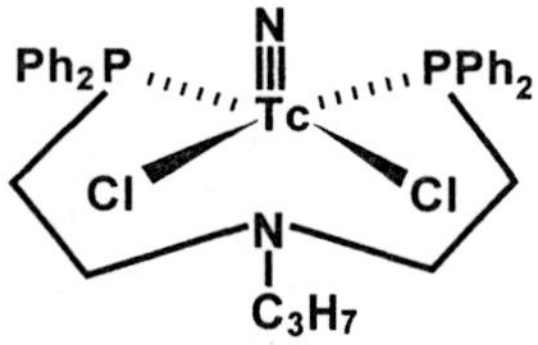

14

FABMS has shown that the bulky Pr$_2^i$PCH$_2$CH$_2$PPr$_2^i$ forms the dimeric [{TcNCl$_2$(P-P)}$_2$]. The ^{31}P NMR spectrum indicates a chlorine bridged structure [57]. The reaction of MePhNNH$_2$/dppe/[TcOCl$_4$]$^-$ in MeOH, however, yields a cationic complex formulated as the oxo-imido *trans*-[TcO(NH)(dppe)$_2$]$^+$. Few details are available, but the crystal structure determination showed marked asymmetry in the bonding of the two axial ligands [74]. A distinction between the [HN=Tc=O]$^+$ core and the tautomeric [N≡Tc–OH]$^+$ core should be possible

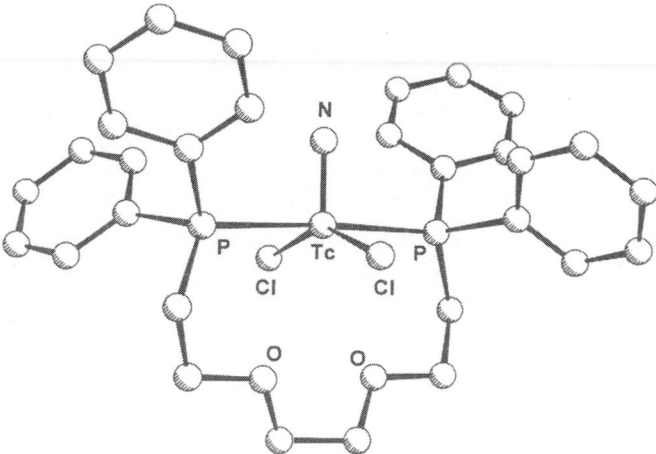

Fig. 1. The structure of [TcNCl₂(Ph₂PCH₂CH₂OCH₂CH₂OCH₂CH₂PPh₂)] (15) [76]

from the IR spectrum. It may be noted that in the distorted octahedral d^0 imido-oxo complex cis-[Mo^{VI}O(NH)Cl₂(OPEtPh₂)₂] the imido and oxo ligands are mutually cis [75] and the formal bond order is 2.5 [6].

Trigonal bipyramidal geometry is observed for **15** (Fig. 1), with the nitrido and chloro ligands in the equatorial plane. The Tc≡N bond length is 1.601(4) Å, Tc–Cl 2.385(1) Å, and the P–Tc–P angle is near linear at 176.46(4)°. The P–Tc–Cl [88.11(3), 90.99(4)°] and P–Tc–N [91.77(8)°] angles are close to 90°. If the ether oxygens are included in the coordination sphere then the geometry may be regarded as distorted pentagonal bipyramidal [76]. However, the long Tc···O contact distances of 3.190(2) Å indicate only a weak interaction and that the oxygen atoms are best regarded as uncoordinated. A related [TcNCl₂L] complex **14** with an amine-bridged diphosphine has square pyramidal geometry with a Tc≡N bond length of 1.60(1) Å and Tc displaced 0.410(2) Å above the P₂Cl₂ plane. The lone pair on the amine nitrogen is directed into the "*trans*" position and the Tc···N_amine distance of 2.70(1) Å is comparable to some *trans*-NTc–OH₂ bond lengths. This indicates incipient coordination and the expansion of the coordination sphere to octahedral [76]. The neutral [TcNCl₂L] [L=MeC(CH₂PPh₂)₃] [76] and the cationic [TcNBrL]A [L = E(CH₂CH₂PPh₂)₃: E = N, A = BPh₄; E = P, A = PF₆] [77] have been prepared by substitution of [TcNX₂(PPh₃)₂]. Electrochemical reduction of the cationic species does not occur until below −1.0 V vs SCE and is irreversible.

2.5 Complexes with TcN{O₂S₂}, TcN{N₂O₂}, TcN{N₂S₂} and Other Mixed-Coordination Cores

The versatility of the [TcN]²⁺ core is indicated by the preparation of a wide variety of mixed-ligand complexes. TcN{O₂S₂} coordination is represented by the thio-β-diketonato complex [TcN{PhC(S)CHC(O)Ph}$_2$], prepared from NBu$_4$ [TcNBr$_4$] [78], and TcN{N$_2$O$_2$} coordination by azomethine complexes such as [TcN{MeCOC(COOEt)CHNCH$_2$CH$_2$NCHC(COOEt)COMe}] [79]. The [TcNL] complex, where L represents a potentially pentadentate Schiff base, may have ONNNO coordination [57]. Structurally characterized TcN{N$_2$S$_2$} complexes are the square pyramidal [TcN(tox)$_2$] (tox = 8-quinolinethiolate), in which the coordination arrangement is NSNS and the Tc≡N bond length 1.623(4) Å [16], the Schiff base complex **16** [Tc≡N 1.621(8) Å] [80] and the dithiocarbazate Schiff base derivative **17** [Tc≡N 1.613(3) Å] [81]. These complexes, and a variety of other dithiocarbazate Schiff base [81] and diaminedithiolato complexes [82] have been prepared from [TcNCl$_4$]$^-$ or [TcNCl$_2$(PPh$_3$)$_2$].

16 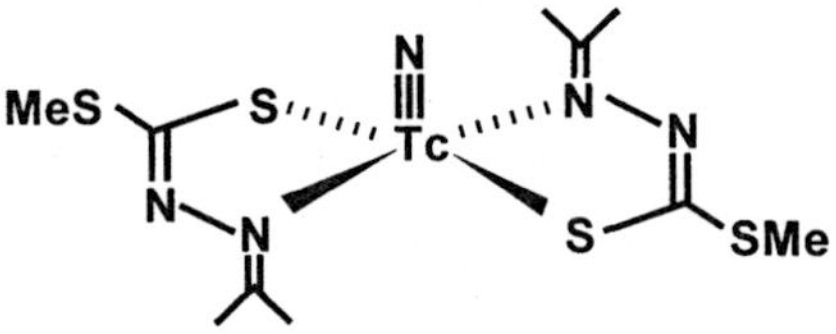   **17**

A novel complex is [TcN(tmbt)$_2$L$_2$] (**18**), where L represents the proton sponge 1,1,2,2-tetramethylguanidine, which is generally regarded as a "non coordinating" ligand. The Tc≡N bond length and v(TcN) are unexceptional at 1.615(6) Å and 1057 cm^{-1} [59]. The robust *trans*-[TcN(tmbt)$_2$(py)$_2$] is prepared in quantitative yield from 2,3,5,6-tetramethylbenzenethiol (tmbtH) and [TcN(OH)(py)$_4$]$^+$ [59].

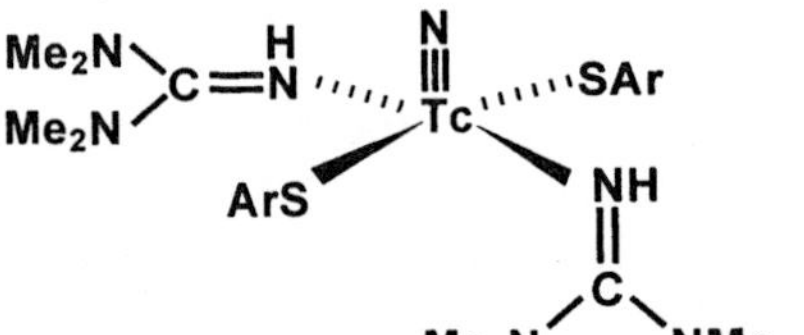

18 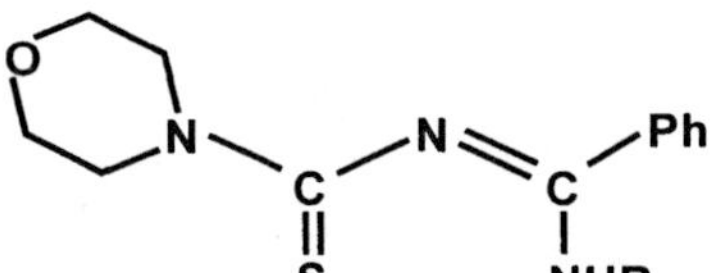   **19**

Neutral *N*-(thiocarbamoyl)benzamidinato complexes [TcNL$_2$] (HL = **19**) have been prepared and show v(TcN) at 1075–1060 cm^{-1} [83, 84]. TcN{P$_2$S$_2$} coordination is found in [TcNL$_2$] [HL = 2-(Ph$_2$P)C$_6$H$_4$SH]. Surprisingly, this complex is electrochemically inactive over the range ±1.5 V [85]. The [TcN(S$_2$CO)L] and [TcN(S$_2$CNEt$_2$)L]BPh$_4$ complexes, where L is MeC(CH$_2$PPh$_2$)$_3$ or the diphosphines in **14** and **15**, are prepared in good yield by the reaction of [TcNCl$_2$L] with K(S$_2$COEt) or K(S$_2$CNEt$_2$) [76].

A considerable number of complexes containing one PPh$_3$ ligand in a mixed coordination sphere are known [79, 86–89]. In **20** the Tc≡N bond length is 1.608(5) Å and the Tc–N amido bond length 2.067(4) Å. The Tc atom is 0.66 Å, and the P atom of the phosphine is 0.33 Å, above the $ON_{amido}O$ plane. Surprisingly, the reaction of **20** with PMe_2Ph results in chlorine abstraction from the $CHCl_3$ solvent to yield $[TcNCl_2(PMe_2Ph)_3]$ [87].

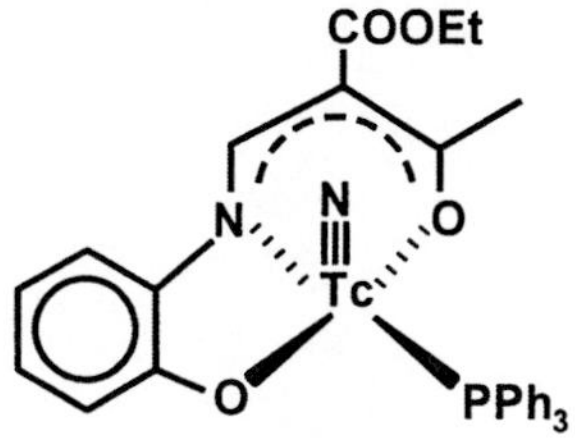

20

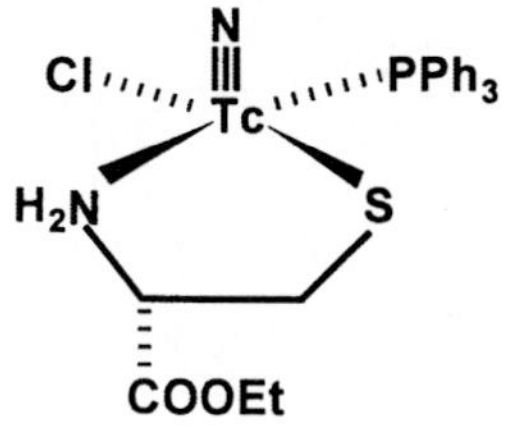

21

The amino acid complexes $[TcNCl(L)(PPh_3)]$ (HL=L-cysteine, L-cysteine ethyl ester, cysteamine) have been prepared from $[TcNCl_2(PPh_3)_2]$ or $AsPh_4$ $[TcNCl_4]/PPh_3$ [88]. The crystal structure of the L-cysteine ethyl ester complex **21** shows a Tc≡N bond length of 1.605(3) Å and the Tc atom displaced by 0.594(1) Å above the square basal plane [88]. Other structurally characterized examples are the square pyramidal **22** with ONSP coordination and Tc≡N 1.611(3) Å [81], and **23** with NSPCl coordination and Tc≡N 1.615(7) Å [89].

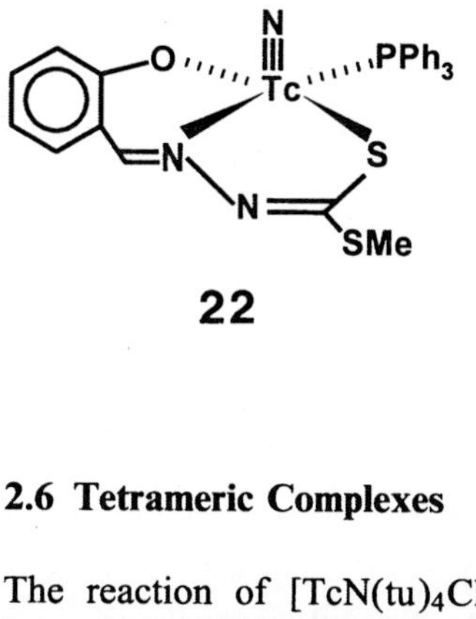

22

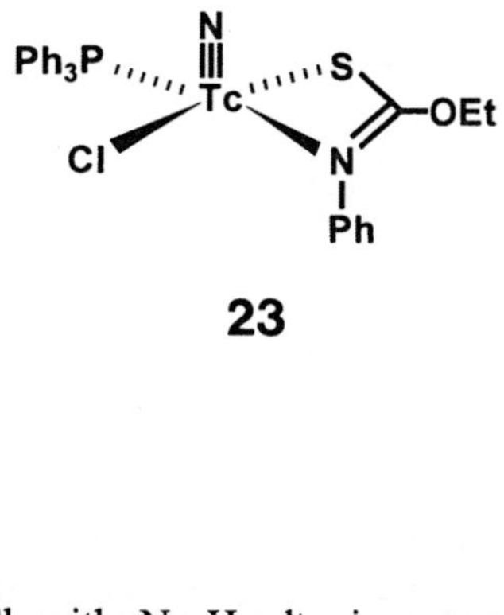

23

2.6 Tetrameric Complexes

The reaction of $[TcN(tu)_4Cl]Cl$ with Na_2H_2edta in aqueous solution gives a low yield of $[\{TcN(tu)\}_4(edta)_2] \cdot 6H_2O$, the first complex containing the cyclic tetrameric $[\{N≡Tc\}_4]^{8+}$ core (Fig. 2) [90]. The complex is insoluble in water and common organic solvents but was found to react slowly with $Na(S_2CNEt_2)$ in dimethylformamide to yield $[TcN(S_2CNEt_2)_2]$. A peak at 984 cm^{-1} in the IR spectrum was assigned to $v(TcN)$. The crystal structure shows that the N≡Tc–NTc bonding is asymmetric, with N≡Tc bond lengths in the range 1.681(7)–1.695(7) Å and the Tc–NTc bond lengths 1.977(7)–2.009(7) Å. The $[TcN]_4$ core is V-shaped, with N≡Tc–N approximately linear at 163.2(4)–178.6(4)°. The equatorial atoms about each Tc are closely planar, with the Tc atoms displaced by 0.247(3)–0.323(3) Å above this plane towards the nitrido

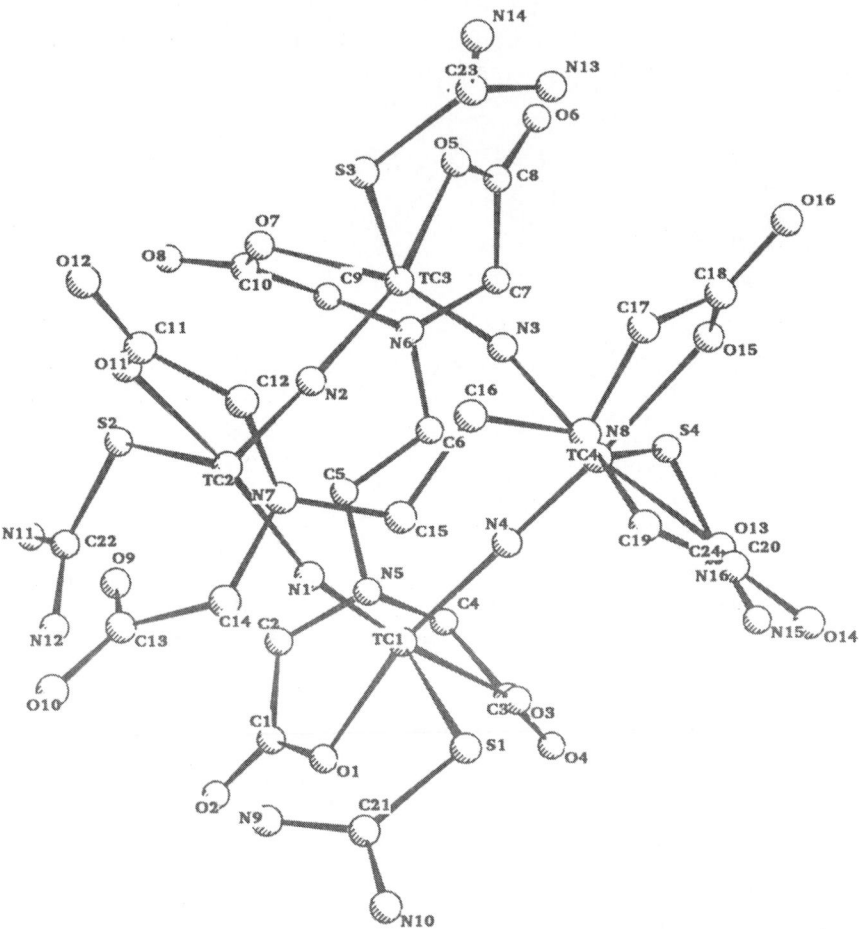

Fig. 2. The structure of [{TcN(thiourea)}$_4$(edta)$_2$] · 6H$_2$O [90]

ligand, and the thiourea ligands are *S*-bonded. A notable feature of the structure is that the hexadentate edta^{4-} ligands bridge diagonal pairs of Tc atoms, one edta above and the other below the [TcN]$_4$ core, to form an open cage-like structure with an extensive hydrogen-bonded network involving the lattice water molecules. The stability of the tetramer may be due to the "clamping effect" of the edta ligands and the need to break four TcN$_{nitrido}$ bonds for dissociation to occur.

2.7 Technetium-99m Nitrido Complexes

In view of the importance to diagnostic nuclear medicine of ^{99m}Tc radiopharmaceuticals containing the [^{99m}Tc^{V}O]$^{3+}$ and [^{99m}TcO$_2$]$^+$ cores [11, 12, 91], there is considerable interest in the preparation of radiopharmaceuticals based on the chemically stable $[^{99m}TcN]^{2+}$ core. It has been shown that $[^{99m}Tc^{VI}NCl_4]^-$ is formed in high yield from the $^{99m}TcO_4^-/NaN_3/HCl$ reaction at the "no carrier added" level ($[Tc]$ 10^{-7}–10^{-9} M) and that, after removal of HCl, the residue may be reacted with ligands to yield ^{99m}TcN radiopharmaceuticals [92–94]. For reducing ligands such as thiols [95] or phosphines [96] it has been established chromatographically that reduction to $^{99m}Tc^VN$ occurs and that the ^{99m}TcN complex is the same as the $^{99}Tc^VN$ complex prepared at the macroscopic level, but the oxidation state with ligands such as gluconate or diethylenetriaminepentaacetate is unclear [92, 93]. In some cases different TcN products are formed at the ^{99m}Tc and ^{99}Tc concentration levels [88, 92]. At pH > 4 and in the absence of added ligands, $[^{99m}Tc^{VI}NCl_4]^-$ is oxidized by air to $^{99m}TcO_4^-$ [97]. Exchange reactions of $^{99m}Tc^VN$ thiolato complexes have been studied by HPLC [43]. A more recent procedure for the preparation of $^{99m}Tc^VN$ radiopharmaceuticals is by the initial reaction of $^{99m}TcO_4^-$ with S-methyl N-methyldithiocarbazate [$H_2NNMeC(=S)SMe$], or derivatives, in the presence of PR_3/HCl as reductant followed by addition of the ligand [64, 98, 99]. A wide variety of ^{99m}TcN complexes with ligands including, dithiocarbamates [100–103], dithioetherdithiols [104], 1-thio-β-D-glucose [105], diaminodithiols [106], NS, N_2S_2 and N_2SO Schiff bases [107, 108], L-cysteine and the ethyl ester [88], nitrogen heterocyclic thiols [109], tropolone [110], β-diketonates [111], macrocyclic amines [112], ethylenediamine-N-N'- diacetic acid [113] and chelating phosphines [96, 114] have been prepared by either one or both of these methods or variations thereof. $[^{99m}TcNCl_4]^-$ has been shown to effectively label monoclonal antibodies with high retention of antibody specificity [115]. In general, the biological distribution of ^{99m}TcN complexes is different to that of the ^{99m}TcO complexes prepared from $^{99m}TcO_4^-/ligand$ and Sn^{2+} as the reducing agent [92, 93]. Mass spectrometry has shown that $[^{99/99m}TcN(S_2CNEt_2)_2]$ is recovered in unmetabolized form from rat liver [116]. The diphosphine cations $[^{99m}TcNX(P-P)_2]^+$ (P-P = dmpe, depe, dppe; X = Cl or Br) showed lack of myocardial retention due to facile in vivo reduction [96]. However, the neutral $[^{99m}TcN\{S_2CN(OEt)R\}_2]$ (R = Et, OEt) complexes, prepared by the dithiocarbazate method, are promising myocardial imaging agents [103].

3 Technetium(VI) Nitrido Complexes

The ability of the nitrido ligand to stabilize the Tc(VI) oxidation state is apparent in the stability of the $[TcN]^{3+}$ core to hydrolysis and disproportionation. This is in marked contrast to the $[Tc^{VI}O]^{4+}$ core, which is highly susceptible to oxidation and to disproportionation according to $3Tc^{VI} \rightarrow Tc^{IV} + 2Tc^{VII}$ and is not readily stabilized by coordination [4]. A characteristic feature is the formation of dimeric $[NTcOTcN]^{4+}$ and $[NTc(\mu\text{-}O)_2TcN]^{2+}$ complexes, which have no analogues for any other transition metal, but show many resemblances to the well-known isoelectronic $[OMo^VOMo^VO]^{4+}$ and $[OMo^V(\mu\text{-}O)_2Mo^VO]^{2+}$ complexes [117, 118]. Monomeric $Tc^{VI}N$ (d^1 configuration) is easily and reliably detected by EPR spectroscopy [7, 8, 119–121]. Dimeric $Tc^{VI}N$ species are EPR silent due to spin pairing [122, 123].

3.1 Monomeric $Tc^{VI}N$ Complexes

3.1.1 Preparation, Structure and Reactions

The key halide complexes, the orange-red $R[TcNCl_4]$ and the intensely blue $R[TcNBr_4]$ ($R = AsPh_4$, NBu_4) are readily prepared in high yield by the re-action of TcO_4^-/NaN_3 in refluxing HX followed by precipitation with the organic cations [119]. These reactions indicate the greater stability of $[TcNX_4]^-$ compared to $[Tc^VOX_4]^-$, which is readily further reduced to $[Tc^{IV}X_6]^{2-}$ by hot HX[124]. If the $TcO_4^-/NaN_3/HCl$ reaction mixture is evaporated to dryness, the residue extracted with MeCN and the extract then evaporated to dryness and dissolved in conc. HCl, the orange-red solution probably contains $H[TcN(OH_2)Cl_4]$. Addition of CsCl to this solution yields red crystals of the six-coordinate $Cs_2[TcNCl_5]$, which is the key starting material for the study of $Tc^{VI}N$ aqueous solution chemistry [16, 122, 123], while addition of NEt_4Cl gives orange crystals of $(NEt_4)trans\text{-}[TcN(OH_2)Cl_4]$ [125]. The aquabromo complex $NEt_4[24]$ may be prepared in high yield directly from the $TcO_4^-/NaN_3/HBr$ reaction mixture [125].

24   **25**

EPR studies of $Cs_2[TcNX_5]$ in HX ($X = Cl, Br$) solution and of $AsPh_4$-$[TcNX_4]$ in organic solvents in the presence of a large molar ratio of X^- show no evidence for the equilibria [126]

$$trans\text{-}[TcN(OH_2)X_4]^- + X^- \rightleftarrows [TcNX_5]^{2-} + H_2O$$

$$\text{or} \quad [TcNX_4]^- + X^- \rightleftarrows [TcNX_5]^{2-}$$

These results indicate that the form in HX solution is most likely $[TcN(OH_2)X_4]^-$, with the *trans* ligand readily lost or exchanged, and the form precipitated dependent on the cation size. A similar cation effect is found in the isolation of R[TcOX$_4$], Cs$_2$[TcOX$_5$] [127], NEt$_4$[TcO(OH)$_2$)Br$_4$] [128], and Group 6–8 oxo and nitridohalo species in general. Raman spectroscopy has shown that in 12 M HCl [TcOCl$_5$]$^{2-}$ is in equilibrium with *trans*-[TcO(OH$_2$)Cl$_4$]$^-$, but that the aqua form predominates by a factor of 60 [127]. A lesser tendency to form [TcNX$_5$]$^{2-}$ (X = Cl, Br) is consistent with the greater *trans* effect of the nitrido ligand. The Tc^{VI}N species in conc. HX solution is generally denoted simply as [TcNX$_4$]$^-$ [123].

The R[TcNCl$_4$] (R = AsPh$_4$ or NBu$_4$) salts are also formed from R[Tc^{V}OCl$_4$] and NaN$_3$ [129], *N,N,N'*-tri(trimethylsilyl)benzamidine or NCl$_3$ [130], by the oxidation of [Tc^{V}NCl$_2$(AsPh$_3$)$_2$] with SOCl$_2$ [67], and [TcNCl$_4$]$^-$ has been detected by EPR in the oxidation of Tc^{V}N thiolato complexes by Cl$_2$ in solution [13, 37]. The reaction of NH$_2$OSO$_3$H with TcO$_4^-$/HCl also yields [TcNCl$_4$]$^-$ and shows that a single amine nitrogen attached to a good leaving group may serve as an N^{3-} precursor, but the product is contaminated with nitrosyl species and [TcCl$_6$]$^{2-}$ [131]. In the preparation of single crystals of NBu$_4$ [(TcNBr$_4$)$_{0.8}$(TcOBr$_4$)$_{0.2}$] from TcO$_4^-$/HBr/aqueous NH$_3$ the nitrido ligand may originate from NBr$_3$ formed under the reaction conditions [132]. The use of ^{15}N labelled NaN$_3$ provides a convenient route to R[Tc^{15}NCl$_4$] and Cs$_2$[Tc^{15}NCl$_5$] [17, 133].

The [TcNX$_4$]$^-$ and [TcOX$_4$]$^-$ (X = Cl, Br) anions in the AsPh$_4^+$ salts have ideal C$_{4v}$ symmetry. Structural data are given in Table 2. The structural features observed for the isoelectronic [Tc^{V}N]$^{2+}$/[Tc^{V}O]$^{3+}$ pairs are again apparent in a comparison of the square pyramidal or octahedral [Tc^{VI}N]$^{3+}$/[Tc^{V}O]$^{3+}$ pairs in Table 2. The Tc≡N and Tc=O bond lengths are comparable but the nitrido ligand exerts the greater *trans* influence, in terms of the *trans*-NTc−OH$_2$ or −Cl bond lengths in the six-coordinate complexes. The O=Tc−X angles are greater than the N≡Tc−X angles and the displacement of Tc above the square basal or equatorial plane is consequently greater for the oxo complexes. Also, the square basal or equatorial Tc−X bond distances are significantly greater for the nitrido complexes. The *trans* influence of the nitrido ligand is apparent in the structure of *mer*-[TcNCl$_3$(bpy)] (**25**), where the Tc−N$_{bpy}$ bond lengths *cis* and *trans* to the nitrido ligand are 2.136(5) and 2.371(4) Å, respectively [137]. The *trans*-aqua ligand in crystals of [Rb(15-crown-5)$_2$][TcN(OH$_2$)Cl$_4$] grown from SOCl$_2$ solution appears to result from the diffusion of atmospheric water into the anhydrous crystal lattice, presumably topotactically, with retention of crystallinity [133]. The weak binding of the *trans*-aqua ligand is shown by the complete dehydration of NEt$_4$[TcN(OH$_2$)Cl$_4$] to NEt$_4$[TcNCl$_4$] under vacuum and the reformation of the aqua complex on exposure to atmospheric moisture. The bromo complex NEt$_4$[**24**], however, does not undergo dehydration under the same conditions, indicating a shorter NTc−OH$_2$ bond length than in the chloro complex [125].

The remarkable resistance of the Tc^{VI}N bond to acid hydrolysis is apparent from the methods of preparation, which involve reflux in conc. HCl or HBr. The halo ligands in R[TcNX$_4$] (X = Cl, Br) are labile and readily undergo exchange, as in the reaction of AsPh$_4$[TcNCl$_4$] with LiBr in acetone to give AsPh$_4$[TcNBr$_4$]

Table 2. Structural and IR data for monomeric $[Tc^{VI}N]^{3+}$ complexes and some $[Tc^{V}O]^{3+}$ analogues

Complex	Tc≡N/ =O (Å)	Tc–X$_{cis}$[a] (Å)	Tc–OH$_2$ or Tc–L$_{trans}$ (Å)	O/ N≡Tc–X (°)	δ[b] (Å)	ν(TcN)/ (TcO) cm^{-1}	Ref.
AsPh$_4$[TcNCl$_4$][c]	1.581(5)	2.3220(9)		103.34(3)	0.54	1076	119
AsPh$_4$[TcOCl$_4$]	1.593(8)	2.309(2)		106.8(1)	0.67	1025	134
AsPh$_4$[TcNBr$_4$][c]	1.596(6)	2.4816(5)		103.04(2)	0.56	1074	135
AsPh$_4$[TcOBr$_4$]	1.613(9)	2.460(1)		106.59(3)	0.70		136
NEt$_4$[TcN(OH$_2$)Br$_4$] (24)	1.599(9)	2.510(1), 2.518(1)	2.443(7)	97.2(2), 98.0(2)	0.33	1063	125
NEt$_4$[TcO(OH$_2$)Br$_4$]	1.618(9)	2.505(1), 2.508(1)	2.317(9)	97.6(2), 99.5(3)	0.37	1000	128
[Rb(15-crown-5)$_2$][TcN(OH$_2$)Cl$_4$]	1.600(3)	2.320(2)	2.43(4)	94.5(2)	–	1074	133
Cs$_2$[TcNCl$_5$]	1.600[d]	2.373(5)	2.740(5) (Cl)	99.73(8)	0.401	1027	125
Cs$_2$[TcOCl$_5$][e]	1.65	2.36	2.50 (Cl)	–	–	954	127
mer-[TcNCl$_3$(bpy)] (25)	1.669(4)	2.308(2)- (Cl) 2.136(5) (cis N$_{bpy}$)	2.371(4) (trans N$_{bpy}$)	96.4(2)- 102.7(2) (Cl) 93.3(2) (N$_{bpy}$)	–	1026	137

[a] Square basal or equatorial ligands

[b] Displacement of Tc above square basal or equatorial plane

[c] The [TcYX$_4$]$^-$ (Y = N,O) anions have ideal C_{4v} symmetry

[d] Value fixed in the refinement due to the statistical disorder of the ligands in the cubic space group

[e] Structural data from solid state EXAFS spectrum

[16] and the formation of $AsPh_4[TcNCl_4]$ on dissolution of the bromo complex in $SOCl_2$ [131]. Mixtures of chloro and bromo $R[TcNX_4]$ exchange in organic solvents to give mixed-ligand species [27] (see Sect. 3.1.2) and $[TcNCl_4]^-$ is rapidly and completely converted to $[TcNBr_4]^-$ on dissolution in conc. HBr [138]. A large number of $[Tc^{VI}NL_{4/5}]^{n-}$ and mixed-ligand species have been detected by EPR spectroscopy in solution and are discussed in Sect. 3.1.2. In dilute aqueous acid solutions $[TcNX_4]^-$ undergoes dimerization to yield the $Tc^{VI}N$ μ-oxo and bis(μ-oxo) complexes described in Sect. 3.3.

Substitution of $R[TcNX_4]$ (X = Cl, Br) in organic solvents generally results in reduction and these compounds are generally useful starting materials for the preparation of Tc^VN complexes [16]. Reduction also occurs on substitution by non-reducing ligands such as 2,2'-bipyridine or 1,10-phenanthroline in methanol solution to yield cis-$[Tc^VNX(bpy)_2]BPh_4$ or cis-$[TcNCl(phen)_2]PF_6$ [57, 58]. However, reaction of $NBu_4[TcNCl_4]$ with 2,2'-bipyridine in MeCN yields the neutral structurally characterized mer-$[Tc^{VI}NCl_3(bpy)]$ (25) [137] and $NBu_4[TcNBr_4]$ reacts with bis(salicylidine)ethylenediamine (salenH$_2$) in acetone to yield $[Tc^{VI}N(salen)]Cl$ [139]. Reaction of $NBu_4[TcNCl_4]$ with $(NBu_4)_4[H_3PW_{11}O_{39}]$ in MeCN results in the incorporation of TcN to give dark crystals of the Keggin polyoxotungstate derivative $(NBu_4)_4[PW_{11}TcNO_{39}]$, where the Tc(VI) oxidation state is thought to be retained [140]. A remarkable reaction is the abstraction of sulfur from $S_2O_4^{2-}$ by $NBu_4[TcNCl_4]$ in neat pyridine, 2-methyl- or 3,5-dimethylpyridine (L) to yield the thionitrosyl complexes cis-mer-$[Tc^I(NS)Cl_2L_3]$ [141]. Loss of the nitrido group is unusual, but $AsPh_4[TcNCl_4]$ reacts with 1,2-benzenedithiol (bdtH$_2$) in acetone at room temperature to give the trigonal prismatic $AsPh_4[Tc(bdt)_3]$ [142] and reaction of $NBu_4[TcNBr_4]$ with 2,2'-bipyridine yields $\{[Tc^VNBr(bpy)_2]\}_2[Tc^{II}Br_4]$ [57]. $AsPh_4[TcNCl_4]$ is unaffected by reflux in $SOCl_2$ for one hour, but in the presence of PPh_3 reduction occurs to $[Tc^{IV}Cl_6]^{2-}$ [67].

3.1.2 Spectroscopic Properties and Studies

The $v(TcN)$ IR absorption is sensitive to the presence and type of $trans$ ligand. For the five-coordinate $R[TcNX_4]$ (R = AsPh$_4$, NBu$_4$; X = Cl, Br) $v(TcN)$ occurs at 1080–1074 cm^{-1} while the presence of $trans$ halide has a marked effect with $v(TcN)$ at 1027 (X = Cl) and 1028 (X = Br) cm^{-1} in $Cs_2[TcNX_5]$ [119, 16, 126]. The $trans$ 2,2'-bipyridine nitrogen in 25 results in absorption at 1026 cm^{-1} [137]. A $trans$ aqua ligand has relatively little effect as is shown by the small change in $v(TcN)$ from 1065 to 1070 cm^{-1} on dehydration of $NEt_4[TcN(OH_2)Cl_4]$ to $NEt_4[TcNCl_4]$ [125]. It may be noted that $v(Tc^{16}O)$ at 1025 cm^{-1} for $AsPh_4[TcOCl_4]$ [134], $v(Tc^{14}N)$ at 1076 cm^{-1} for $AsPh_4[TcNCl_4]$ [119], and calculation by the simple diatomic oscillator model yields stretching force constants of 8.52 mdyn $Å^{-1}$ for Tc^VO and 8.37 mdyn $Å^{-1}$ for $Tc^{VI}N$, implying a greater bond strength for TcO in this pair. However, for $Cs_2[TcOCl_5]$ [$v(TcO)$ 954 cm^{-1}] [127] and $Cs_2[TcNCl_5]$ [$v(TcN)$ 1027 cm^{-1}] this approximation gives 7.38 mdyn $Å^{-1}$ for TcO and 7.62 mdyn $Å^{-1}$ for TcN. The multiply bonded ligand may be regarded as effectively "energy factored" from the low energy TcCl vibrations as indicated by $NBu_4[TcOCl_4]$ [v(TcO) 1020 cm^{-1}], for which normal coordinate analysis gives a TcO stretching force constant of 8.41 mdyn $Å^{-1}$ [143] and the diatomic oscillator model a good approximation of 8.44 mdyn $Å^{-1}$.

The electronic spectra of $[TcNX_4]^-$ and the isoelectronic $[Mo^VOX_5]^{2-}$ (X = Cl, Br) in conc. HX solution show a striking similarity in shape and intensity for the technetium and molybdenum pairs (Table 3, Fig. 3). The $[TcNX_4]^-$ spectra are essentially red-shifted versions of the $[MoOX_5]^{2-}$ spectra [118]. In the axially compressed tetragonal field of $[TcNX_4]^-$, $[MoOX_5]^{2-}$ or $[MoOX_4]^-$ (C_{4v} symmetry) the d orbital energy ordering is $d_{xy}(b_2) < d_{xz,yz}(e) < d_{x^2-y^2}(b_1) < d_{z^2}(a_1)$, with the unpaired electron located in the low energy, essentially nonbonding, $d_{xy}(b_2)$ orbital. Three d-d transitions may thus be expected. For $[MoOCl_5]^{2-}$, the weak $^2E \leftarrow {}^2B_2 (d_{xy} \rightarrow d_{xz}, d_{yz})$ and $^2B_1 \leftarrow {}^2B_2(d_{xy} \rightarrow d_{x^2-y^2})$ transitions have been identified at 14 000 ($\varepsilon = 12$ M^{-1} cm^{-1}) and 22 500 (14) cm^{-1}, respectively [144].

However, the spectra of $[TcNCl_4]^-$ at high concentration in HCl and of $NBu_4[TcNCl_4]$ in MeCN did not reveal any additional weak peaks which could be attributed to d-d transitions [118]. That the bands in the $[TcNCl_4]^-$ spectrum in Fig. 3 are ligand-to-metal charge transfer (LMCT) in nature is shown by the intensity, the red shift on replacement of the Cl^- ligands by Br^- and comparison with the Mo^VO analogues. The lowest energy LMCT band is at 20 800 cm^{-1} for $[TcNCl_4]^-$ and at 14 250 cm^{-1} for $[TcNBr_4]^-$. The difference of 6550 cm^{-1} is in good agreement with the 6000 cm^{-1} shift predicted from the theory of charge

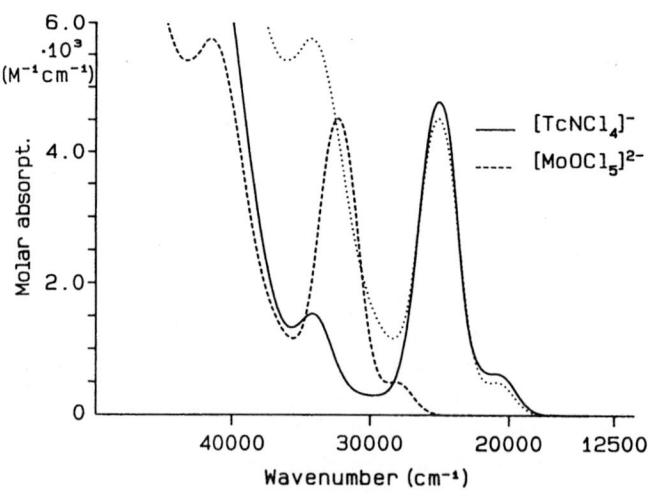

Fig. 3. Electronic spectra of $[TcNCl_4]^-$ in 7.5 M HCl and $[MoOCl_5]^{2-}$ in 10.7 M HCl. The *dotted line* represents the spectrum of $[MoOCl_5]^{2-}$ displaced by 7240 cm^{-1} to lower energy in order to demonstrate near coincidence with the spectrum of $[TcNCl_4]^-$

Table 3. LMCT electronic spectral data for $[TcNX_4]^-$ and $[MoOX_5]^-$ (X=Cl, Br) [118]

Complex	$\tilde{v}_{max}$ $(\varepsilon)^a$	Complex	$\tilde{v}_{max}$ $(\varepsilon)^a$
$[TcNCl_4]^-$ in 7.5 M HCl	20 800 (625) 25 060 (4780) 34 100 (1540)	$[TcNBr_4]^-$ in 7.5 M HBr	14 250 (470) 17 600 (3290) 20 080 (3030) 25 000sh (930)
$[MoOCl_5]^{2-}$ in 11 M HCl	28 200 (500) 32 300 (4520) 41 500 (5750)	$[MoOBr_5]^{2-}$ in 7.5 M HBr^b	21 200 (590) 24 150 (3625) 26 600 (2860)
NBu$_4$[TcNCl$_4$] in MeCN	19 000 (244) 21 600 (1930) 22 300 (1910) 24 750 (4950) 33 700 (1560) 43 300 (7200)	NBu$_4$[TcNBr$_4$] in MeCN	13 100 (225) 16 390 (3680) 17 850 (2620) 19 610 (3220) 24 600 (670) 33 600 (5700) 36 800 (4900)

^{a} Absorption maxima are given in cm^{-1} and molar absorptivities (ε) in $M^{-1}cm^{-1}$
^{b} Data from Ref. 131

transfer spectra and the optical electronegativities of $Cl^-(\pi)$ and $Br^-(\pi)$ [118, 145] and is consistent with the transition originating from π rather than σ halide orbitals (in which case a difference of ~ 3000 cm^{-1} would be expected [145]). The spectra of NBu$_4$[TcNX$_4$] (X = Cl, Br) in MeCN (Table 3) are significantly red shifted and show additional structure to the spectra of $[TcNX_4]^-$ in HX [118]. For the lowest energy LMCT band this shift is 1800 cm^{-1} for the chloro and 1150 cm^{-1} for the bromo complex. While it is likely that $[TcNX_4(MeCN)]^-$ is the major form in MeCN solution, the red shift does not appear to be the result of coordination in the *trans* position since essentially the same spectra are observed in "non-coordinating" solvents such as CH_2Cl_2 or benzene [118]. By analogy with magnetic circular dichroism studies of $[MOCl_4]^-$ (M = Cr, Mo, W) [146], the first three low energy features in the spectrum of NBu$_4$[TcNCl$_4$] have been assigned as, 19 000 cm^{-1} to 3b$_1$ (Cl p_z) $\rightarrow$ 2b$_2(d_{xy})$, 21 600 and 22 300 cm^{-1} to 5e $\rightarrow$ 2b$_2(d_{xy})$, and the intense band at 24 750 cm^{-1} to the 4e $\rightarrow$ 2b$_2(d_{xy})$ transition [147]. In the case of **25** in dmso, absorption maxima are observed at 21 500 sh ($\varepsilon = 1470$ M^{-1} cm^{-1}) and 24 750 (2460) cm^{-1} [137].

The low energy LMCT spectra of $[TcNX_4]^-$ (X = Cl, Br) in organic solvents and the availability of the half-filled low energy d_{xy} orbital are consistent with the ease of reduction of the $[Tc^{VI}N]^{3+}$ core to $[Tc^V N]^{2+}$. For $[TcNI_4]^-$ in organic solvents, the first LMCT transition is calculated to occur at ~ 4000 cm^{-1} and explains why attempts to prepare $[TcNI_4]^-$ salts by $[TcNCl_4]^-/I^-$ exchange result in the oxidation of I^- to I_2 [148]. As expected, $[TcNF_4]^-$ in HF does not show significant absorption at energies less than 36 000 cm^{-1} due to the high optical electronegativity of $F^-(\pi)$ [118]. Spectroelectrochemical studies at $-60\,^\circ C$ have shown that $[TcNX_4]^-$ (X = Cl, Br) is reversibly reduced to the colourless d^2

species $[Tc^VNX_4]^{2-}$ ($\tilde{\nu}_{max}$ 40 700 cm^{-1} for X = Br) [18]. This "optical bleaching" is expected since low energy LMCT transitions are unavailable due to the filled d_{xy} orbital and the large energy gap to the next d orbital(s). The colourless monomeric $[TcN(HSO_4)_4]^-$ and $[TcN(H_2PO_4)_4]^-$ anions have been identified by EPR in the concentrated acid solutions and show a single intense absorption at 33 800 and 37 600 cm^{-1}, respectively, which may be due to a $\pi N \rightarrow Tc$ LMCT transition [123].

Magnetic susceptibility measurements on single crystals of AsPh$_4$[TcNCl$_4$] from 300 to 4.5 K indicate a well-behaved S 1/2 system following the Curie law with a Weiss constant $\theta - 0.13(5)$ K [149]. The d^1 configuration of Tc(VI) and the nuclear spin of ^{99}Tc ($I = 9/2$) result in readily observed EPR spectra at room temperature. At lower temperatures the 10 hyperfine lines due to ^{99}Tc are clearly resolved together with ligand superhyperfine splitting. As a result, Tc^{VI}N complexes have been extensively investigated theoretically, in the study of ligand exchange reactions and to identify new species in solution [7, 8]. The spectra of $[TcNX_4]^-$ (X = Cl, Br) in particular, have been examined in great detail [13, 119, 126, 150, 151], including single-crystal EPR, electron nuclear double resonance (ENDOR) and electron spin echo envelope modulation (ESEEM) studies of ^{15}N-enriched AsPh$_4$[TcNCl$_4$] doped into the diamagnetic AsPh$_4$[TcOCl$_4$] host [152, 153] and single-crystal EPR and ^{15}N powder ENDOR studies of NBu$_4$[TcNBr$_4$]/[TcOBr$_4$] [154]. Analysis of the hyperfine data for AsPh$_4$[TcNCl$_4$] indicates that 20% of the spin density is localized in mainly the $3p$ orbitals of the Cl atoms [152], and the localization is greater on the bromine atoms in $[TcNBr_4]^-$ [154]. A polarized neutron diffraction study of a single crystal of AsPh$_4$[TcNCl$_4$] has, however, found 46(5)% of the spin density to be located on the Cl atoms, corresponding to exceptionally high covalence of the Tc–Cl bonds [155]. The $[TcNX_4]^-$ (X = Cl, Br) anions in HX solution show $g_\parallel > g_\perp$ $\{[TcNCl_4]^-, g_\parallel = 2.0075(5), g_\perp = 2.0020(5); [TcNBr_4]^-, g_\parallel = 2.145(1), g_\perp = 2.032(2)\}$ [126], whilst for $[TcNF_4]^-$ $g_\perp > g_\parallel$ $[g_\parallel = 1.895(2), g_\perp = 1.990(3)]$ [138]. The effect of the increase in the separation of the Tc^{VI} ions in polycrystalline samples of $[TcNCl_4]^-$, $[TcN(OH_2)Cl_4]^-$ and $[TcNCl_5]^{2-}$ salts at 130 K on the Tc$\cdots$Tc interactions has been studied [133].

EPR spectroscopy has proven particularly useful for the identification of $[TcN]^{3+}$ species in solution and for the monitoring of ligand exchange reactions. Mixed-ligand $[TcNBr_{4-n}Cl_n]^-$ ($n = 1$–3) species have been identified in mixtures of $[TcNCl_4]^-$ and $[TcNBr_4]^-$ [27, 150, 156]. The mixed species are readily assigned as there is a linear dependence of $g_\parallel$, $A_\parallel$ and g_0 on the averaged sum of the spin-orbit coupling constants of the basal donor ligands [150]. The exchange of $[TcNCl_4]^-$ and $[TcNBr_4]^-$ in HCl/HBr mixtures has been studied. For the exchange of $[TcNCl_4]^-$ with Br$^-$ the stepwise equilibrium constants K_1–K_4 are 0.78, 0.47, 0.30, and 0.12, respectively [156]. Studies of $[TcNX_4]^-$ (X = Cl, Br) in conc. HX or in the presence of a large molar ratio of Cl$^-$ or Br$^-$ in organic solvents show no evidence for the formation of $[TcNX_5]^{2-}$ [126]. A 0.002 M solution of Cs$_2$[TcNCl$_5$] in 28.6 M HF shows the presence of the five $[TcNF_{4-n}Cl_n]^-$ ($n = 0$–4) species, presumably due to the low activity of fluoride ion in HF solution. The $[TcNF_4]^-$ anion may be prepared in solution by the dissolution of "$TcN(OH)_3$" in 50% HF, but has not been isolated [157]. For the $[TcNF_nCl_{4-n}]^-$ ($n = 1-4$) species, $A_\parallel$ is linearly dependent on the effective basal ligand spin-orbit coupling, but for $g_\parallel$ a linear relationship does not hold [138]. The addition of NBu_4F to $[TcNX_4]^-$ ($X = Cl, Br$) in MeCN, however, results in essentially stoichiometric replacement of X^- by the "naked" fluoride ion. After addition of 4 equivalents of NBu_4F, $[TcNF_4]^-$ is the only species observed by EPR. Further addition results in the formation of $[TcNF_5]^{2-}$, which is the major species after addition of 8 equivalents [147].

The mixed-ligand species $[TcNCl_3(CN)]^-$ and $[TcNCl_2(CN)_2]^-$ have been identified by EPR in the cleavage of $[\{TcN(CN)_2\}_2(\mu-O)_2]^{2-}$ with HCl [19, 20], $[TcNBr_3(NCS)]^-$ and $[TcNBr_2(NCS)_2]^-$ in the reaction of $[TcNBr_4]^-$ with NCS^- [27] and $[TcNX_n(N_3)_{4-n}]^-$ ($X = Cl, Br$; $n = 1-3$) and $[TcN(N_3)_4]^-$ in the reaction of $NBu_4[TcNX_4]$ with NaN_3 in acetone [158]. A variety of $[TcN]^{3+}$ oxygen-coordinated species such as $[TcN(HSO_4)_4]^-$ and $[TcN(H_2PO_4)_4]^-$ have been identified in the concentrated acid solution and analogous species in anhydrous $MeSO_3H$, CF_3SO_3H [123], $ClSO_3H$ [159], and aqueous solutions of H_3PO_3 and $H_4P_2O_7$ [123]. These species show $g_\parallel$ 1.89–1.91, $g_\perp$ 1.99–2.00, $A_\parallel$ $344-363\times10^{-4}$ cm^{-1}, and $A_\perp$ $162-174\times10^{-4}$ cm^{-1}. The oxidation of $[Tc^VNCl_2(EPh_3)_2]$ ($E = P, As$) to $[TcNCl_4]^-$ by $SOCl_2$ has been shown by EPR to proceed via the Tc(VI) species $[TcNCl_3(EPh_3)]$ [67] and intermediate species have been observed in the oxidation of Tc^VN thiolato complexes by Cl_2 or Br_2 [14, 37].

3.2 Dimeric and Polymeric Halide Complexes

The very moisture sensitive $TcNCl_3$ may be prepared by the reaction of $TcCl_4$ with IN_3 or of $NBu_4[TcNCl_4]$ with $GaCl_3$. The Tc≡N IR stretch at 1018 cm^{-1} and $v(TcCl)$ at 400–320 and 240 cm^{-1} indicate a polymeric structure with TcNTc and $TcCl_nTc$ bridges. $TcNCl_3$ is insoluble in CH_2Cl_2 but dissolves on addition of $AsPh_4Cl$ due to the formation of $AsPh_4[TcNCl_4]$ [130].

Addition of 18-crown-6 to a suspension of $Cs_2[TcNCl_5]$ in $SOCl_2$ results in the formation of an orange-red solution, which on slow evaporation of the solvent yields crystals of $[Cs(18-crown-6)][TcNCl_4]$ [Fig. 4(a)] [133, 160]. The structure consists of the unprecedented "infinite sandwich" M^+/crown ether configuration with ordered and disordered infinite chains of $[TcNCl_4]^-$ anions arranged in an antiparallel fashion between columns of $[Cs(18-crown-6)]^+$ cations. In the ordered Tc≡N···Tc≡N··· chain the Tc≡N and N···Tc bond lengths are 1.561(36) and 2.714(36) Å, respectively. The tetragonal ($P4/n$) unit cell $[a = b = 22.459(13)$, $c = 4.275(4)$ Å, $Z=4]$ results in the unusual situation that the nearest neighbours of each Cs^+ cation are two Cs^+ cations at 4.275 Å, while the Tc atoms of the four nearest anions are at 7.95 Å and the nearest $Cs^+···Cl$ contacts are 6.4–6.6 Å. Also, while each Cs^+ cation has neighbours at 4.275, 8.55 and 12.825 Å along each vertical column, in the horizontal plane the

John Baldas

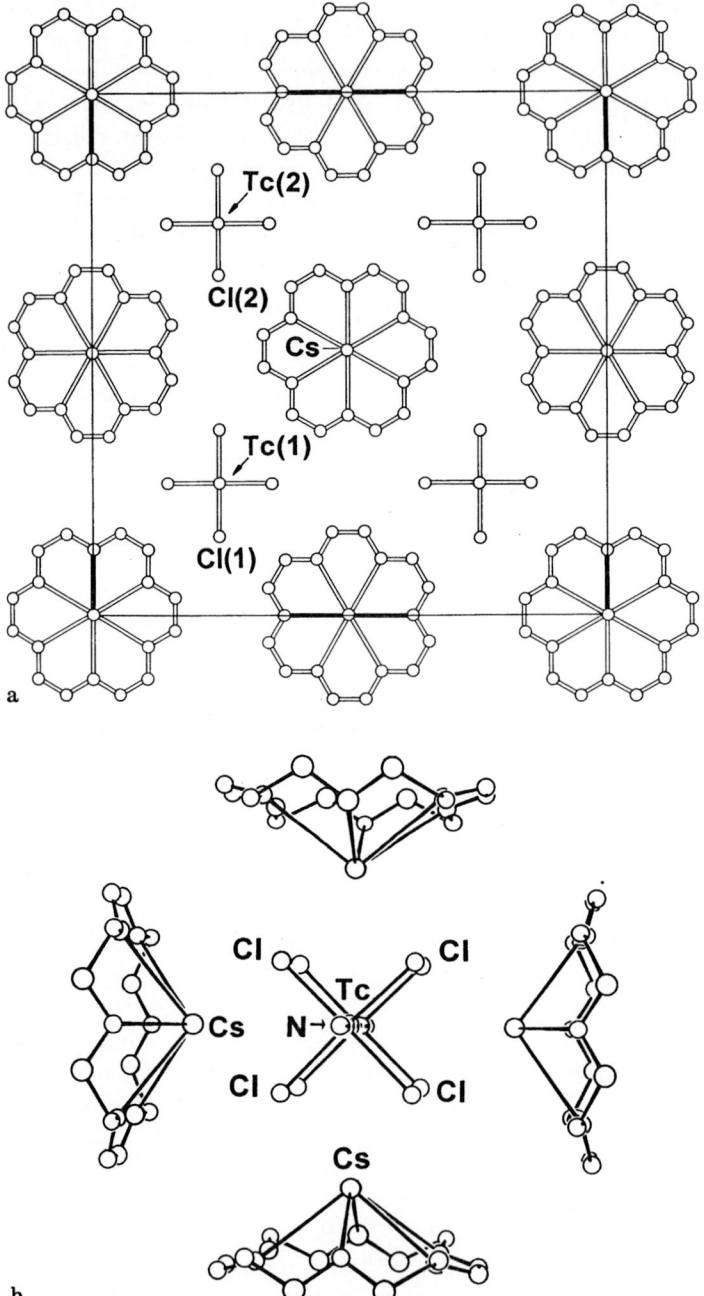

Fig. 4. The structure of **a** the infinite sandwich [Cs(18-crown-6)][TcNCl₄] viewed down the *c*-axis [133] and **b** portion of the structure of [Cs(18-crown-6)]₄[(TcNCl₄)₄(H₂)₃] (the two lattice [TcN(OH₂)Cl₄]⁻ anions are not shown) [161]

nearest Cs^+ neighbours are at 11.23 Å. The IR spectrum shows a single $v(TcN)$ absorption at 1041 cm^{-1}, but partial ^{15}N labelling results in complex behaviour due to the coupling of TcN oscillators in the infinite $[TcNCl_4]^-$ chains and is diagnostic for this arrangement. EPR spectra over the temperature range 130–290 K indicate the presence of exchange interactions along the $\cdots TcN \cdots TcN \cdots$ chains [133]. Recrystallization of the infinite sandwich $[Cs(18\text{-crown-}6)][TcNCl_4]$ from MeCN, acetone or ethanol, or the reaction of $Cs_2[TcNCl_5]$ with 18-crown-6 in 6 M HCl, yields $[Cs(18\text{-crown-}6)]_4[(TcNCl_4)_4(OH_2)_3]$. For this complex two $v(TcN)$ absorptions at 1055 and 1046.5 cm^{-1}, which are shifted to 1023 and 1016 cm^{-1} on ^{15}N labelling, indicate the presence of two types of TcN centres [133]. The crystal structure [Fig. 4(b)] shows the novel arrangement of a dimeric $[N\equiv TcCl_4 \cdots N\equiv Tc(OH_2)Cl_4]^{2-}$ unit surrounded by a square cage formed by four $[Cs(18\text{-crown-}6)]^+$ cations, with two monomeric $[TcN(OH_2)Cl_4]^-$ units present in the lattice [161]. These two forms appear to differ little in energy since ether diffusion into an MeCN solution of the aqua complex may result in the crystallization of either the infinite sandwich or a mixture of the infinite sandwich and the aqua complex. Recrystallization of the aqua complex in the anhydrous conditions of $SOCl_2$ gives only the infinite sandwich. Addition of 18-crown-6 to $[TcNCl_4]^-$ (prepared from "$TcN(OH)_3$", and free of Cs^+) in conc. HCl yields $[(H_3O)(18\text{-crown-}6)]_2[(TcNCl_4)_2(OH_2)]$, which has been shown by crystallography to contain only dimeric $[N\equiv TcCl_4 \cdots N\equiv Tc(OH_2)Cl_4]^{2-}$ units and $[(H_3O)(18\text{-crown-}6)]^+$ cations [28]. The infinite sandwich form, $[(H_3O)(18\text{-crown-}6)][TcNCl_4]$, has also been isolated and structurally characterized [28]. The cell height of 4.501(2) Å shows a looser packing than for the Cs^+ analogue.

3.3 Oxo-bridged Complexes

3.3.1 Preparation, Structure and Spectra

A characteristic and novel feature of the chemistry of $[TcN]^{3+}$ is the formation of dimeric complexes based on the $[NTc\text{-}O\text{-}TcN]^{4+}$ and $[NTc(\mu\text{-}O)_2TcN]^{2+}$ cores [20, 122, 123, 162, 163]. The chemistry and structural aspects of these $Tc^{VI}N$ dimers have some $Re^{VI}O$ analogues, but most closely parallel those of the well-known isoelectronic $[OMo^V\text{-}O\text{-}Mo^VO]^{4+}$ and $[OMo^V(\mu\text{-}O)_2Mo^VO]^{2+}$ dimers [117]. The only known $Tc^{VI}O$ oxygen-bridged dimer is the organometallic $[(Me_2TcO)_2(\mu\text{-}O)_2]$ [164].

Hydrolysis of $Cs_2[TcNCl_5]$ in an ample quantity of water gives a brown precipitate of "$TcN(OH)_3$", which has been formulated as the bis(μ-oxo) dimer $[\{TcN(OH)(OH_2)\}_2(\mu\text{-}O)_2]$ (26) on the basis of its reactions and the presence of $v(TcOTc)$ absorptions in the IR spectrum [163]. This precipitate is isoelectronic with the well-known "$MoO(OH)_3$", for which recent solution EXAFS studies have indicated the bis(μ-oxo) structure $[\{MoO(OH)_2\}_2(\mu\text{-}O)_2]^{2-}$ for the anionic form [165]. Addition of ethanol to a solution of 26 in aqueous CsOH precipitates the yellow $Cs_2[\{TcN(OH)_2\}_2(\mu\text{-}O)_2]$ with $v(TcN)$ at 1046 cm^{-1} and

v(TcOTc) at 734 cm^{-1} [163]. Solutions of **26** in 7.5 M CF$_3$SO$_3$H (a very weakly coordinating medium) are orange ($\lambda_{max} = 474$ nm) and EPR silent, showing the absence of monomeric species. The monomeric aqua cation [TcN(OH$_2$)$_5$]$^{3+}$ is thus not a viable species even in strongly acid solution and appears to spontaneously dimerize to the μ-oxo aqua cation [{TcN(OH$_2$)$_4$}$_2$(μ-O)]$^{4+}$ (**27**) [123, 166].

27   **28**

Solutions of **26** in the weakly-coordinating 1 M p-toluenesulfonic acid, CF$_3$SO$_3$H or MeSO$_3$H are pale yellow and have been shown by paper electrophoresis to contain a single cationic species [123, 163]. Also, dilution of a solution of **27** in 7.5 M CF$_3$SO$_3$H by the addition of water leads to the slow formation of the pale yellow species. This species has been formulated as the bis(μ-oxo) aquanitrido cation [{TcN(OH$_2$)$_3$}$_2$(μ-O)$_2$]$^{2+}$ (**28**) on the basis of the similarity of the electronic spectrum to that of the well-established [{MoO(OH$_2$)$_3$}$_2$(μ-O)$_2$]$^{2+}$ cation, the absence of EPR signals, and the isolation of [{Tc^{VI}N(S$_2$CNEt$_2$)}$_2$(μ-O)$_2$] on reaction with Na(S$_2$CNEt$_2$) [122]. Solution EXAFS studies have confirmed the presence of the bis(μ-oxo) ring, and structure **28** thus seems secure except for the exact number of coordinated water molecules [167].

The only μ-oxo complex to have been isolated and structurally characterized is the cyclic tetramer (AsPh$_4$)$_4$[Tc$_4$N$_4$(O)$_2$(ox)$_6$] (**29**) prepared by the reaction of AsPh$_4$[TcNCl$_4$] with oxalic acid in aqueous acetone [162]. The centrosymmetric structure, shown in Fig. 5, consists of two [(ox)TcN-O-TcN(ox)] units joined by two tetradentate oxalates. The Tc$\equiv$N bond lengths are 1.639(17) and 1.606(17) Å and the Tc–O–Tc bridges are only approximately linear with angles of 150.4(8)° and Tc–O$_{bridge}$ bond lengths of 1.840(13) and 1.869(13) Å. The nitrido ligands are in the *syn* conformation **30**. A marked asymmetry due to the *trans* influence of the nitrido ligand is apparent in the Tc–O bond lengths of the bridging oxalates, with NTc–O$_{trans}$ 2.410(11), 2.369(12) Å and NTc–O$_{cis}$ 2.076(11), 2.061(11) Å. The μ-oxo structure of the oxalato complex with the nitrido ligands *cis* to the oxygen bridge is in contrast to the linear (or near linear) [O=Tc^V-O-Tc^V=O]$^{4+}$ d^2-d^2 dimers where the oxo ligands are *trans* to the oxygen bridge [168, 169]. This difference in geometry is explained in terms of closed-shell electronic structures [170]. In the *syn* conformer **30** (C_{2v} symmetry), and also the *anti* conformer **31** (C_{2h} symmetry), the π interaction between the t_{2g} sets of the two Tc atoms and the three (p_x, p_y) orbital sets on the two nitrogen and the bridging oxygen atoms results in only one nonbonding metal orbital being available for the Tc d electrons. The d^1-d^1 configuration with occupancy of this nonbonding orbital

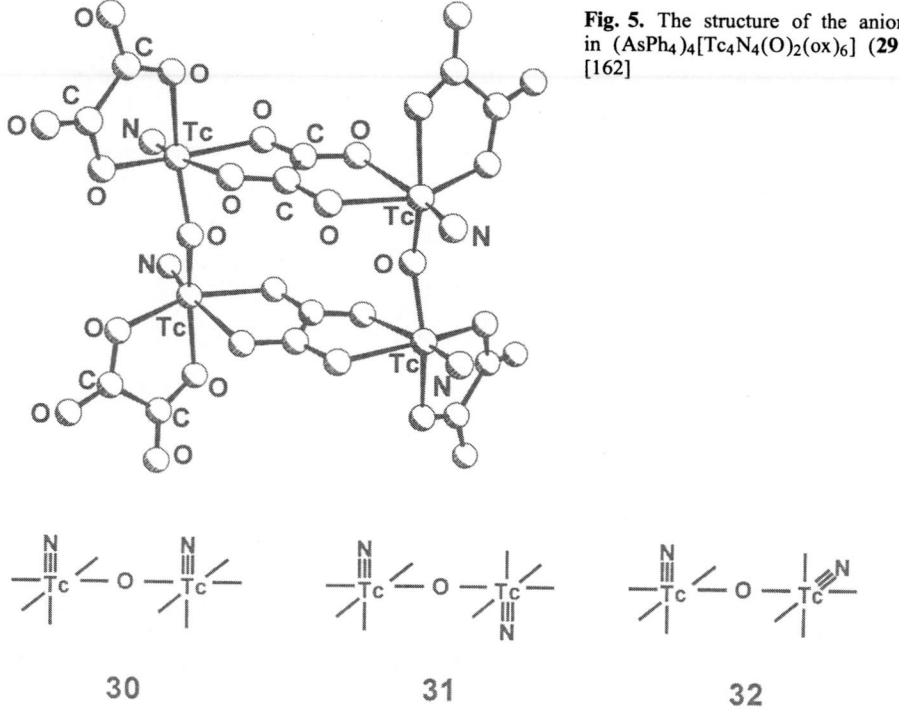

Fig. 5. The structure of the anion in $(AsPh_4)_4[Tc_4N_4(O)_2(ox)_6]$ **(29)** [162]

30   31   32

(a_2 in C_{2v} and b_g in C_{2h} symmetry) satisfies the closed-shell requirement [131, 170]. Skew conformations (C_2 symmetry), with the angle between the two Tc–N vectors ranging between 0 and 180° are possible. In the 90° skew conformer **32**, the d electrons are not paired and the complex would be expected to be paramagnetic. Skew conformations have not been observed for the large number of known $[Mo^V_2O_3]^{4+}$ complexes [117]. The electronic spectrum of **29** is dominated by an intense visible absorption at 493 nm ($\varepsilon = 39\ 200$ M^{-1} cm^{-1} for the tetramer) arising from a transition within the three-centre Tc⋯O⋯Tc π-bond system [20, 118]. Intense visible absorptions are characteristic of $[Mo^V_2O_3]^{4+}$ complexes [117]. In the IR spectrum of **29**, v(TcN) occurs at 1050 cm^{-1}, but v_{asym}(TcOTc) has not been identified [131].

Addition of $AsPh_4Cl$ and then dropwise addition of HCl to a solution of $Cs_2[TcNCl_5]$ in water with sufficient $MeSO_3H$ added to dissolve the initial precipitate gives a high yield of the yellow $(AsPh_4)_2[(TcNCl_2)_2(\mu\text{-}O)_2]$. The yellow bromo complex may be similarly prepared by use of $AsPh_4Br \cdot HBr/HBr$ [118, 166]. Dithiocarbamato, cyano and ethanedithiolato complexes have been prepared by the addition of the ligand to solutions of $Cs_2[TcNCl_5]$ in aqueous $Na_4P_2O_7$ [20]. The $[(TcNCl_2)_2(\mu\text{-}O)_2]^{2-}$ anion in Fig. 6 shows the general structural features of the $[NTc(\mu\text{-}O)_2TcN]^{2+}$ core [118, 166]. Structural data are summarized in Table 4.

John Baldas

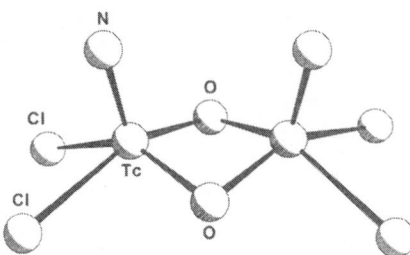

Fig. 6. The structure of the anion in $(AsPh_4)_2$ $[(TcNCl_2)_2(\mu\text{-}O)_2]$ [118, 166]

Table 4. Structural data for $[NTc(\mu\text{-}O)_2TcN]^{2+}$ dimers and some $[OMo(\mu\text{-}O)_2MoO]^{2+}$ analogues

Complex	Tc≡N/Mo=O av. (Å)	M–M (Å)	M–O$_{bridge}$ (Å)	δ-sbp[a] (Å)	N···N/ O···O[b] (Å)	Ref.
$(AsPh_4)_2[(TcNCl_2)_2(\mu\text{-}O)_2]$	1.649(8)	2.579(1)	1.896(7)–1.953(5)	0.53(1)	3.27	118
$(AsPh_4)_2[(MoOCl_2)_2(\mu\text{-}O)_2]$	1.69(1)	2.591(2)	1.90(1)–1.94(1)	0.615(2)	3.26	172
$(AsPh_4)_2[(TcNBr_2)_2(\mu\text{-}O)_2]$	1.609(13)	2.575(2)	1.93(1)–1.95(1)	0.59(1)	3.30	118
$(AsPh_4)_2[\{TcN(CN)_2\}_2(\mu\text{-}O)_2]$	1.640(8)	2.560(1)	1.933(5)–1.937(5)	0.56(1)	3.49	118
$[\{TcN(S_2CNEt_2)\}_2(\mu\text{-}O)_2]$	1.623(4)	2.543(1)	1.935(3)–1.942(3)	0.65(1)	3.47	20
$[\{MoO(S_2CNEt_2)\}_2(\mu\text{-}O)_2]$	1.679(2)	2.580(1)	1.940(2)–1.943(2)	0.729(1) 0.738(1)	3.55	171
$[\{TcN(S_2CNC_4H_8)\}_2(\mu\text{-}O)_2]$	1.65(2) 1.59(2)	2.542(2)	1.934(13)–1.947(12)	0.66(1)	3.47	20

[a] Displacement of M atom above the square basal plane
[b] Non bonded contact distance between the nitrido or terminal oxo (Mo) ligands. Data from Ref. 118

A comparison of the isostructural $[\{TcN(S_2CNEt_2)\}_2(\mu\text{-}O)_2]$ and $[\{MoO(S_2CNEt_2)\}_2(\mu\text{-}O)_2]$ [20] again shows the greater structural effect of the oxo ligand, with the Mo atoms displaced by 0.73 Å above the square basal planes [171] compared to 0.65 Å for the Tc atoms. The Tc–Tc distances of 2.542(2)–2.591(1) Å in the bis(μ-oxo) dimers correspond to a d^1_{xy}-d^1_{xy} single bond and account for the absence of EPR spectra [20]. These distances may be compared to the unsupported Tc^{VI}–Tc^{VI} bond length of 2.744(1) Å in the imido dimer $[(ArN)_3Tc\text{-}Tc(NAr)_3]$ (Ar = 2,6–diisopropylphenyl) [173]. The two nitrogen atoms in $[Tc_2N_2O_2]^{2+}$ complexes are bent back from each other to a N···N contact distance of 3.3–3.5 Å. In the absence of this bending the N···N distance would be the same as the Tc–Tc distance and rather shorter than sum of the N van der Waals radii of about 2.9 Å [118]. All $[Tc_2N_2O_2]^{2+}$ complexes isolated to date are yellow and do not show significant absorption at wavelengths longer than 450 nm [118].

The TcO_2Tc ring system is readily detected in the IR spectrum by the presence of a strong asymmetric stretching mode at 710–700 cm^{-1} and a weaker symmetric mode at 515–450 cm^{-1}. These assignments have been confirmed by ^{18}O labelling. All known dimers have the *syn* stereochemistry in Fig. 6 and show two ν(TcN) absorptions as a result of the in-phase (A_1 in C_{2v} symmetry) and out-of-phase (B_1) vibrations of the coupled TcN oscillators. For $(AsPh_4)_2[(TcNX_2)_2(\mu$-$O)_2]$ the higher energy peak is the more intense and the splitting is small with absorptions at 1063, 1054 (X = Cl) and 1059, 1051 cm^{-1} (X = Br) [118]. This coupling has been demonstrated for $(AsPh_4)_2[\{TcN(CN)_2\}_2(\mu$-$O)_2]$ by varying the level of ^{15}N labelling of the nitrido ligands. Surprisingly, neither $(AsPh_4)_2[\{TcN(CN)_2\}_2(\mu$-$O)_2]$ or the 50% ^{13}CN or C^{15}N labelled compounds (where there are species with C_{2v}, C_s and C_2 symmetry) show significant ν(CN) absorptions. The effect of 99% ^{13}CN and 98% C^{15}N labelling on a TcCN bending mode confirms that the cyano ligands are C-bonded and the crystal structure shows the C≡N bond lenghts to be normal at 1.12–1.18 Å. The absence of significant ν(CN) absorptions appears to be due to the localization of the d^1 electrons in the Tc–Tc bond and hence negligible Tc–CN π-bonding. The asymmetric ν(TcO$_2$Tc) mode at 723 cm^{-1} has been confirmed by ^{18}O labelling but the symmetric mode has not been identified [147].

The formation of bis(μ-oxo) dimers greatly reduces the susceptibility of Tc^{VI}N to reduction. Thus, reaction of $(AsPh_4)_2[(TcNCl_2)_2(\mu$-$O)_2]$ with Na(S$_2$CNEt$_2$) in MeCN gives $[\{TcN(S_2CNEt_2)\}_2(\mu$-$O)_2]$ in good yield while reaction of $[TcNCl_4]^-$ in the same solvent gives only the reduced $[Tc^VN(S_2CNEt_2)_2]$ [20, 118]. The greater sensitivity of the TcO$_2$Tc ring to cleavage by HCl in organic solvents than to aqueous acid indicates that cleavage is initiated by protonation of the oxygen bridge [118]. Cleavage occurs via the intensely coloured [NTc-O-TcN]$^{4+}$ species and the process is readily monitored spectrophotometrically [20, 118]. Cleavage reactions may be used to prepare novel monomeric Tc^{VI}N species such as $[TcN^{VI}Cl_2(S_2CNEt_2)]$, which are not accessible by partial substitution of $[TcNCl_4]^-$.

3.3.2 Monomer, μ-Oxo Dimer, Bis(μ-Oxo) Dimer Interconversions in Solution

The interconversions and equilibria of $[TcN]^{3+}$ species in solutions of inorganic and organic acids are described by Scheme 1. Monomeric species are readily identified by their EPR spectra, without interference from the presence of the spin-paired EPR silent dimers. The μ-oxo dimers are readily distinguished from bis(μ-oxo) dimers by the presence of an intense visible absorption at 470–580 nm. The interconversions may occur over a period of hours and are conveniently followed by UV-vis spectrophotometry. The reaction sequence in Scheme 1 may occur in either direction depending on whether 26 or Cs$_2$[TcNCl$_5$] is dissolved in the acid. High acidity and the presence of coordinating anions such as Cl$^-$ favour the monomeric species [118, 123, 166]. Solutions of Cs$_2$[TcNCl$_5$] in 3.33 M HCl show only the presence of $[TcNCl_4]^-$ while in 0.5 M HCl a pink species ($\lambda_{max} = 538$ nm), probably the μ-oxo dimer 34a (L = Cl) with water coordinated *trans* to the nitrido ligands, is formed. If Cs$_2$[TcNCl$_5$] is first hydrolysed to 26 and then HCl added to 3.33 M, an intensely blue species ($\lambda_{max} = 566$ nm), probably the μ-oxo dimer 34b (L = Cl), is formed which converts to $[TcNCl_4]^-$ by

Scheme 1. (L = monoanionic ligand)

first order kinetics [118]. Spectrophotometric studies of $Cs_2[TcNCl_5]$ in $MeSO_3H$ solutions indicate the presence of μ-oxo dimers with varying levels of aqua and chloride substitution [123]. EXAFS studies have confirmed the presence of μ-oxo and bis(μ-oxo) dimers in HCl solution [167]. Magnetic susceptibility studies in solution by the Evans NMR method indicate that both the pink and blue species are diamagnetic [174]. Pink solutions of $Cs_2[TcNCl_5]$ in 1 M HCl show reversible thermochromic behaviour. On heating, conversion to the orange-yellow $[TcNCl_4]^-$ occurs and the pink colour slowly returns on cooling, thus showing that the monomer $\rightarrow$ μ-oxo dimer conversion is an exothermic process [118]. The behaviour of **26** and $[TcNBr_4]^-$ in HBr solution is similar to that in HCl [118].

The sequence of interconversions is illustrated by the spectra of a solution of **26** in 5 M H_3PO_4 shown in Fig. 7 [123]. During the first hour the fast reaction **33** $\rightarrow$ **34a** or **34b** ($L = H_2PO_4^-$ or $L_2 = HPO_4^{2-}$) predominates, and well defined "isosbestic" points are observed. After about 19 h the main reaction is the slower conversion of **34a** or **34b** $\rightarrow$ $[TcN(H_2PO_4)_4]^-$ and new "isosbestic" points appear. The increase and then decrease of the μ-oxo dimer absorption at 501 nm is characteristic of A $\rightarrow$ B $\rightarrow$ C consecutive reactions and the decrease of the bis(μ-oxo) dimer at 342 nm and increase in the monomer at 279 nm are also evident. When **26** is dissolved in 17.5 M H_3PO_4 the reaction sequence is rapid and complete conversion to $[TcN(H_2PO_4)_4]^-$ occurs ($\lambda_{max} = 266$ nm). Dissolution of $Cs_2[TcNCl_5]$ in the alkaline 1 M K_2HPO_4 (pH 9.5) results in rapid conversion of $[TcNCl_4]^-$ to the μ-oxo dimer ($\lambda_{max} = 508$ nm) followed by slower complete conversion to the bis(μ-oxo) dimer ($\lambda_{max} = 327$ nm). Similar behaviour is observed in H_2SO_4 solutions of **26**, where the orange μ-oxo dimer absorbs at 485

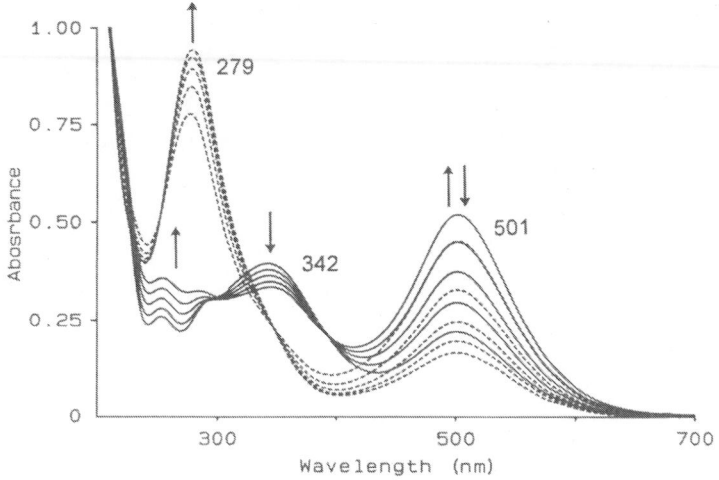

Fig. 7. UV-vis spectra of **26** in 5 M H$_3$PO$_4$ ([Tc] = 3.2 × 10^{-4} M). *Solid lines* at preparation, 15, 30, 45, 60 min; *dashed lines* at 19, 21, 23, 25, 27 hours

nm and the colourless [TcN(HSO$_4$)$_4$]$^-$ at 296 nm. At low acid concentrations the interconversions may be slow and paper electrophoresis is a useful technique for the separation of species [123].

The formation of dimeric species on dissolution of Cs$_2$[TcNCl$_5$] in a wide variety of 1 M aqueous carboxylic acids, including citric, tartaric, oxalic, glycolic, D,L-malic, gluconic, glucuronic, α-hydroxyisobutyric, and glycine has been studied spectrophotometrically. The retention of the Tc^{VI}N moiety is shown by the formation of [TcNCl$_4$]$^-$ on addition of conc. HCl. For solutions of Cs$_2$[TcNCl$_5$] in 1 M oxalic acid, addition of AsPh$_4$Cl results in the precipitation of **29** [131]. The reaction of [TcNCl$_4$]$^-$ with Na$_2$H$_2$edta in aqueous solution has been reported to yield the red-violet monomeric [TcN(Hedta)]·3H$_2$O on the basis of microanalytical and IR data [175]. However, the absorption at 505 nm in the electronic spectrum indicates that the product obtained may be a μ-oxo dimer.

3.4 Technetium-99m Nitrido Complexes

The formation of [^{99m}TcNCl$_4$]$^-$ on reaction of ^{99m}TcO$_4^-$/NaN$_3$/HCl has been established [92, 93]. After removal of HCl the residue is stable to oxidation under acidic conditions but undergoes oxidation to ^{99m}TcO$_4^-$ at pH > 4 [97]. While reaction of [^{99m}TcNCl$_4$]$^-$ with thiolato or phosphine ligands undoubtedly leads to reduction to ^{99m}Tc^{V}N [95, 96], the oxidation state with ligands such as gluconate or phosphonates [92, 93] has not been established, but may well be ^{99m}Tc^{VI}N. However, due to the low $^{99m/99}$Tc concentration of 10^{-8}–10^{-7} M in these preparations, dimer formation, which follows second order kinetics, will only be significant if dimerization is a very fast process [166]. Dimer formation has been demonstrated for 4×10^{-4} M solutions of $Cs_2[TcNCl_5]$ in a variety of organic acids [131]. Paper electrophoresis of $[^{99m}TcNCl_4]^-$ in 0.5 M p-toluenesulfonic acid shows largely the presence of a cationic species, which may be $[^{99m}Tc_2N_2O_2(OH_2)_6]^{2+}$.

4 Technetium(VII) Nitrido Complexes

The aqueous solution chemistry of Tc(VII) is dominated by the stability of the TcO_4^- anion [4]. Nitrido complexes are few and are limited to peroxides and one dimeric nitrido-hydrazido example. The peroxo complexes based on the $[Tc^{VII}N(O_2)_2]$ core are analogous to the well-known isoelectronic $[Mo^{VI}O(O_2)_2]$ complexes [117] and are the only examples of nitridoperoxo complexes and rare examples of peroxo complexes of a metal in the $+7$ oxidation state.

Slow evaporation of a solution of $Cs_2[TcNCl_5]$ in 10% aqueous H_2O_2 yields yellow-orange crystals of the explosive nitridoperoxo complex Cs[**35**]. The coordination geometry is a distorted pentagonal pyramid with the nitrido ligand in the apical position [Tc≡N 1.63(2) Å] and η^2 peroxo ligands with O–O bond distances of 1.41(2) and 1.46(2) Å [176]. The $AsPh_4[TcN(O_2)_2X]$ (X = Cl, Br) salts are prepared from $AsPh_4[TcNX_4]/H_2O_2$ in MeCN and are thermally more stable [176, 177]. Addition of 2,2′-bipyridine,1,10-phenanthroline or oxalic acid to the pale-yellow solution of **26** in 10% H_2O_2 yields $[TcN(O_2)_2(L-L)]$ (L-L = bpy, phen) and the dimeric $(AsPh_4)_2[\{TcN(O_2)_2\}_2(ox)]$, respectively [177]. The stability of the $[Tc^{VII}N(O_2)_2]$ core may be contrasted with the immediate oxidation of $[TcOCl_4]^-$ to TcO_4^- by H_2O_2, with no evidence for the formation of transitory peroxo species [176].

35   **36**

The crystal structure of the oxalate dimer (Fig. 8) shows the anion to consist of two $TcN(O_2)_2$ units bridged by a tetradentate sideways-bound oxalate with distorted pentagonal bipyramidal geometry about each Tc atom [178]. The bond lengths are Tc≡N 1.61(4), 1.69(3) Å, O–O 1.43(4)–1.50(4) Å, and the *trans* influence of the nitrido ligand is apparent in Tc–O$_{oxalate}$ bond lengths of 2.09(2), 2.11(2) Å *cis* and 2.40(2), 2.47(2) Å *trans*.

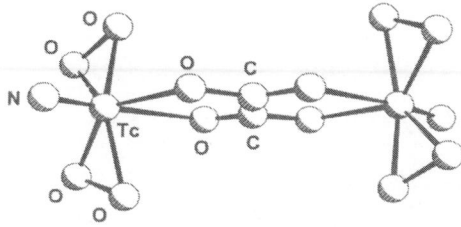

Fig. 8. The structure of the anion in $(AsPh_4)_2$
$[\{TcN(O_2)_2\}_2(ox)]\cdot 2Me_2CO$ [178]

The structure of the bpy and phen complexes is also most probably pentagonal bipyramidal with the nitrido ligand in an axial position and the heterocyclic ligand spanning axial and equatorial positions [178]. In the IR spectra of the nitridoperoxo complexes $v(TcN)$ occurs at 1069–1035, $v(O–O)$ at 912–894 and $v_{sym}(TcO_2)$ at 665–647 cm^{-1} [177].

The reaction of $NBu_4[TcOCl_4]$ with Ph_2NNH_2 and 2,4,6-triisopropylbenzenethiol (ArSH) yields yellow crystals of the novel nitrido-hydrazido(2-) formally Tc(VII) binuclear complex [36]·0.5Et$_2$O. The nitrido ligands result from N–N bond cleavage of the organohydrazine. The geometry about each Tc atom is distorted square pyramidal with Tc≡N bond lengths of 1.64(1) Å, long Tc=NNPh$_2$ bonds of 1.88(1) av. Å, and Tc–N–NPh$_2$ angles of 140.2(11) and 141.7(11)°. The Tc–S$_{bridging}$ bond lengths of 2.470(7) Å are significantly longer than the average Tc–S$_{terminal}$ lengths of 2.379(6) Å [179]. The long Tc=NNPh$_2$ bonds and the bent Tc–N–NPh$_2$ arrangement indicate that the bound N atom of the hydrazido-(2-) ligand is substantially sp^2 hybridized and that the formal Tc–N bond order is close to 2. A similar effect is observed for the nonequivalent imido ligands in *cis*-$[Mo^{VI}(NPh)_2(S_2CNEt_2)_2]$ where one of the Tc–N–C bond angles is 139.4(4)° and the other 169.4(4)° [180].

5 Abbreviations

15-crown-5	1,4,7,10,13-pentaoxacyclopentadecane
18-crown-6	1,4,7,10,13,16-hexaoxacyclooctadecane
14S4	1,4,8,11-tetrathiacyclotetradecane
16S4-(OH)$_2$	1,5,9,13-tetrathiacyclohexadecane-3,11-diol
18S6	1,4,7,10,13,16-hexathiacyclooctadecane
av.	average value
bpy	2,2′-bipyridine
Bu	*n*-butyl
depe	1,2-bis(diethylphosphino)ethane
dmpe	1,2-bis(dimethylphosphino)ethane
dmso	dimethylsulfoxide
dppe	1,2-bis(diphenylphosphino)ethane

ε	molar absorptivity $(M^{-1} cm^{-1})$
edtaH$_4$	ethylenediaminetetraacetic acid
en	1,2-ethanediamine
EPR	electron paramagnetic resonance
EXAFS	extended X-ray absorption fine structure
FABMS	fast atom bombardment mass spectrometry
HPLC	high performance liquid chromatography
LMCT	ligand-to-metal charge transfer
mntH$_2$	maleonitriledithiol
ox	oxalate(2-)
phen	1,10-phenanthroline
py	pyridine
SCE	saturated calomel electrode
tu	thiourea

Acknowledgements. The author thanks Dr. S.F. Colmanet for preparing the PLUTO figures.

6 References

1. Griffith WP (1972) Coord Chem Rev 8: 369
2. Baldas J, Bonnyman J, Pojer PM, Williams GA, Mackay MF (1981) J Chem Soc, Dalton Trans 1798
3. Kaden L, Lorenz B, Schmidt K, Sprinz H, Wahren M (1981) Isotopenpraxis 17: 174
4. Baldas J (1994) Adv Inorg Chem 41: 1
5. Cowie M, Lock CJL, Ozog J (1970) Canad J Chem 48: 3760
6. Lin Z, Hall MB (1993) Coord Chem Rev 123: 149
7. Kirmse R, Abram U (1990) Isotopenpraxis 26: 151
8. Raynor JB, Kemp TJ, Thyer AM (1992) Inorg Chim Acta 193: 191
9. Dehnicke K, Strähle J (1981) Angew Chem Int Ed Engl 20: 413
10. Dehnicke K, Strähle J (1992) Angew Chem Int Ed Engl 31: 955
11. Nicolini M, Bandoli G, Mazzi U (eds) (1986) Technetium in chemistry and nuclear medicine 2, Cortina International, Verona/Raven, New York
12. Nicolini M, Bandoli G, Mazzi U (eds) (1990) Technetium and rhenium in chemistry and nuclear medicine 3, Cortina International, Verona/Raven, New York
13. Kirmse R, Stach J, Abram U (1985) Polyhedron 8: 1403
14. Abram U, Kirmse R, Stach J, Lorenz B (1985) Z Chem 25: 153
15. Rummel S, Schnurpfeil D, Willecke L (1985) Z Chem 25: 150
16. Baldas J, Bonnyman J, Williams GA (1986) Inorg Chem 25: 150
17. Lorenz B, Schmidt K, Kaden L, Wahren M (1986) Isotopenpraxis 22: 444
18. Baldas J, Heath GN, Raptis RG, Williams GA (unpublished)
19. Baldas J, Boas JF, Colmanet SF, Mackay MF (1990) Inorg Chim Acta 170: 233
20. Baldas J, Boas JF, Colmanet SF, Williams GA (1992) J Chem Soc, Dalton Trans 2845
21. Purcell W, Potgieter IM, Damoense LJ, Leipoldt JG (1992) Transition Met Chem 17: 387
22. Leipoldt JG, Basson SS, Roodt A, Purcell W (1992) Polyhedron 11: 2277
23. Baldas J, Colmanet SF, Ivanov Z, Williams GA (unpublished)
24. Cotton FA, Kraihanzel CS (1962) J Am Chem Soc 84: 4432
25. Krasser W, Bohres EW, Schwochau K (1972) Z Naturforsch 27a: 1193

26. Abram U, Spies H, Abram S, Kirmse R, Stach J (1986) Z Chem 26: 140
27. Abram U, Abram S, Spies H, Kirmse R, Stach J, Köhler K (1987) Z Anorg Allg Chem 544: 167
28. Baldas J, Colmanet SF, Williams GA (unpublished)
29. Williams GA, Baldas J (1989) J Nucl Med Allied Sci 33: 327
30. Baldas J, Bonnyman J, Williams GA (1984) J Chem Soc, Dalton Trans 833
31. Hathaway BJ, Holah DG (1964) J Chem Soc 2400
32. Mutalib A, Sekine T, Omori T, Yoshihara K (1993) Radiochim Acta 63: 123
33. Abram U, Spies H (1984) Inorg Chim Acta 94: L3
34. Abram U, Spies H, Görner W, Kirmse R, Stach J (1985) Inorg Chim Acta 109: L9
35. Williams GA, Baldas J (1989) Aust J Chem 42: 875
36. Colmanet SF, Mackay MF (1988) Inorg Chim Acta 147: 173
37. Abram U, Münze R, Kirmse R, Köhler K, Dietzch W, Golič L (1990) Inorg Chim Acta 169: 49
38. Abram U, Abram S, Stach J, Dietzsch W, Hiller W (1991) Z Naturforsch 46b: 1183
39. Pasqualini R, Duatti A (1992) J Chem Soc, Chem Comm 1354
40. Rossi R, Marchi A, Magon L, Casellato U, Graziani R (1990) J Chem Soc, Dalton Trans 2923
41. Stach J, Abram U, Münze R (1990) in ref 12, p 79
42. Stach J, Dietzsch W, Abram U (1989) Z Chem 29: 295
43. Baldas J, Bonnyman J (1990) in ref 12 p 429
44. Baldas J, Bonnyman J, Mackay MF, Williams GA (1984) Aust J Chem 37: 751
45. Baldas J, Colmanet SF, Williams GA (1991) Aust J Chem 44: 1125
46. Colmanet SF, Mackay MF (1988) Aust J Chem 41: 151
47. Bandoli G, Mazzi U, Abram U, Spies H, Münze R (1987) Polyhedron 6: 1547
48. Baldas J, Bonnyman J (1988) Inorg Chim Acta 141: 153
49. Pietzsch H-J, Spies H, Leibnitz P, Reck G (1993) Polyhedron 12: 2995
50. Parkin G (1993) Coord Chem Rev 93: 887
51. Marchi A, Garuti P, Duatti A, Magon L, Rossi R, Ferretti V, Bertolasi V (1990) Inorg Chem 29: 2091
52. Marchi A, Rossi R, Magon L, Duatti A, Casellato U, Graziani R, Vidal M, Riche F (1990) J Chem Soc, Dalton Trans 1935
53. Duatti A, Marchi A, Bertolasi V, Ferretti V (1991) J Am Chem Soc 113: 9680
54. Bernardi R, Zanotti M, Bernardi G, Duatti A (1992) J Chem Soc, Chem Comm 1015
55. Bertolasi V, Ferretti V, Gilli P, Marchi A, Marvelli L (1991) Acta Crystallogr Sect C 47: 2535
56. Archer CM, Dilworth JR, Kelly JD, McPartlin M (1989) J Chem Soc, Chem Comm 375
57. Archer CM, Dilworth JR, Griffiths DV, McPartlin M, Kelly JD (1992) J Chem Soc, Dalton Trans 183
58. Clarke MJ, Lu J (1992) Inorg Chem 31: 2476
59. de Vries N, Costello CE, Jones AG, Davison A (1990) Inorg Chem 29: 1348
60. Helm L, Deutsch K, Deutsch EA, Merbach AE (1992) Helv Chim Acta 75: 210
61. Rummel S, Hermann M, Schmidt K (1985) Z Chem 25: 152
62. Dilworth JR, Griffiths DV, Hughes JM, Morton S, Hiller W, Archer CM, Kelly JD, Walton G (1992) Inorg Chim Acta 192: 59
63. Nicholson T, Davison A, Jones AG (1991) Inorg Chim Acta 187: 51
64. Duatti A, Marchi A, Pasqualini R (1990) J Chem Soc, Dalton Trans 3729
65. Abram U, Lorenz B, Kaden L, Scheller D (1988) Polyhedron 7: 285
66. Abram U, Mäding P, Kirmse R, Köhler K (1989) Z Chem 29: 183
67. Baldas J, Boas JF, Colmanet SF, Williams GA (1991) J Chem Soc, Dalton Trans 2441
68. Abrams MJ, Larsen SK, Shaikh SN, Zubieta J (1991) Inorg Chim Acta 185: 7
69. Batsanov AS, Struchkov YuT, Lorenz B, Olk B (1988) Z Anorg Allg Chem 564: 129
70. Kaden L, Lorenz B, Kirmse R, Stach J, Abram U (1985) Z Chem 25: 29
71. Hiller W, Hübener R, Lorenz B, Kaden L, Findeisen M, Stach J, Abram U (1991) Inorg Chim Acta 181: 161
72. Kaden L, Lorenz B, Kirmse R, Stach J, Behm H, Beurskens PT, Abram U (1990) Inorg Chim Acta 169: 43
73. Archer CM, Dilworth JR, Kelly JD, McPartlin M (1989) Polyhedron 8: 1879
74. Archer CM, Dilworth JR, Jobanputra P, Thompson RM, McPartlin M, Povey DC, Smith GW, Kelly JD (1990) Polyhedron 9: 1497
75. Chatt J, Choukroun R, Dilworth JR, Hyde J, Vella P, Zubieta J (1979) Transition Met Chem 4: 59

76. Marchi A, Marvelli L, Rossi R, Magon L, Uccelli L, Bertolasi V, Ferretti V, Zanobini F (1993) J Chem Soc, Dalton Trans 1281
77. Dilworth JR, Griffiths DW, Hughes JM, Morton S, Archer CM, Kelly JD (1992) Inorg Chim Acta 195: 145
78. Uhlemann E, Spies H, Pietzsch H-J, Herzschuh R (1992) Z Naturforsch 47b: 1441
79. Abram U, Abram S, Münze R, Jäger E-G, Stach J, Kirmse R (1990) in ref 12 p 69
80. Tisato F, Mazzi U, Bandoli G, Cros G, Darbieu M-H, Coulais Y, Guiraud R (1991) J Chem Soc, Dalton Trans 1301
81. Marchi A, Duatti A, Rossi R, Magon L, Pasqualini R, Bertolasi V, Ferretti V, Gilli G (1988) J Chem Soc, Dalton Trans 1743
82. Marchi A, Marvelli L, Rossi R, Magon L, Bertolasi V, Ferretti V, Gilli P (1992) J Chem Soc, Dalton Trans 1485
83. Abram U, Hartung J, Beyer L, Kirmse R, Köhler K (1987) Z Chem 27: 101
84. Abram U, Hartung J, Beyer L, Stach J, Kirmse R (1990) Z Chem 30: 180
85. Dilworth JR, Hutson AJ, Morton S, Harman M, Hursthouse MB, Zubieta J, Archer CM, Kelly JD (1992) Polyhedron 11: 2151
86. Abram U, Münze R, Jäger E-G, Stach J, Kirmse R (1989) Inorg Chim Acta 162: 171
87. Abram U, Abram S, Münze R, Jäger E-G, Stach J, Kirmse R, Admiraal G, Beurskens PT (1991) Inorg Chim Acta 182: 233
88. Marchi A, Rossi R, Marvelli L, Bertolasi V (1993) Inorg Chem 32: 4673
89. Rossi R, Marchi A, Aggio S, Magon L, Duatti A, Casellato U, Graziani R (1990) J Chem Soc, Dalton Trans 477
90. Baldas J, Colmanet SF, Ivanov Z, Williams GA (1994) J Chem Soc, Chem Comm 2153
91. Jurisson S, Berning D, Jia W, Ma D (1993) Chem Rev 93: 1137
92. Baldas J, Bonnyman J (1985) Int J Appl Radiat Isot 36: 133
93. Baldas J, Bonnyman J (1985) Int J Appl Radiat Isot 36: 919
94. Baldas J, Bonnyman J, Kanellos J (1986) in ref 11 p 103
95. Baldas J, Bonnyman J, Ivanov Z (1989) J Nucl Med 30: 1240
96. Dilworth JR, Archer CM, Latham IA, Kelly JD, Griffiths DV, York DC, Mahoney PM, Higley B (1991) Nucl Med Biol 18: 547
97. Alagui A, Apparu M, du Moulinet d'Hardemare A, Riche F, Vidal M (1989) Appl Radiat Isot 40: 813
98. Duatti A, Marchi R, Pasqualini R (1991) J Labelled Comp Radiopharm 30: 13
99. Pasqualini R, Comazzi V, Bellande E, Duatti A, Marchi A (1992) Appl Radiat Isot 43: 1329
100. Baldas J, Bonnyman J, Pojer PM, Williams GA (1982) Eur J Nucl Med 7: 187
101. Abram S, Abram U, Spies H, Münze R (1986) J Radioanal Nucl Chem, Articles 102: 309
102. Abram S, Beyer R (1990) Isotopenpraxis 26: 107
103. Pasqualini R, Duatti A, Bellande E, Comazzi V, Brucato V, Hoffschir D, Fagret D, Comet M (1994) J Nucl Med 35: 334
104. Drouillard S, Alagui A, Apparu M, Mathieu JP, Pasqualini R, Vidal M (1992) Appl Radiat Isot 43: 1227
105. Giganti M, Duatti A, Uccelli L, Cittanti C, Colamussi P, Piffanelli A (1994) Eur J Nucl Med 21: 806
106. Apparu M, Drouillard S, Mathieu JP, du Moulinet d'Hardemare A, Pasqualini R, Vidal M (1992) Appl Radiat Isot 43: 597
107. Coulais Y, Cros G, Darbieu MH, Gantet P, Tafani JAM, Vende D, Pasqualini R, Guiraud R (1993) Nucl Med Biol 20: 263
108. Coulais Y, Cros G, Darbieu MH, Tafani JAM, Belhadj-Tahar H, Bellande E, Pasqualini R, Guiraud R (1994) Nucl Med Biol 21: 263
109. Baldas J, Bonnyman J (1988) Nucl Med Biol 15: 451
110. Baldas J, Bonnyman J (1986) J Radioanal Nucl Chem, Letters 105: 267
111. Mutalib A, Sekine T, Omori T, Yoshihara K (1993) Radiochim Acta 63: 117
112. Borel M, Rapp M, Pasqualini R, Madelmont JC, Godeneche D, Veyre A (1992) Appl Radiat Isot 43: 425
113. Chappuis PP, Truffer S, Ianoz E, Lerch P (1993) J Radioanal Nucl Chem, Letters 175: 113
114. Marchi A, Marvelli L, Rossi R, Magon L, Uccelli L (1994) Int J Appl Radiat 45: 397
115. Kanellos J, Pieterz GA, McKenzie IFC, Bonnyman J, Baldas J (1986) J Nat Cancer Inst 77: 431
116. Stach J, Abram S, Abram U (1988) J Radioanal Nucl Chem, Letters 128: 131

117. Stiefel EI (1977) Progr Inorg Chem 22: 1
118. Baldas J, Colmanet SF, Ivanov Z, Williams GA, James BD (unpublished)
119. Baldas J, Boas JF, Bonnyman J, Williams GA (1984) J Chem Soc, Dalton Trans 2395
120. Abram U, Kirmse R (1988) J Radioanal Nucl Chem, Articles 122: 311
121. Abram U, Kirmse R (1993) Radiochim Acta 63: 139
122. Baldas J, Boas JF, Bonnyman J, Colmanet SF, Williams GA (1990) J Chem Soc, Chem Comm 1163
123. Baldas J, Boas JF, Ivanov Z, James BD (1993) Inorg Chim Acta 204: 199
124. Davison A, Jones AG (1982) Int J Appl Radiat Isot 33: 875
125. Baldas J, Colmanet SF, Williams GA (1991) Inorg Chim Acta 179: 189
126. Baldas J, Boas JF, Bonnyman J (1987) J Chem Soc, Dalton Trans 1721
127. Thomas RW, Heeg MJ, Elder RC, Deutsch E (1985) Inorg Chem 24: 1472
128. Mantegazzi D, Ianoz E, Lerch P, Tatsumi K (1990) Inorg Chim Acta 167: 195
129. Lorenz B (1990) Isotopenpraxis 26: 452
130. Abram U, Wollert R (1993) Radiochim Acta 63: 149
131. Baldas J, Ivanov Z (unpublished)
132. Mantegazzi D, Ianoz E, Lerch P, Nicolo F, Schenk K, Chapuis G (1990) in ref 12, p 153
133. Baldas J, Boas JF, Colmanet SF, Rae AD, Williams GA (1993) Proc R Soc Lond A 442: 437
134. Baldas J, Colmanet SF (1989) Aust J Chem 42: 1155
135. Baldas J, Bonnyman J, Williams GA (1985) Aust J Chem 38: 215
136. Hübener R, Abram U (1992) Z Anorg Allg Chem 617: 96
137. Lorenz B, Kränke P, Schmidt K, Kirmse R, Hübener R, Abram U (1994) Z Anorg Allg Chem 620: 921
138. Baldas J, Boas JF, Bonnyman J (1989) Aust J Chem 42: 639
139. Pietzsch H-J, Abram U, Kirmse R, Köhler K (1987) Z Chem 27: 265
140. Abrams MJ, Costello CE, Shaikh SN, Zubieta J (1991) Inorg Chim Acta 180: 9
141. Lu J, Clarke MJ (1990) Inorg Chem 29: 4123
142. Colmanet SF, Williams GA, Mackay MF (1987) J Chem Soc, Dalton Trans 2305
143. Baran EJ, Cabello CI (1983) Z Naturforsch 38a: 563
144. Winkler JR, Gray HB (1981) Comments Inorg Chem 1: 257
145. Lever ABP (1984) Inorganic electronic spectroscopy. 2nd edn. Elsevier, Amsterdam
146. Sabel DM, Gewirth AA (1994) Inorg Chem 33: 148
147. Baldas J, Boas JF, Ivanov Z, James BD (unpublished)
148. Lever ABP (1974) J Chem Ed 51: 612
149. Figgis BN et al, cited in ref 133
150. Kirmse R, Stach J, Abram U (1986) Inorg Chim Acta 117: 117
151. Köhler K, Kirmse R, Abram U (1986) Z Chem 26: 339
152. Kirmse R, Köhler K, Abram U, Böttcher R, Golič L, de Boer E (1990) Chem Phys 143: 75
153. Köhler K, Kirmse R, Böttcher R, Abram U, Gribnau MCM, Keijzers CP, de Boer E (1990) Chem Phys 143: 83
154. Köhler K, Kirmse R, Böttcher R, Abram U (1991) Chem Phys 160: 281
155. Figgis BN, Reynolds PA, Cable JW (1993) J Chem Phys 98: 7743
156. Köhler K, Kirmse R, Abram U (1991) Z Anorg Allg Chem 600: 83
157. Baldas J, Boas JF, Bonnyman J, Colmanet SF (1990) in ref 12 p 63
158. Abram U, Köhler K, Kirmse R, Kalinichenko NB, Marov IN (1990) Inorg Chim Acta 176: 139
159. Baldas J, Boas JF, Ivanov Z (unpublished)
160. Baldas J, Colmanet SF, Williams GA (1991) J Chem Soc, Chem Comm 954
161. Baldas J, Colmanet SF, Craig DC, Rae AD, Williams GA (unpublished)
162. Baldas J, Colmanet SF, Mackay MF (1988) J Chem Soc, Dalton Trans 1725
163. Baldas J, Boas JF, Bonnyman J, Colmanet SF, Williams GA (1991) Inorg Chim Acta 179: 151
164. Herrmann W, Alberto R, Kiprof P, Baumgärtner F (1990) Angew Chem Int Ed Engl 29: 189
165. Startsev AN, Klimov OV, Shkuropat SA, Fedotov MA, Degtyarev SP, Kochubey DI (1994) Polyhedron 13: 505
166. Baldas J, Boas JF, Colmanet SF, Ivanov Z, Williams GA (1993) Radiochim Acta 63: 111
167. Williams GA, Martin LJ (personal communication)
168. Pillai MRA, John CS, Lo JM, Schlemper EO, Troutner DE (1990) Inorg Chem 29: 1850
169. Pietzsch H-J, Spies H, Leibnitz P, Reck G, Beger J, Jacobi R (1993) Polyhedron 12: 187
170. Lin Z, Hall MB (1991) Inorg Chem 30: 3817
171. Ricard L, Martin C, Wiest R, Weiss R (1975) Inorg Chem 14: 2300

172. Moynihan KJ, Boorman PM, Ball JM, Patel VD, Kerr KA (1982) Acta Crystallogr Sect B 38: 2258
173. Burrell AK, Bryan JC (1993) Angew Chem Int Ed Engl 32: 94
174. Baldas J, Boas JF (1988) J Chem Soc, Dalton Trans 2585
175. Takayama T, Sekine T, Yoshihara K (1993) J Radioanal Nucl Chem, Letters 176: 325
176. Baldas J, Colmanet SF, Mackay MF (1989) J Chem Soc, Chem Comm 1890
177. Baldas J, Colmanet SF (1990) Inorg Chim Acta 176: 1
178. Baldas J, Colmanet SF, Williams GA (1991) J Chem Soc, Dalton Trans 1631
179. Abrams MJ, Chen Q, Shaikh SN, Zubieta J (1990) Inorg Chim Acta 176: 11
180. Haymore BL, Maatta EA, Wentworth RAD (1979) J Am Chem Soc 101: 2063

Technetium(V) Chemistry as Relevant to Nuclear Medicine

Bernd Johannsen and Hartmut Spies

Institut für Bioanorganische und Radiopharmazeutische Chemie, Forschungszentrum Rossendorf, P.O.B. 510119, D-01314 Dresden, Germany

Table of Contents

Topics in Current Chemistry, Vol. 176
© Springer-Verlag Berlin Heidelberg 1996

This review covers the coordination chemistry of technetium(V) and – to some extent – rhenium(V). In this survey, the complexes are arranged according to donor-atom sets in the coordination sphere.

Acknowledging the important part that ^{99m}Tc(V) radiopharmaceuticals are playing in nuclear medicine, advances in their coordination chemistry, as attained by new syntheses and better structural characterization, are discussed with respect to the design of Tc(V) radiotracers. In the light of current interest in making technetium complexes active in vivo, a chapter considering several aspects of reactivity is included.

1 Introduction

Until the mid 1970s, technetium(V) had been widely ignored and its chemistry misinterpreted. Then, the development of the chemistry of technetium(V) was prompted by the trends and needs of nuclear medicine, which predominantly uses technetium-99m radiopharmaceuticals for a broad range of diagnostics.

The better known rhenium(V) chemistry is of great interest, not only because it resembles the second-row congener technetium and easily permits non-radioactive model studies for technetium. ^{186}Re and ^{188}Re are attractive isotopes for therapeutic radiopharmaceuticals. ^{99m}Tc and ^{186}Re can be considered to be a "matched pair" for diagnostics and therapy.

Aside from ^{99m}Tc tagged radiopharmaceuticals, in which technetium is attached to large molecules without significantly interfering with the properties of the native compound, the majority of Tc complexes have to possess special molecular configurations and properties, in order to allow their application as tracers in modern single-photon emission tomography (SPECT) or other specific functional studies. Among the various oxidation states of technetium explored in radiotracer design, Tc(V) has proven to be the most suitable for synthesis of well-defined monomeric complexes that are sufficiently stable in aqueous solution and permit great variability in molecular structure and properties. Modern technetium-99m radiopharmaceuticals for brain perfusion imaging or renal function studies make use of the predominant Tc(V)oxo core and the mostly square-pyramidal configuration of the complexes. The focus on tomography and on the specific functions to be assessed in nuclear medicine diagnostics has initiated a quest for new technetium complexes, which are not inert in vivo. Basically, molecular interactions with the target organ are essential, either to trap blood flow tracers in the brain, myocardium, or other targets, or to handle the tracer or bind it specifically. Technetium tracers that bind preferentially to a specific site, and respond to a biochemical change of the binding site as a function of a specific disease, would bring routine nuclear medicine closer to characterizing the patient's problems in regiospecific biochemical terms. Positron emission tomography (PET) with organic tracer molecules, which is mainly restricted to research, paves the way for such an in vivo biochemistry. It is an important task to extend this approach to coordination chemistry from PET to SPECT. Receptor binding molecules, peptides and specific antibodies are among the important challenges for the technetium chemist. At first glance, the nonphysiological element, technetium, does not appear very suitable for such a purpose. The challenging question is whether reactivity can be obtained by appropriate molecule design and will persist under the mild conditions prevailing in vivo.

Although technetium(V) complexes have been studied extensively in order to design new radiotracers, investigations of in vivo reactive complexes are still in the formative stages and their potential for "mimicking" biological substrates is not yet predictable. It is the intention of this review to give an introduction into the chemistry of technetium(V) and to some extent rhenium(V) and to outline this relatively new and expanding interdisciplinary area of research.

2 Cores

Owing to its central position in the d-block elements, technetium exhibits a wide variety of oxidation states, coordination numbers, and coordination geometries. As far as radiopharmaceutically relevant chemistry in aqueous solution is concerned, technetium(V) chemistry is governed by oxotechnetium species. (The chemistry of nitridotechnetium(V) complexes is dealt with in preceding chapter of this monograph [Chapt. 3]). The reason for the appearance of oxo species is the need for neutralization of the high formal charge on technetium(V). Whether the Tc(V) center will contain one or two oxo atoms depends on the ability of the coordinated ligands to donate negative charge to the metal [1, 2]. The *oxo* chemistry of technetium(V) resembles that found for the isoelectronic molybdenum(VI) [3]. Oxotechnetium species are most conveniently categorized on the basis of their cores (Fig. 1). One dominant structural element is the monooxotechnetium core, TcO^{3+}. The presence of the *oxo* ligand has a significant effect on the structure and reactivity of the derived complexes. One such

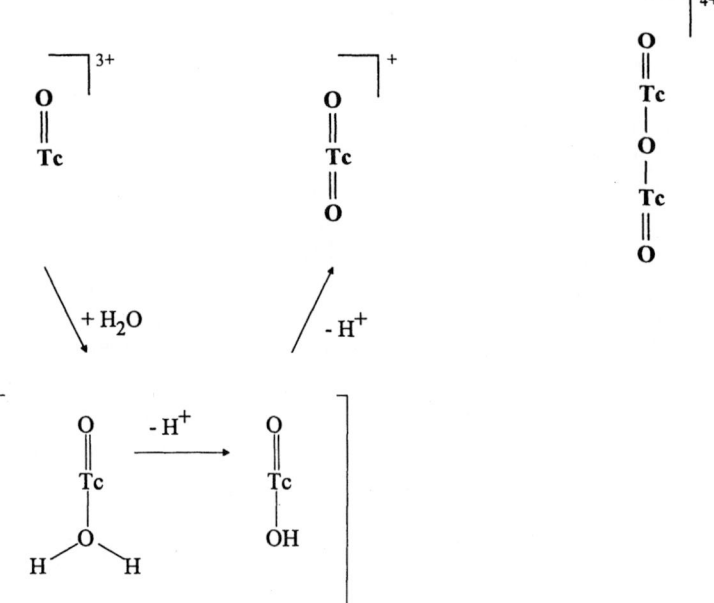

Fig. 1. Oxotechnetium cores

consequence is that the TcO group induces a large *trans* effect, which favors formation of five-coordinated complexes by making the *trans* position labile. Furthermore, the large steric requirement related to the short Tc=O distance causes the Tc center to be located significantly above the plane of the four ancillary ligand atoms, and results in a square-pyramidal arrangement of the donor atoms. This is represented schematically in Fig. 2 and is realized in a large variety of complexes, ranging from the most simple one – five coordinated [TcOCl$_4$]$^-$ containing four monodentate ligands – to those containing more complicated ligands having two or more donor groups.

As observed from X-ray crystal data, which are available for a variety of complexes [4, 5, 6], *oxo*technetium complexes show the following general feature: the metal lies above the equatorial plane of the four basal ligand atoms and the oxo group is at the apex. Tc=O bond lengths correlate with the displacement of the technetium atom from the basal plane towards the oxygen at the apical position.

A good indicator of the presence of the Tc=O group in the complex is its stretching vibration, observed in the ir spectra as a sharp, intensive band. Its values range from 890 to about 1020 cm^{-1} and are related to the electronegativities of the ligands in the equatorial plane as well as of the presence and nature of a *trans* ligand.

In [TcOCl$_4$]$^-$, there are different labilities in the molecule. While the *oxo* position is inert to substitution, the four chlorine ligands may be replaced by other donor atoms to give a variety of TcO complexes. The position *trans* to the oxo group is very labile and is free in the special [TcOCl$_4$]$^-$ complex. Coordination of the sixth position may occur with ligands which occupy equatorial positions and force an additional donor group of the ligand into the position *trans* to the oxo group. That is the case, for example, with the complexes of penicillamine [7] or the tripodal ligand hydrotris(1-pyrazolyl)borate, HBPz3 [8].

Because it appears that the major factor of the formation of oxo complexes is neutralization of the high formal charge on technetium(V), negatively charged ancillary ligands favour monooxospecies as shown above. Neutral ligands, however, such as amines, or those with efficient π-back-bonding properties,

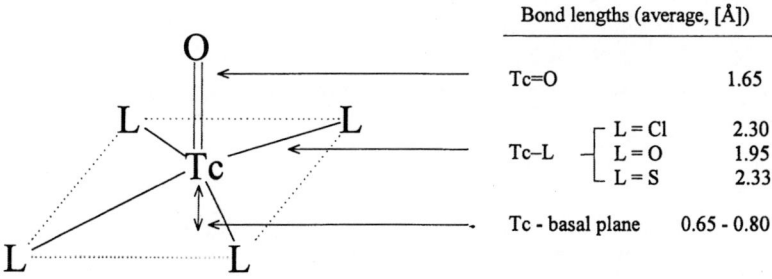

Fig. 2. Schematical representation of the square-pyramidal TcOL$_4$ unit

favour the formation of another important type of complexes, the bisoxotechnetium(V) species containing the TcO_2^+ core. The complexes $[MO_2L_4]$ are six-coordinated and adopt octahedral geometry. This type of Tc(V) complexes is exemplified by amine complexes such as $[TcO_2(py)_4]^+$ [9, 10, 11]. The mean values of the Tc $=$ O and the Tc–L_{eq} bond lengths in hexacoordinated technetium complexes are greater than in penta-coordinated compounds, obviously due to lesser steric hindrance in the latter [5].

A third group of oxo complexes contains a $Tc_2O_3^{4+}$ core. A structure of this type can be thought as being derived from the bisoxo core. The oxo group can function as a ligand to a second Tc center, thus generating a nearly linear oxygen bridged structure. While bridge-formation converts a double-bonded oxo ligand into a bis-monodentate ligand, the Tc–O bond lengthens considerably. Examples of complexes derived from this structural element are tetradentate N_2O_2 Schiff base ligands [12] and bidentate neutral thioethers [13].

Kinetic and structural studies, mainly with cyanide and pyridine ligands, have shown [14, 15, 16] that the chemistry is very complicated, involving several kinds of transitions between the above-mentioned structures. As the sixth position is either free or substitutionally labile, enabling hydrolysis or alcoholysis, to generate trans-$TcO(OR)^{2-}$ or trans-TcO_2^+ species, the trans-TcO_2^+ species can undergo successive protonation reactions quite reversibly, to form the corresponding hydroxo oxo and aqua oxo species. Additionally, coordinated water may be displaced by monodentate ligands. An illustrative example – showing the versatility of the oxotechnetium core – is $[TcO(CN)_5]^{2-}$, which can be recrystallized from water unchanged in the presence of excess cyanide. In the absence of cyanide however, trans-$[TcO_2(CN)_4]^{3-}$ is formed quantitatively [17].

3 Donor Atoms

3.1 Halide and Pseudohalide Donors

Halides form with Tc(V) both neutral compounds, such as $TcOX_3$ (X $=$ F, Cl, Br), and oxohalide anions. Fluoride differs from the other halides by forming $KTcF_6$ [18]. For X $=$ Cl and Br, simple oxohalotechnetates $[TcOX_4]^-$ are accessible in a reaction of pertechnetate with concentrated acids HX according to

$$[TcO_4]^- + 6HX \rightarrow [TcOX_4]^- + 3H_2O + X_2 .$$

Further reduction to the Tc(IV) species, $[TcX_6]^{2-}$, is slow compared with their rapid formation from pertechnetate. This fact, as well as trapping by precipitation with large organic cations, allows isolation of the thermodynamically less favoured $TcOX_4^-$ anions.

Tetrachlorooxotechnetate(V) results from action of conc. HCl on pertechnetate at ambient temperatures and is preferably isolated as the tetrabutylammonium salt [19]. Tetrabromooxotechnetate(V) was similarly obtained with hydrobromic acid at 0 °C [8]. The molecular structures of both compounds are reported in [20, 21]. The analogous iodo complex, tetraiodooxotechnetate(V), was synthesized by ligand exchange of the chloro compound with sodium iodide in acetone [22]. However, it suffers from considerable decomposition during isolation.

All $[TcOX_4]^-$ species hydrolyse in aqueous solution and then disproportionate into pertechnetate and Tc(IV) oxide hydrate according to

$$3[TcOX_4]^- + 7H_2O \rightarrow [TcO_4]^- + 2(TcO(OH)_2)_n + 10HX + 2X^-.$$

The tendency to hydrolyse increases from the chloro to the iodo analogues. Oxohalide technetates are often used as precursors for oxotechnetium complexes, thus making use of their reactivity and good solubility of their tetraalkyl ammonium salts in various organic solvents. Particularly, salts of tetrachlorooxotechnetate(V) have proven to be such effective starting materials.

3.2 O-Donors

The technetium(V) and rhenium(V) oxo core, although soft rather than hard, binds the hard donor atom oxygen. Actually, there is a large variety of technetium and rhenium complexes with O-donor ligands. The assortment of ligands, with or without carboxylic acid groups, comprises simple polyols, diphenols, di- and tripodal catechols, as well as OH/COOH-containing ligands, such as hydroxy carbonic acids or carbohydrate derivatives. OH-ligands act in their deprotonated forms, e.g., as alkoxy or carboxylate groups.

A general mode of access to polyhydric complexes of Tc(V) is reduction of pertechnetate with two equivalents of stannous chloride in aqueous solution of the excess O-donor ligand, e.g.:

$$TcO_4^- + Sn^{2+} + Na \text{ gluconate (excess)} \rightarrow [TcO(gluc)_2]^-.$$

The reaction can be easily monitored by UV spectroscopy (Fig. 3). A common feature of this type of complex with polyhydroxy alcohols and hydroxycarbonic acids is a low intensive absorption (lg ε about 2) near 500 nm [23]. Alternatively, the complexes can be prepared from $[TcOCl_4]^-$ in organic solvents. Apart from a few cases, the complexes have not been structurally characterized, because of the difficulty of obtaining pure compounds in crystalline form. The Tc glycolato complex [24, 25] has been identified as $[TcO(OCH_2CH_2O)_2]^-$. Analogous complexes with catechol [25], tetrachlorocatechol [26] and 4-nitrocatechol [27] have been prepared and characterized by X-ray crystal-structure analysis. The anion shows the square-pyramidal arrangement typical of five-coordinated oxotechnetium(V) complexes with the short Tc-O(oxo) bond and longer Tc-O bonds between the metal and the diol oxygen atoms (Fig. 4).

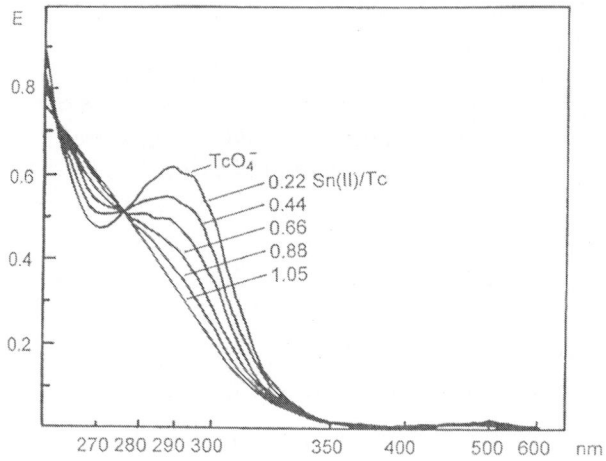

Fig. 3. UV spectra of the formation of $[TcO(gluc)_2]^-$ from pertechnetate

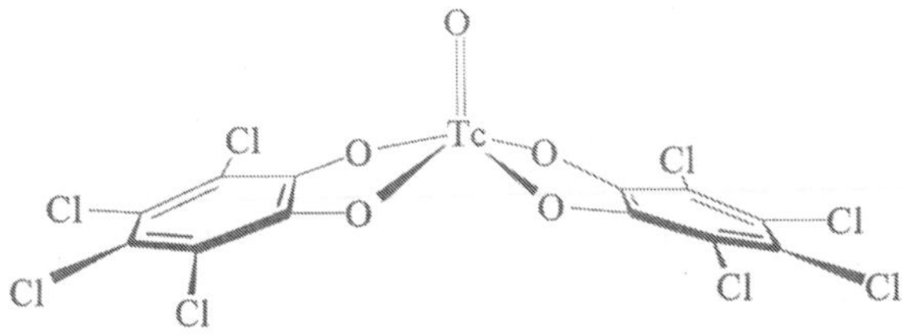

Fig. 4. Tc complex of tetrachloro catechol [26]

Hydroxycarbonic acids, such as citric, tartaric, malic, and hydroxyisobutyric acid, but also gluconate or glucoheptonate, are assumed to form the same kind of complex as the polyols do [23, 28, 29].

The O-donor complexes of Tc(V) exhibit moderate and differential stability in aqueous solution. In the presence of reducing agents, such as stannous chloride, they are reduced to mainly undefined products of Tc in a lower oxidation state. However, at the low technetium concentration of ^{99m}Tc that is used in nuclear medicine, the rate of the reduction process is very low. This makes it possible to prepare Tc(V) radiopharmaceuticals with O-donor ligands by the usual procedure, in which an excess of reducing agent over technetium is unavoidably used. The Tc(V) complexes also tend either to be easily oxidized or to disproportionate [23].

Another aspect of stability is ligand exchange, which can easily occur with stronger ligands, such as mercaptides, according to the rank order of donor atoms toward Tc(V). This qualifies them as a convenient starting material for preparing a great number of metal complexes in the given oxidation state $+5$.

O-donor complexes that are currently used as Tc(V) precursors in the chemical and radiopharmaceutical synthesis are listed in Table 1.

The complexes of the ligands listed above differ widely in their stability, which makes it possible to select the most suitable candidate for the given attacking ligand and the appropriate reaction conditions. The short-chained compounds, like the ethylene glycol complex, are more reactive in water than the longer chained ones. This also implies a greater tendency to decompose, which should be recognized in the interpretation of analytical procedures.

Tc(V) gluconate and glucoheptonate are often used when the reaction should be carried out in a neutral aqueous- or mixed aqueous/organic medium, and a rapid exchange reaction and high radiochemical purity of the product is required. In radiopharmaceutical preparations, Tc(V) tartrate is also quite often used. For labelling modified antibodies Tc(V) tricine has recently been particularly recommended [33].

Monobasic bidentate O-donor ligands such as 3-oxy-4-pyronates and 3-oxy-4-pyridinonates form neutral complexes [TcOL$_2$Cl] and [TcOLCl$_3$] [34]. Technetium and rhenium(V) chemistry was enriched by the introduction of tripodal O-donor ligands (Fig. 5). The di- and tricatechole ligands DIPACE and TRIPACE reduce ^{99m}TcO$_4^-$ to form stable species. The resulting complexes are assumed to contain trans-TcO$_2^+$ units [35, 36].

Table 1. TcO(ligand)$_2$]$^{n-}$ precursors used in the chemical and radiopharmaceutical synthesis of Tc(V) complexes

Ligand	References
Glycol	[23, 30]
Catechol	[26, 30]
Tartrate	
α-Hydroxyisobutyrate	
Citrate	[28, 29]
Gluconate	[31]
Glucoheptonate	[32]
Tricine, (trishydroxymethyl)methylglycine	[33]

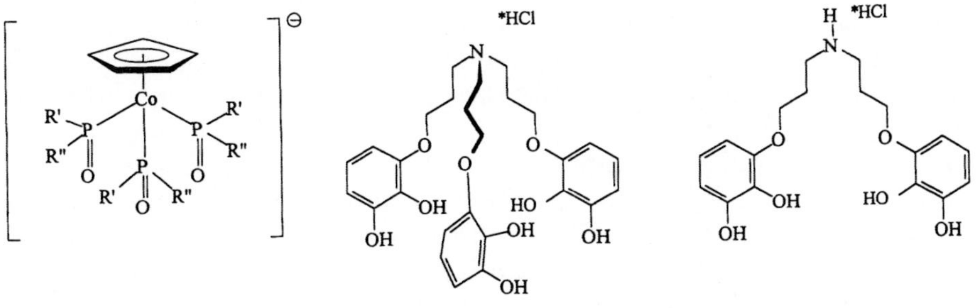

Fig. 5. Tripodale ligands L$_{OR}$ (a), TRIPACE (b), and the analog DIPACE (c)

Another type of ligand is the monoanionic, tridentate oxygen donor $[(C_5H_4R)Co-(P(O)R'R'')_3]^-$ (L_{OR}), which has been used to prepare the complexes of technetium [37] and rhenium: [38] $[MO_3L]$ and $[MOX_2L]$ (X Cl, Br). These complexes are stable in organic solvents but hydrolyse slowly in water. In order to evaluate their usefulness in radioimmunotherapy, the corresponding compounds were also prepared with radioactive rhenium isotopes.

3.3 N-Donors

N-donor ligands for Tc(V) and Re(V) are amines, amides, imines and oximes and combinations of these groups. Among these various groups significant differences exist in their ability to form stable complexes, and in the nature of these complexes, mainly because of the different basicity of the groups.

Aliphatic, aromatic and heterocyclic **amines** are all capable of serving as ligands in complexing the metal, either alone or in combination with other donor atoms. As amines are basic, the N-donor atoms remain protonated on pure-amine-N-coordination to Tc(V) or Re(V). Amines coordinate as neutral donors. Therefore, the positive charge of the metal has to be diminished by additional charge-donating groups. This often results in dioxo complexes of the general type $[MO_2(N_4)]^+$ containing $trans$-$[MO_2]^+$. When amine ligands undergo ligand exchange with oxotechnetium(V) precursors such as $[TcOCl_4]^-$, the formation of $[TcO(N)_4]^{3+}$ or deprotonated species is not to be expected. TcO_2-species are formed, which means that water coordinates $trans$ to the oxo group, which is sufficiently acid so that a second oxo group is formed by double deprotonation. This underlines the high-charge requirements of the Tc(V) center, that are poorly satisfied by the σ-donating amine ligands. In this respect, the amine-N donor ligands differ from π-donor ligands such as halides, thiolates, and alcoholates, which tend to form complexes containing the oxo metal group $[M=O]^{3+}$. One exception is 1,2-diaminobenzene. The highly delocalized, unsaturated ligand acts in its deprotonated form to give the anionic complex $[TcO(N_2)_2]^-$ [39].

The Tc-pyridine complex has been known since the early 70s [9, 11]. Cationic Tc complexes with aliphatic amines and N-containing heterocyclic compounds were described in the early 80s [40, 41, 42]. The structure of two very similar Tc/Re complexes, ethylenediamine [43] and the macrocyclic tetra-aza ligand 1, 4, 8, 11-tetraazacyclotetradecane (cyclam) [44, 45], are shown in Fig. 6. The Tc–O distances in the characteristic $trans$-TcO_2 centers are about 0.1 Å longer than the average Tc–O distance in monooxo complexes [4, 5, 6]. While in monooxo complexes without a $trans$ ligand the single oxygen donor forms a stronger bond, utilizing more of the metals σ- and π-orbitals, lengthening of the Tc–O bond distance in the dioxo complexes is the result of competition between the two $trans$ oxygen atoms for the same σ- and π-orbitals of the metal. The average Tc–N bond lengths are 2.15 Å for $[TcO_2(en)_2]^+$ and 2.125 Å for $[TcO_2(cyclam)]^+$.

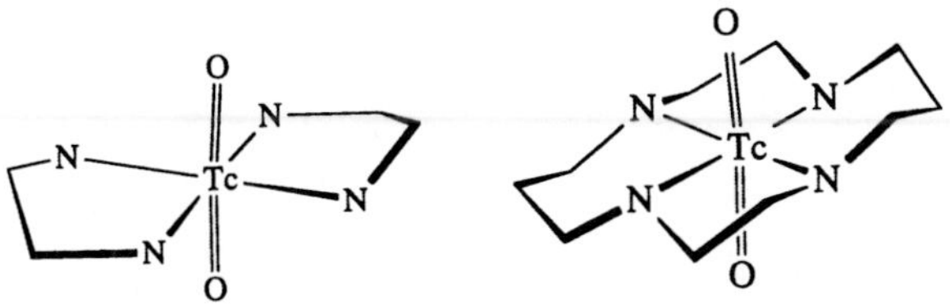

Fig. 6. $[TcO_2(en)_2]^+$ [43] and $[TcO_2(cyclam)]^+$ cations [44]

The N_4 donor set, with its potential to produce cationic complexes, has been of limited interest in nuclear medicine. A cyclam-based bifunctional chelating agent has been recommended for the preparation of antibodies specifically labelled with ^{99m}Tc [46]. To avoid nonspecific binding of the metal to the protein, the kinetically inert complex with the cyclam derivative is formed and subsequently coupled to the antibodies. Recently [47], a bifunctional tetraamine chelating group has been used successfully to balance the lipophilicity/hydrophilicity of an octreotide derivative, which has a high affinity for somatostatin receptors and will be used for the scintigraphic localization of receptor-positive tumours. The N_4 group is coupled to octreotide through a *thiourea bond* (Fig. 7). For the same reason, the approach has also been used to synthesize a new biotin conjugate, ^{99m}TcN_4-Lys-biotin [48], which retains its affinity to avidin. Recently, a modification of the cyclam ligand has been recommended (Fig. 8) in which one or two amide group(s) are substituted for one or two amine group(s) [49, 50]. Monooxocyclam and dioxocyclam still seem to form complexes with the *trans dioxo* metal core, although structural proof has not been given [50]. The deprotonable amide in this metal/ligand system seems to be unable to make the N_4 complex with the *trans oxo* ligand. The amide-N operates however at pH > 3.5 as anionic donor to give the neutral [TcO$_2$oxocyclam] or [TcO$_2$dioxocyclam] complex, at pH > 11.5 even the anionic complex [TcO$_2$dioxocyclam] $^-$.

N-donor groups in the form of ε-amino groups of lysine as nonspecific binding sites for technetium may play an undesired role in Tc-labelling of monoclonal antibodies for tumour imaging [51]. Nonspecifically bound Tc has a poor in vivo stability and appears to increase the undesired liver uptake and reduces tumour uptake. The metal oxidation state and coordination is not yet known.

Going from aliphatic to aromatic amines, complexes with N-heterocycles such as pyridine [9, 10], bipyridine [52], phenanthroline [52], and imidazoles [53], have been described. The complex $[TcO_2(py)_4]^+$ has long been known [9, 11]. Pyridines, mainly those having electron-withdrawing substituents (CN, NO$_2$) in the *para*-position – and thus decreased donor ability of the ring nitrogen – give rise to complexes with halogen atoms as co-ligands. Imidazole complexes [53] resemble aliphatic amine complexes [43] in their structure.

The stability of the complexes differs significantly, depending on the nature of the amine. Complexes with chelating amines persist in water for some time.

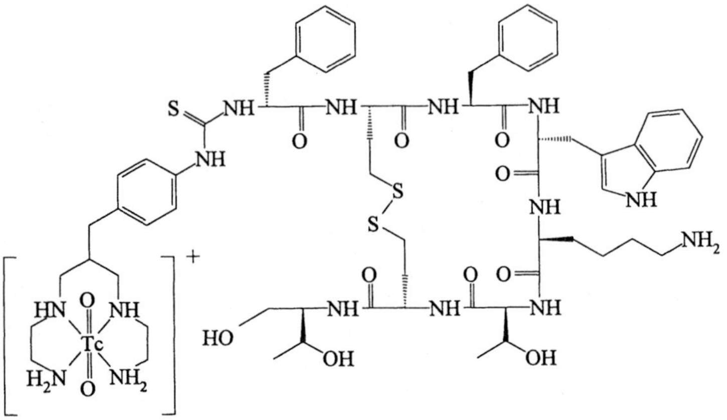

Fig. 7. ^{99m}Tc [N$_4$-(D)Phe^1]-octreotide [47]

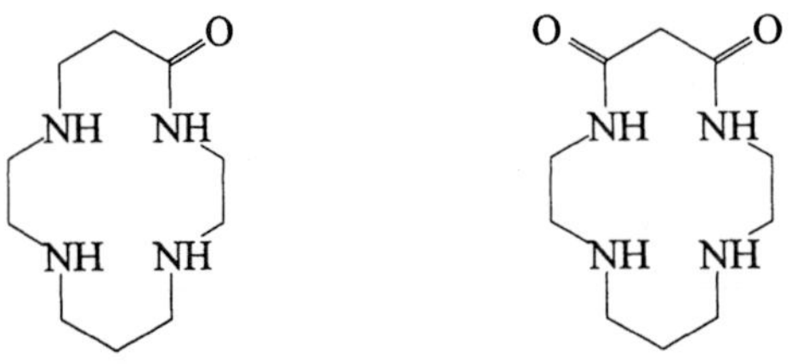

Fig. 8. Monooxo- and dioxocyclam

Diamine chelate complexes are more stable than the monodentate amine hetero-cycles and, therefore, can be studied under physiological conditions. The imidazole complexes are unstable in aqueous solution and decompose rapidly to technetium oxide hydrate. Six-membered ring chelates are significantly less stable than five-membered ones. Lesser flexibility of the ligand, such as 1.2-diamino-cyclohexane, parallels somewhat lower stability of the complex [53]:

1.2-diaminoethanes > 1.2-diaminocyclohexane > 1.3-diaminopropane.

In acid media, the complexes undergo rapid hydrolysis to yield free diamines, as shown for [TcO$_2$(en)$_2$]$^+$ [43]. The stability of the complexes is between that of O-coordinated and S-coordinated complexes, as shown in the reaction sequence of Tc(V) gluconate with pyridine, ethylenediamine or o-phenylenediamine to give the cationic N-coordinated species; these, in turn, are able to exchange the nitrogen ligands by dithiols, such as DMSA [40].

The inability of amines to deprotonate upon coordination, and thus to compensate the charge of the MO^{3+} core, can be overcome by combination with other types of donors – including N-donor groups – in the ligand, as has already been discussed above for oxocyclam. A prominent example is provided by technetium complexes of tetradentate amine oxime ligands.

Oximes initially played a minor role in Tc(V) chemistry. A Tc(V) complex of dimethylglyoxime with uncommon seven-coordination was synthesized and characterized. The complex [Tc(dimethylglyoxime)$_3$SnCl$_3$(OH)]·3H$_2$O involves Sn, which was introduced as stannous ion to reduce pertechnetate in the synthesis of the complex [54]. Oximes came into the limelight when the neutral Tc(V) complexes of amine oximes, PnAO and analogs (Fig. 9), proved to be very useful in nuclear medicine [55, 56]. On complexation, the open-chain dioxime ligand loses two protons from the two NH groups and – unlike cyclam – forms a complex with the monooxo core [Tc=O]$^{3+}$ (Fig. 10). The +3 charge is compensated by the two deprotonated N atoms and the one OH group of the dioxime, which forms an intramolecular hydrogen bond. The four nitrogen atoms in the complex are sp^2-hybridized, with all bonds in one plane [57]. The Tc=O bond length of 1.676 Å is in the range expected for TcO complexes, although at its long end, thus indicating a weakening of the Tc=O bond by the multiple-bond character of the two Tc–N amine bonds (average 1.9 Å in the basal plane [57, 58]).

An interesting insight into the nature of this kind of complex has been provided by variation of the ring size, going from an ethylene hydrocarbon backbone between the two amines to a pentyl bridge [58]. The ethylene analog of PnAO, with its propylene bridge, has the same crystal structure as the Tc PnAO complex. The pentyl analog, however, gives a six-coordinate dioxo complex of slightly distorted octahedral geometry, in which the technetium atom lies in the plane of the four nitrogens. The two amines are no longer deprotonated, only one oxime proton is lost, resulting in a neutral complex. Thus, both *monooxo* and *dioxo* complexes exist within this class of N-donor

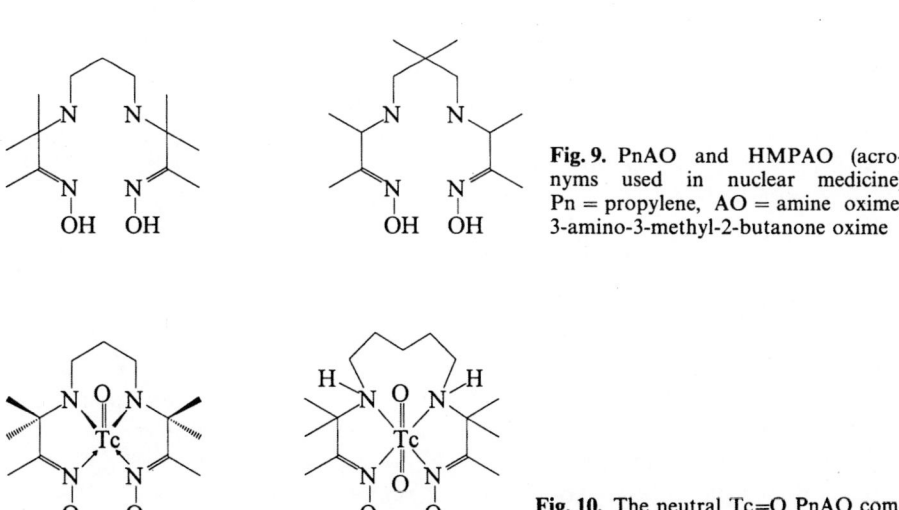

Fig. 9. PnAO and HMPAO (acronyms used in nuclear medicine) Pn = propylene, AO = amine oxime; 3-amino-3-methyl-2-butanone oxime

Fig. 10. The neutral Tc=O PnAO complex and the six-coordinate pentyldioxo analog [58]

ligands, with the amines donating either an anionic nitrogen or hard neutral donor N atom. The monooxo Tc(V) core is favoured, but obviously inhibited by steric constraints in the pentyl analog.

Systematic studies have been carried out to optimize the biological behaviour of the Tc-PnAO complex, particularly its retention in the brain, by variation of the number and position of methyl substituents on the carbon backbone of the ligand. On this basis the radiopharmaceutical ^{99m}Tc- D,L-HMPAO [59] for brain perfusion studies has been introduced, as well as PnAO derivatives used as bifunctional chelating agents to prepare technetium- and rhenium-labelled biomolecules [60, 61, 62, 63]. The optically active ligand HMPAO gives Tc- D,L- and *meso*-HMPAO complexes, which are able to penetrate the blood-brain barrier and remain trapped within the brain due to a rapid conversion into a nondiffusible species. The D,L complex is more unstable than the *meso* complex; the instability and the mechanism of the conversion that takes place in vivo is not understood. In the case of the PnAO complex, impurities present in the ^{99m}TcO$_4^-$ eluate have recently been blamed for the instability. The secondary complex formed was converted back to the primary complex by heating [64]. Glutathione (GSH) within cells is considered to be responsible for the rapid conversion of the lipophilic complex to an anionic species of unknown structure. This hypothesis is based on the reactivity of GSH with ^{99m}Tc-D,L-HMPAO in aqueous solution and supported by the in vivo correlation observed between GSH concentration and ^{99m}Tc- D,L-HMPAO uptake in brain tumours [65]. The reaction rate is much slower with ^{99m}Tc-*meso*-HMPAO relative to the D,L-isomer, the second-order rate constants differ by a factor of *ca.* 20 [66]. The cerebral uptake of ^{99m}Tc-*meso*-HMPAO is also lower, roughly half of ^{99m}Tc-D,L-HMPAO. Surprisingly, the analogous TMPAO complexes, without the two methyl groups at the sixth position on the ligand backbone, do not fit into the simple GSH-governed trapping mechanism. The complex with the *meso*-ligand has a similar slow in vitro GSH decomposition rate as ^{99m}Tc *meso*-HMPAO, but its cerebral uptake is as high as that of ^{99m}Tc-D,L-HMPAO [66].

Another interesting aspect of reactivity within this class of complexes is interconversion of *syn* and *anti* isomers, as has been observed with Tc-PnAO complexes substituted at the central carbon of the propylene bridge [67]. The same phenomenon has been described for DMSA complexes and will be discussed in the section on S-donor groups.

Another important N-donor group is the **amide** group. Contrary to the basic amino groups, the more acidic amide functions tend to be deprotonated in the complex and therefore operate as a monoanionic donor. Alkaline conditions promote the deprotonation and subsequent complex formation. The amide group is a very useful component of mixed donor sets, such as N$_2$S$_2$ or N$_3$S, as discussed in the next chapter. Whether a pure amide coordination may occur in M(V) complexes has not yet been proved. Tetrapeptides do form Tc(V) complexes [68], apparently without involvement of the carboxyl group. The N-donor atom provided by Schiff bases plays only a role in mixed donor sets and will be discussed below.

Hydrazine ligands, that exhibit electronic flexibility, are known to bind transition metals. Proper rhenium(V) oxo complexes react with aromatic hydrazides and hydrazines to form rhenium diazenido species with a Re=N double bond [69]. Technetium-hydrazido chemistry is complicated by a tendency toward N–N bond cleavage and formation of Tc-nitrido and Tc-imido species [70]. An unusual binuclear Tc(V)/Tc(VI) catecholate complex with two bridging hydrazido ligands has also been described [26]. For Tc(V), the hydrazino nicotinamide (HYNIC) group seems to be very promising in labeling biomolecules such as peptides or proteins [33, 71, 72, 73]. The active ester of hydrazinonicotinic acid is used to derivatize the ε-amino groups of lysine residues in peptides or proteins. These conjugates form Tc complexes by ligand exchange with Tc(V) gluconate. Using this procedure, it is possible to achieve specific activities above 10 000 mCi/μmol at peptide concentrations of appr. 10 μg/ml without purification [71]. The first step is the synthesis of an aromatic hydrazine linker, the next step is conjugation of this linker to the protein (Fig. 11), then, this modified protein is "labeled" by ligand exchange with a Tc(V) precursor, possibly forming a Tc-diazenido linkage. The exact structure of the chelate moiety is not yet known.

3.4 S-Donors

Due to the high affinity of sulphur as a soft donor atom to the borderline metals technetium and rhenium, and in view of the multitude of potential sulphur-containing ligands, a large variety of coordination compounds of technetium and rhenium with sulphur donor ligands are known. S-donor ligands coordinate the metals in various oxidation states, ranging from + I to + VI. At least for SH ligands, however, the preferred oxidation state is + V. Actually, SH ligands brought the once ignored + 5 oxidation state of Tc into the limelight [74, 75, 76] and have made sulphur-coordinated metal (V) complexes the most intensively studied compounds in Tc and Re chemistry. S-donor ligands for

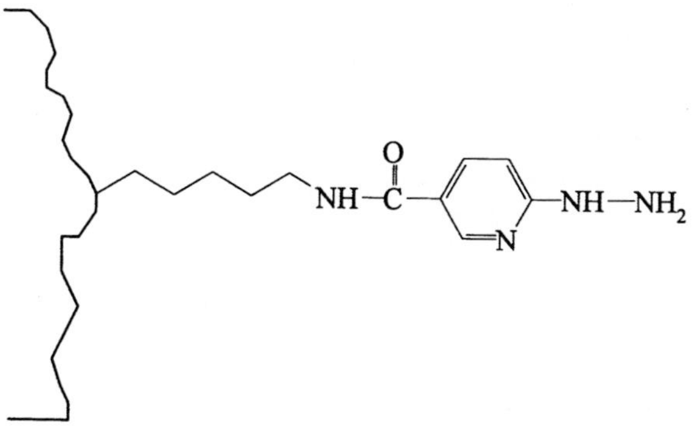

Fig. 11. Hydrazino nicotinamide linked to a protein [73]

Tc(V) and Re(V) are thiols, thioethers and thiocarbonyl compounds and molecules containing combinations of their functional groups. There exist significant differences in the ability of the various groups to form stable complexes and in the kind of complexes formed. Unlike thiols, a neutral S-donor shows a preference for the metals in oxidation states lower than +5.

Monodentate or multidentate **thiols** as the ligand coordinate are anionic donors in form of thiolates. The metal core of their complexes is normally $[M{=}O]^{3+}$. The charge requirements of this core are easily met by the ligands. In the case of pure thiolato coordination, anionic complexes result. In a few cases, neutral binuclear oxo complexes and oxo-free species also occur.

The reaction of complexes with monothiols to give $[TcOS_4]^-$, as described in [77, 78], e.g. for sterically hindered benzenethiols with phosphines, alkyl isocyanides or alkyl cyanides, led to partial loss of the thiol ligands and reduction of the oxotechnetium core by O abstraction [78]. Dithiols such as aliphatic, unsaturated ("dithiolene"), aromatic and heteroaromatic ligands give the very stable salts of anionic bis(dithiolato)oxotechnetates listed in Table 2. The same type of complex is obtained with O,S-donor ligands such as mercaptoethanol [79]. Sulphur in dithiol ligands can also be replaced by selenium. All of these compounds have the general formula $[TcO(S_2)_2]^-$ and the typical square-pyramidal arrangement [74, 75, 80, 81, 82, 83, 84]. The complex of *meso*-2,3-dimercaptosuccinic acid dimethylester (Fig. 12) is one of the first structures identified [85].

Nuclear medicine has taken manifold advantage of the thiophilic nature of technetium and uses thiol ligands in radiopharmaceutical design. An outstanding ligand is *meso*-dimercaptosuccinic acid (DMSA), which forms – in addition to a kidney imaging agent at the metal oxidation +4 or possibly +3 – well-defined bis(dithiolato)-oxotechnetate(V). It was originally only described as one of the first model complexes to study structure-biodistribution relationships of technetium complexes [86]. The introduction of a tumour imaging agent [87], which proved to be identical with the model complex [88], stimulated renewed studies of the chemical and biological properties of the complex. Technetium is also easily bound by metallothioneins, because of the high cysteine content of the protein [89]. In a few cases, thiols also produce binuclear species [30, 90, 91]. This can occur at an excess of the metal over the ligand during complexation

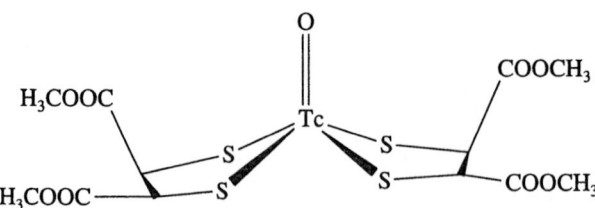

Fig. 12. Tc complex of *meso*-2,3-dimercaptosuccinic acid dimethylester [85]

Table 2 Dithiol ligands used to prepare oxotechnetium complexes

Ligand	R	Ref.
(structure: R R on C–C, $^\ominus$S S$^\ominus$)	R = H R = COOH R = COOMe	[30, 75, 76, 94, 95, 96] [76, 97, 98] [76, 85, 97]
(structure: propylene dithiol ring, $^\ominus$S S$^\ominus$)		[74, 94]
(structure: O R, $^\ominus$S S$^\ominus$)		[83, 94]
(structure: O O, $^\ominus$S S$^\ominus$)		[30, 96]
(structure: R R on C=C, $^\ominus$S S$^\ominus$)	R = CN	[76, 84, 94]
(structure: $^\ominus$S R / $^\ominus$S R)	R = CN (RR) = N-CN selenium analogues	[76] [76] [76, 81]
(structure: benzene ring with R, $^\ominus$S S$^\ominus$)	R = H R = (···)-CH_3	[82, 84] [76, 94]
(structure: S, S S ring, $^\ominus$S S$^\ominus$)		[76]
(structure: S S Mo, $^\ominus$S S$^\ominus$)		[80]

and is illustrated in Fig. 13a, for ethylene dithiol as ligand. Binuclear species are similarly favored when the ligand is used in S-protected form, and the protecting group leaves on complexation under the influence of a reactive precursor, such as $[TcOCl_4]^-$ [90].

Another type of binuclear thiolato complex is derived from certain tridentate dithiols, as illustrated in Fig. 13b [92]. Oxo-free species, which are exceptions,

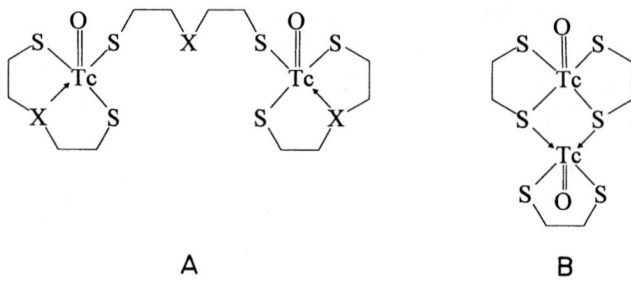

Fig. 13. Types of binuclear complexes $[(TcO)_2(S_2)_3]$

are exemplified by the anionic complex $[Tc(tdt)_3]^-$ (H_2tdt = toluene-3,4-dithiol) [93]. $AsPh_4[Tc(bdt)_3]$ was obtained unexpectedly by the reduction of $AsPh_4[TcNCl_4]$ in the presence of benzene-1,2-dithiol H_2bdt. The formation of this compound by removal of the nitrogen atom is surprising, since the TcN bond in such complexes is exceptionally stable. The compound has a trig-onal-prismatic coordination geometry [93].

Apart from pure S_4 coordination, as in the $[TcO(DMSA)_2]^-$ complex, modern radiopharmaceuticals have the SH-group in combination with other groups forming N_3S, N_2S_2 donor sets. If not a radiopharmaceutical per se, such as ^{99m}Tc mercaptoacetyl-triglycine (^{99m}Tc-MAG_3), the chelating unit is used bifunctionally to synthesize ^{99m}Tc or $^{186/188}Re$- biomolecules. In the labelling of antibodies it plays a particularly important role. To make use of the good donor properties of mercaptide sulphur, SH-groups are generated in the protein molecule or introduced into it. Free thiol groups in proteins, for example up to about 30 sulphhydryl groups per antibody molecule, can be generated by cleavage of its disulphide bonds with 2-mercaptoethanol or other mercaptides. To keep the structure of the antibody molecule intact, only the intra-chain disulphide bonds should be reduced, not the inter-chain bridges that hold the four chains of the immunoglobulin molecule together. The original method [99] has been evaluated and optimized in order to produce a higher labelling yield with minimum trauma to the antibody [100, 101]. The precise site of attachment of technetium to the antibody and to the donor set in the coordination sphere are virtually unknown. Alternatively, free mercapto groups can easily be intro-duced into proteins, e.g. by modification of lysine amino groups with 2-iminothiolane, to produce 1-imino-4-mercaptobutyl derivatives of the protein [102] as determined by the number of available lysine residues in the protein, three to four SH-groups can be introduced into albumin, or six to nine into immunoglobulins. Again, it is unknown which donor groups in the vicinity of the SH-groups are involved in coordination, occupying the necessary number of coordination sites. The in vitro stability of such technetium-labelled modified immunoglobulin proved to be very high. Competitive experiments with the strong competing ligand DMS dimethyl ester showed stability over some hours.

Cysteine, however, is able to strip ca. 50% of the technetium from the labelled protein after 4 hours of incubation [102]. Methods for complexation of antibodies with technetium must be modified for rhenium, because of its lower redox potential and – consequently – its greater tendency to reoxidize [103].

Thioether ligands offer soft, neutral S-donor atoms. It can be expected that such soft donor atoms, with their tendency to prefer the lower oxidation state of the metal, are well suited to binding technetium or rhenium. It can be further expected that when the metal is in oxidation state + 5, additional ligands must participate in coordination in order to compensate for the high charge density of the metal. Complexes containing oxo metal-, bisoxo metal-, or nitrido metal cores, or mixed ligand complexes, should thus be formed. The formation of Tc(V) complexes with thioether ligands will require sufficiently reactive precursors, such as $[TcOCl_4]^-$.

Several classes of technetium compounds, in which thioether groups are involved in the coordination sphere have been reported so far. There are Tc compounds in which a single thioether group is involved in coordination, compounds derived from macrocyclic N_2S_2 ligands, as well as complexes which contain neutral open-chain and crown thioethers. Thioether complexes may occur, at least as intermediates, in the reaction of S-alkyl-protected thiols with Tc(V) or Re(V) precursors [104, 105]. Thioether sulphur plays a constitutive role in tridentate dithiol ligands when used in the convenient [3 + 1] concept of tracer design [106] described below. Another combination of thioether and thiol sulphur atoms has recently been recommended in radiopharmaceutical development [107]. Neutral complexes were obtained, presumably with the Tc oxo core and one halide in the *trans* position, or the nitrido core [108]. Partial substitution of thioether sulphur for nitrogen in macrocyclic tetraaza ligands, e.g. N_2S_2, does not change the original coordination mode significantly. These ligands form cationic bisoxotechnetium(V) complexes [109, 110] (Fig. 14a) which show structural and stability analogies to the bisoxotechnetium tetraamine complexes.

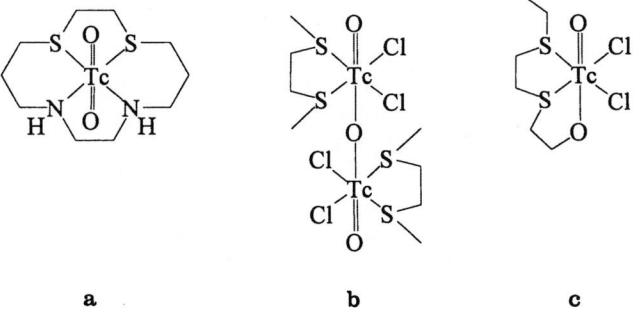

Fig. 14. Technetium complexes of neutral thioethers a) $[TcO_2(S_2N_2)]^+$ [110], b) $([TcO("S_2")Cl_2]_2O$, c) $[TcO("S_2O")Cl_2]$ [114]

Technetium complexes with thioethers in the strict sense, i.e., those without other donor groups in the ligand molecule, comprise homoleptic thioether nitridotechnetium(V) complexes [111], cationic mixed thioether/thiolate complexes of Tc(III) [112], and a cationic Tc(I) complex [113]. However, these latter compounds do not properly fall within the scope of Tc(V) compounds and are excluded from review.

Neutral bidentate thioethers of the type $R-CH_2CH_2-S-CH_2CH_2-S-CH_2-CH_2-R$ ("S_2" $R=H$, alkyl, O alkyl) react with tetrachlorooxotechnetate in acetone with partial exchange of the chlorine ligand by the sulphur atom, to produce binuclear oxotechnetium complexes $[MO("S_2")Cl_2]_2O$ [13] (Fig. 14b). The molecule $[TcO("S_2")Cl_2]_2O$ consists of two independent $[TcO("S_2")Cl_2]$ units bridged by an oxygen atom. Two chloride atoms from the precursor remain coordinated in the complex; they are obviously necessary for compensation of the positive charge of the metal core. Mononuclear complexes $[MO("S_2O")Cl_2]$ [114] are formed instead of binuclear species by providing an intramolecular O-donor anion to occupy the *trans* position, as shown in $HO-CH_2CH_2-S-CH_2CH_2-S-CH_2CH_2-OR$ ("S_2OH" $R=H$, alkyl). The terminal oxygen atom is coordinated in the *trans* position to the Tc=O bond. The presence of the second non-coordinated hydroxyl group and chloride atoms enable the versatile molecules to be modified as required for radiotracer design.

A further group of S-donor ligands are compounds with sulphur in the form of the thiocarbonyl group. Sulphur in the C=S-group has the ability to participate in π-bonding in addition to σ-donation, allowing stabilization of the metal in unusual oxidation states. The thiocarbonyl group is not stable per se. It is available in thioureas, thiosemicarbazones, dithiocarbamates, xanthates, and dithiocarboxylates. Thioureas form complexes at the metal(III) oxidation state. Unlike thiourea itself, as well as dimethylthiourea, both of which form the cationic Tc(III) complexes $[Tc(tu)_6]Cl_3$ and $[Tc(dmtu)_6]Cl_3$, tetramethyl-thiourea (tmtu) gives an oxotechnetium(V) complex [115]. The preference of Tc(V) over Tc(III) in this case is obviously due to steric constraints of the bulky tetramethyl ligand in a six-coordinated arrangement. The complex $[TcO(tmtu)_4]^{3+}$ has been used as a convenient precursor to synthesize the known Tc(V) complex $[trans\text{-}TcO_2(py)_4]^+$ and $[TcO(ema)]^{-1}$ [116]. In the concept of bifunctional chelating agents for labelling of proteins, the di(N-methylthiosemicarbazone) (DTS) moiety has been successfully used (Fig. 15). Various DTS derivatives with diverse spacer lengths were synthetized and tested, in order to obtain stable complexation of the metal with minimum steric

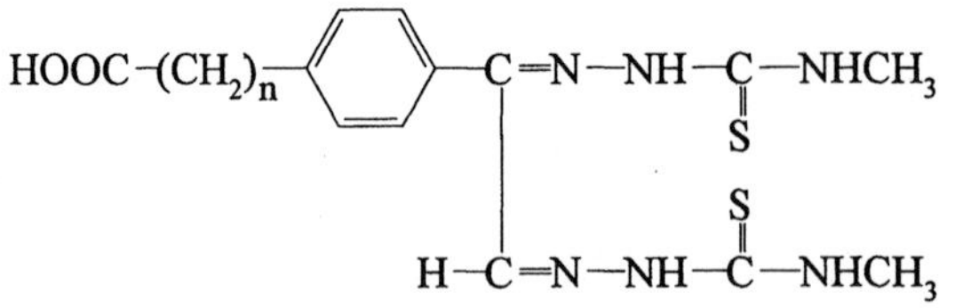

Fig. 15. DTS derivatives as used in the bifunctional chelating approach to label proteins with technetium

interference with the protein [101]. Further complexes of C=S-group-containing ligands are derived, for example, from dithiocarbazates [117], morpholino-*N*-carbodithioate [118], and *N*-(thiocarbamoyl)benzamidines [119].

3.5 P-Donors

Tertiary phosphines – and the analogous arsines – are able to stabilize transition metals in a variety of oxidation states and coordination geometries. Investigations of complexation with P-ligands were promoted by the high stabilization of metal by P ligands, which is mainly due to π-back bonding.

Complexation studies with bidentate phosphine ligands showed that stable cationic complexes of Tc(V), Tc(III), and Tc(I) are easily accessible. The influence of reaction conditions on reaction route and products is well demonstrated by the reaction of pertechnetate with the prototype 1,2-bis(dimethylphosphino)-ethane (dmpe) (Fig. 16). Careful control of reduction conditions allows the synthesis of $[TcO_2(dmpe)_2]^+$, $[TCl_2(dmpe)_2]^+$, and $[Tc(dmpe)_3]^+$, with the metal in the oxidation states V, III, and I [120, 121]. This series illustrates the variety of oxidation states available to technetium and their successive generation by the action of a 2-electron reducing agent.

Owing to the neutrality of phosphine ligands and their soft character, the domain of Tc/P chemistry comprises complexes with Tc in oxidation states lower than +5. To prepare Tc(V) complexes, the positive charge of the metal has to be diminished by additional charge-donating groups, which are located at the fifth position. This results in dioxo complexes containing a trans-$[TcO_2]^+$ core. In this respect, the phosphino donor groups behave similarly to amino donor groups, and – to a certain extent – to thioether sulphur.

Great interest in diphosphino complexes of technetium arose when Deutsch and co-workers found that lipophilic cations are able to be accumulated in heart tissue and can thus be candidates for myocardium perfusion agents [120]. Both the Tc(V) DMPE complex, $[TcO_2L_2]^+$, and the $[TcL_2X_2]^+$ and $[TcL_3]^+$ species in oxidation states +3 and +1 have been evaluated. Whereas the $[TcO_2L_2]^+$

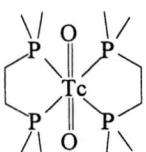 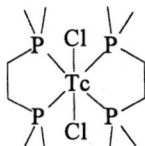 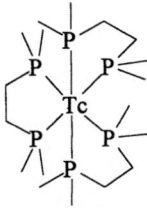

Fig. 16. Formation of Tc(V), (III), and (I) complexes with 1,2-bis(dimethylphosphino)ethane (dmpe) [120, 121]

cation initially did not appear to be useful in nuclear medicine, recent structure – biodistribution relationship studies of complexes with functionalized diphosphines [122] have shown that Tc(V) also enables the development of heart imaging agents. A promising candidate is the *trans*-dioxotechnetium(V) complex with two bidentate phosphine ligands (1,2-bis(bis-ethoxyethyl)phosphino)-ethane (tetrofosmin), shown in Fig. 17 [123, 124]. In the cationic complex the ligand obviously matches the dioxo core, with its significantly higher polarity. Ether functions in the ligand successfully improve the clearance of the lipophilic tracer from non-target tissues.

4 Combination of Donor Groups in the Ligand

4.1 Amine/Thiol

Although simple aliphatic aminethiols, such as the bidentate cysteamine, were shown to give yellow 2:1 Tc(V) oxo complexes, they have not been studied in detail or used. Carboxyl substitution on amine/thiol ligands leads to biochemically relevant substances, such as cysteine and its derivatives and analogues [125]. The complex with penicillamine has been characterized by X-ray crystal structure analysis [7] (Fig. 18). In this complex, prepared in strong acid solution, technetium is six-coordinate, being bound via S,N donor groups of the ligand molecules, and with one carboxylic group in the *trans* position. This structure illustrates the typical propensity of amino groups to coordinate as neutral

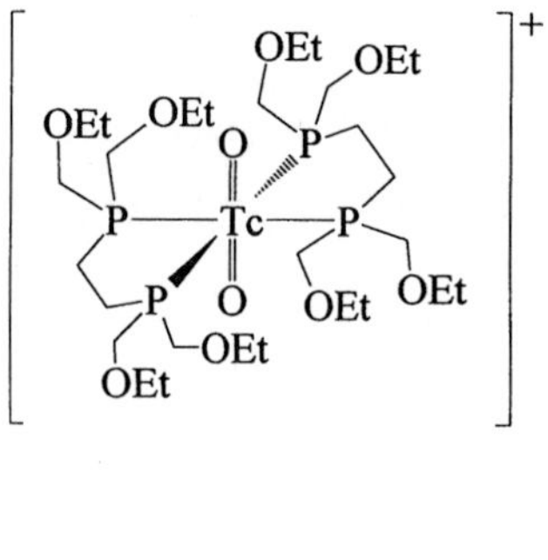

Fig. 17. TcO$_2$ complex with(1,2-bis(bis-ethoxyethyl)phosphino)ethane (Tc-Tetrofosmin) [123, 124]

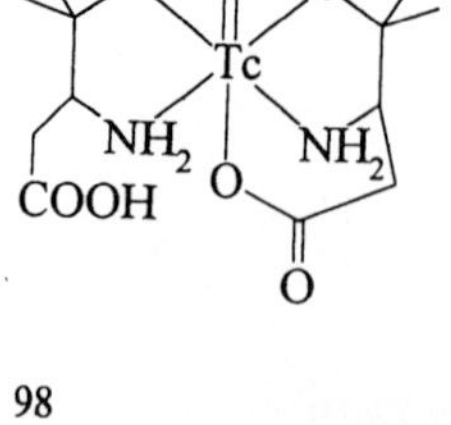

Fig. 18. [TcO(Hpen) (pen)], H$_3$pen = penicillamine [7]

donors, alone or in combination with thiol groups. This implies a tendency to accept the carboxylic group in the *trans* position as an additional anionic donor group to compensate for the charge of the core.

The simple aromatic S, N-ligand 2-aminobenzenethiol (H_2abt) behaves differently, insofar as it may act as a dithiolene-like ligand with mono-deprotonated amino groups. In neutral or acidic solution it forms anionic $[Tc(abt)_3]^-$, which is easily converted to the neutral Tc(VI) complex $[Tc(abt)_3]$, one of the rare Tc(VI) complexes [126, 127]. Another interesting aspect is the relation of oxo- to tris-ligand species, as influenced by the pH of the reaction medium. Under alkaline conditions, oxotechnetium complexes $[TcO(abt)_2]^-$ are formed [128]. Conversion of $[TcO(abt)_2]^-$ to $[Tc(abt)_3]^-$ occurs even on standing in acidic solution. Addition of H_2abt to the reaction mixture results in complete and clean conversion [129]. Treatment with 12 M hydrochloric acid in methanol converts $[TcO(abt)_2]^-$ to $[TcCl_4(abt)]^-$ [130].

Tetradentate N_2S_2 ligands such as **diaminodithiols (DADT)** form neutral lipid-soluble technetium complexes [131]. These complexes are uncharged because of the unique behaviour of the four ionizable atoms in the ligand molecule; the two SH groups and one of the two NH groups are deprotonated in the complex, thus neutralizing the positive charge of the $[TcO]^{3+}$ core. This qualifies them to be used in tracer design for various brain function studies, because – as neutral complexes – they are able to diffuse rapidly through the blood brain barrier into the brain tissue. Series of derivatives with varied backbone ligand structures have been synthesized and characterized [132, 133]. The complexes formed are very stable. Contrary to the DADS system, an increase in the size of the ligand backbone appears to generate a mixture of mono- and dioxo cores [134].

Before complexes can be used in nuclear medicine, especially for brain function studies, in vivo reactivity has to be imposed upon them. Retention in the brain is necessary for perfusion imaging. This can be achieved by substituents on the ligand that bind them to binding sites in the brain. N-alkylation with methyl and other alkyl groups led to the class of complexes shown in Fig. 19. Upon complexation to technetium, the N-alkyl substituent can assume a *syn* or *anti* configuration with respect to the oxo ligand, as proved by X-ray crystal

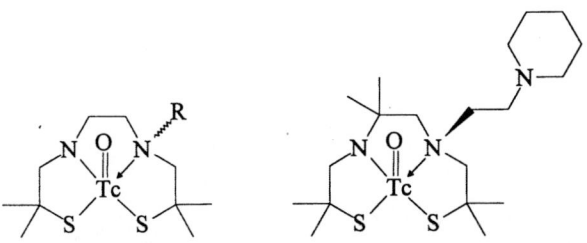

Fig. 19. Technetium complexes of N-alkyl DADT and NEP-DADT [135, 136]

structure analysis for R = methyl [135]. The *syn* configuration seems to be thermodynamically more stable than the *anti* form, and is therefore the dominant product. On this structural basis, (*N*-piperidinylethyl hexamethyl-diaminothiolat)oxotechnetium(V), Tc NEP-DADT, [136] has been developed as a complex that is retained in the brain for a short time, because of the amine side chain. Subclasses within the same type of *N*-alkyl diaminedithiol Tc completes show a structure/in vivo activity relationship for pulmonary accumulation; there is a parabolic correlation between percent lung uptake and the logarithm of the partition coefficient [137]. Studies of the uptake mechanism led to the hypothesis that these Tc complexes interact with the saturable amine-uptake system of the lung. Within a series of neutral, lipophilic ^{99m}Tc complexes of DADT ligands exhibiting pulmonary accumulation in rodents, [TcO(NEt-tmdadt)] (tmdadt = tetramethyl diaminodithiol) showed the highest lung uptake. The ethyl substituent adopts the *syn* configuration with respect to the oxo-Tc group, and the unsubstituted nitrogen is deprotonated, resulting in a neutral molecule [138].

A very promising new brain-perfusion tracer is also based on the DADT ligand system. The technetium complex of *N,N'*-1,2-ethylenediyl-bis-L-cysteine diethyl ester, ECD, shows excellent retention in the brain due to enzymatic cleavage of its ester group(s) to give ionic products that are trapped in the human brain. The structure of the Tc complex (Fig. 20) is very similar to that observed in other DADT complexes. There may be a partial interaction between a carbonyl oxygen and the hydrogen atom of the protonated amine. The interatomic distance is 2.37 Å, shorter than the sum of the van der Waals radii [139]. (The structure of the analogous Re complex is assumed to be very similar.) Within a systematically varied series of analogous Tc-DADT diesters, the relationship between the structure of the complex and its ability to cross the blood brain barrier and be retained in the brain has been studied [140]. In accordance with the assumption that diffusion mediates the penetration of the complexes through the blood brain barrier, the uptake in the brain is related to complex lipophilicity and to the size of the ester groups. The stereochemistry of the complexes is irrelevant. Once in the brain, the complexes undergo ester hydrolysis to the monoacid monoester metabolite and possibly the diacid product (Tc–EC), however, this is restricted to primates. Because of their charge, they are unable to re-diffuse out of the brain. For this retention, however, the stereochemistry of the ligand has shown to be crucial.

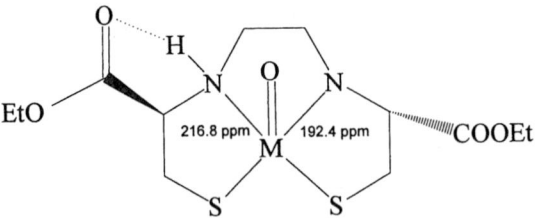

Fig. 20. Tc complex of N,N'-1,2-ethylenediyl-bis- L-cysteine diethyl ester, ECD [139]

The advantage offered by the ease with which amine/thiol ligands react with technetium, even at ambient temperature, has been used for incorporation of technetium and rhenium into biomolecules – particularly antibodies and other proteins [141, 142]. For this purpose, the ligands are modified into Bifunctional Chelating Agents (BCAs) [142, 143]. The N-alkylated ligand forms two, presumably isomeric, cationic products [144]. Going from diaminedithiols (DADT) to triaminedithiols (TADT) makes it possible to use the third nitrogen atom for a more flexible derivatization in the bifunctional concepts [145].

^{99m}TcO-L,L-ethylenedicysteine (LL-EC) was formerly derived from the neutral brain perfusion agent TcO(ECD) by hydrolysis of the two pendant ester bonds. With its free carboxyl groups, this ligand can also be considered as a derivative of the mercapto amino acids discussed above. In the penicillamine complex as the only representative of known molecular structure, a special kind of charge reduction takes place, by involving carboxyl oxygen in *trans* position. Whereas ECD cannot provide this kind of stabilization and operates instead as NH/N$^-$ nitrogen donor, EC offers both possibilities so it is interesting to see how it actually reacts [139, 146]. At neutral pH, Tc and Re complexes have a mixture of mono and diprotonated EC ligands. Initially, the structure of the compounds were proposed to be square-pyramidal rather than similar to the octahedral penicillamine complex. X-ray crystal structures were then reported of the neutral species ReO(ECH$_3$), that exists at low pH, as well as of the trianionic species [NH$_4$]$_3$ [ReO(EC)] [147]. The coordination geometry of ReO(ECH$_3$) has been confirmed quite recently to be distorted octahedral, with a deprotonated carboxylate coordinated *trans* to the oxo ligand [148]. Both sulphur atoms and one carboxylate group undergo ionization during the complexation, so that the complex is neutral [148]. The [NH$_4$]$_3$[ReO(EC)] complex, however, is distorted square-pyramidal. The dynamic processes taking place as a function of pH are best explained by assuming a coupling of carboxyl-group coordination with amine protonation [146, 147]. The ^{1}H NMR shift of the Re complex of D,L-EC indicates that there is only one isomer of the complex in which both carboxylate groups are *syn* to the oxo ligand. ^{99m}TcO-EC is of special interest in nuclear medicine as a renal function agent. The most promising stereo isomer is the complex of D,D-EC [149].

An N$_2$S$_4$ diaminotetrathiol [51], N, N, N', N'-tetrakis(2-mercaptoethyl)-ethylene-diamine [150], has been used to label antibodies with ^{99m}Tc and ^{186}Re. The chelating agent (Fig. 21) is conjugated to the protein via disulphide exchange with one of the thiol groups.

4.2 Amine/Thioether

Technetium compounds with amine/thioether coordination are the cationic *trans*-dioxotechnetium(V) complexes [TcO$_2$(N$_2$S$_2$)]$^+$. The complex in which N$_2$S$_2$ is 1,4-dithia-8,11-diazacyclotetradecane was prepared via an exchange reaction of NBu$_4$[TcOBr$_4$] with the ligand and fully characterized by X-ray crystal structure determination [109, 110]. The coordination around technetium

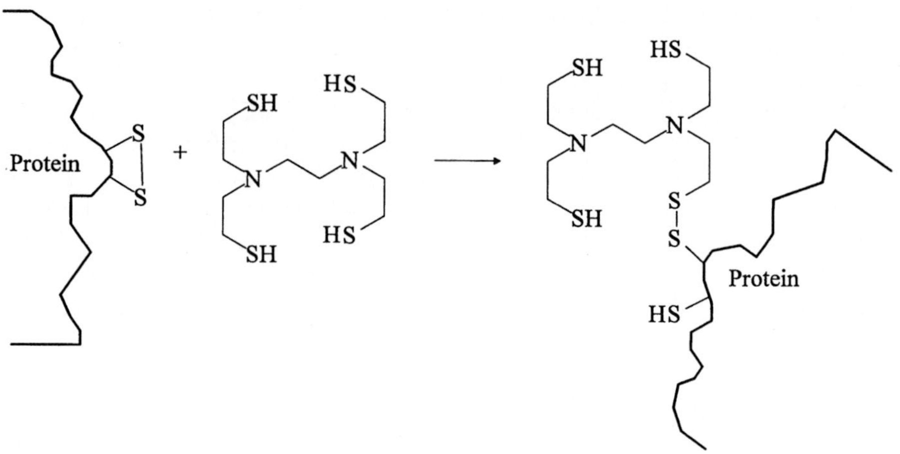

Fig. 21. The N$_2$S$_4$ ligand diaminotetrathiol N,N,N',N'-tetrakis-(2-mercaptoethyl)ethylene-diamine used to label antibodies

can be approximately described as a compressed octahedron, with the ligand heteroatoms on the equatorial plane and a slightly displaced metal atom. The two oxygens are situated on the axis. The two Tc–O bonds are of equal length and in agreement with the average of 1.75 Å, usually found for dioxotechnetium compounds.

4.3 Amine/Phosphine

Bidentate (o-aminophenyl)diphenylphosphine produces TcOL$_2$X (X = mono-negative ligand) complexes of poor stability. However, stable M(V) species (M = Tc, Re) of the type MOLX can be obtained with the corresponding tetradentate P$_2$N$_2$ ligand with a propylene bridge between the two nitrogen atoms [151]. The X ligand is *trans* to the Tc-oxo moiety and the two N and two P atoms of the tetradentate ligand lie on the equatorial plane. The tetradentate NP$_3$ ligand 2-diphenylphosphino-*N*, *N*-bi(2-diphenylphosphinoethyl)-ethaneamine yields a cationic complex [TcOCl$_2$(NP$_3$)]$^+$ [152].

4.4 Amide/Thiol

Tetradentate **diamidedithiol (DADS)** ligands form very stable five-coordinate oxotechnetium anions [104, 153, 154, 155, 156, 157]. Several of them are shown in Fig. 22. The complexes do not undergo substitution reactions with other thiol ligands, such as ethanedithiol. Increasing the size of the ligand backbone: going from ethanediamine-1,2, which forms a 5,5,5-membered chelate ring, to butanediamine-1,4 which forms a 5,7,5-membered ring, has no negative consequences for complex formation. This enables a much more flexible derivatization for tracer design in nuclear medicine [158].

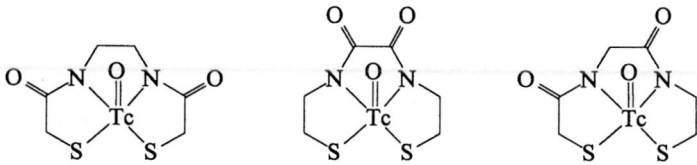

Fig. 22. Some DADS complexes [104]

The complex with the ligand N, N'-ethylenebis(2-mercaptoacetamide), [TcO(ema)]$^-$, showed rapid renal clearance, which initiated a very successful rational design of a radiopharmaceutical for renal function studies. A milestone in this approach was the synthesis of isomeric carboxylate analogues of DADS complexes [159]. This type of stable complex can also be used in the BCA approach. The DADS ligand system requires higher temperatures for complex formation. For labelling proteins in which the DADS chelating moiety has been incorporated, such forcing conditions should be avoided. This can be done with pre-formed complexes that are subsequently conjugated to the protein.

In a next stage, a triamide monothiol ligand was introduced instead of DADS ligands [156]. The N_3S donor ligand mercaptoacetyltriglycine MAG$_3$ forms the radiopharmaceutical [TcO(MAG$_3$)]$^-$ (Fig. 23), which is in routine use in nuclear medicine. The ligand MAG$_3$ combines donor atoms of different reactivity, giving rise to subtle interplay among these donor atoms and diverse reactions. Due to the good donor quality of the mercaptide group, it first attacks to give – hypothetically – 4:1 and then 2:1 complexes under excess ligand conditions [160]. The process which ultimately leads to the 1:1 complex can be facilitated by activation of the amide-N at pH > 11. The carboxylate group is not coordinated to the metal, but when one of the three amide groups is omitted in the ligand molecule, as in MAG$_2$, the carboxylate group is necessary to stabilize the complex and is therefore involved. [TcO(MAG$_2$)]$^-$ is one of the rarely described complexes in which a carboxylic group is coordinated in the equatorial plane [161]. The Tc–O(carboxylate) bond distance is relatively short (2.010 Å) compared to that of the *trans*-coordinated carboxylic group in [TcO(Hpen)(pen)]$^-$ (Hpen = penicillamine) (2.214 Å) [7]. Because of the important part which [TcO(MAG$_3$)]$^-$ plays in medicine as a renal function agent, series of analogous complexes with a systematically varied peptide moiety have been synthesized and studied [162, 163, 164, 165]. The very stable MAG$_3$ complex with its carboxyl group is also very useful in the BCA concept, e.g. for the introduction of ^{99m}Tc and rhenium isotopes into antibodies [166, 167, 168]. Bz-MAG$_3$ has also been attached to a biotin derivative, in order to form the corresponding Tc(V) complex and then to label avidin with it [169].

Instead of the amide thiol ligands themselves, their S-protected precursors are sometimes used. If the protecting groups are not removed before or during complexation, **thioether sulphur** is present in the donor set. This confers different

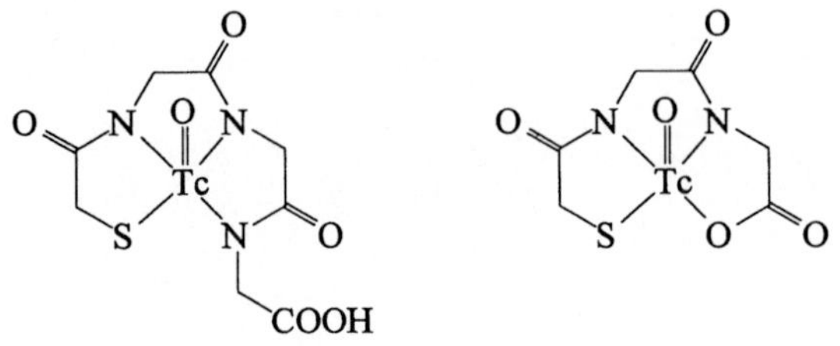

Fig. 23. Tc/Re complexes of MAG$_3$ and MAG$_2$ [156, 161]

characteristics on the ligand, as this sulphur is no longer anionic. Such a variation may be helpful when neutral complexes are to be synthesized instead of the anionic ones. For instance, a stable alkyl substituent of one of the thiol groups in the chelate ema gives the neutral complex [TcO(emaR)] [105]. In some instances, when R = tert. aminoalkyl, these complexes react further to give the original anionic ema complex [TcO(ema)]$^-$.

4.5 Amine/Amide/Thiol

In order to combine the advantages of the two mainly used N$_2$S$_2$ ligand systems, DADTs and DADSs, the combination of an amine-N and amide-N in an N$_2$S$_2$ arrangement has been proposed [170]. The new combination is particularly useful if it is linked to proteins and should, therefore, form Tc or Re complexes under mild conditions. A number of these ligands (Fig. 24), monoaminemono-amide dithiols (MAMA), have been described [170, 171]. The *syn* and *anti* N-benzyl-MAMA complexes show the typical square-pyramidal crystal structure of Tc(V) complexes. In the *syn* isomer the bulky aromatic ring extends away from the central core, while the *anti* isomer is forced to position the bridging methylene group under the Tc=O bond, in order to minimize the interactions of the aromatic ring with the rest of the complex [171]. The MAMA chelate is more polar than the diaminedithiol system. It is therefore particularly useful when less lipophilicity has to be imposed upon biomolecule conjugates, e.g. substituted progestins [172].

4.6 Amine/Amide/Thiol/Thioether

Tc-thioether-thiol-MAMA complexes may occur as intermediates in the synthesis of the above described MAMA complexes [173]. Ligand exchange of Tc(V) gluconate and S-protected MAMA ligands produces the complex of the unprotected ligand, at acid or neutral pH – depending on the nature of the labile protecting groups. If two different labile protecting groups are used for the two thiol groups in the ligand, the relative position of the amino group with respect to the protected groups influences the rate of complex formation and its reaction mechanism. With S-protected ligands, intermediate thioether complexes are produced. Upon heating or in aqueous solution, the intermediates quickly convert into the S-dealkylated species [105].

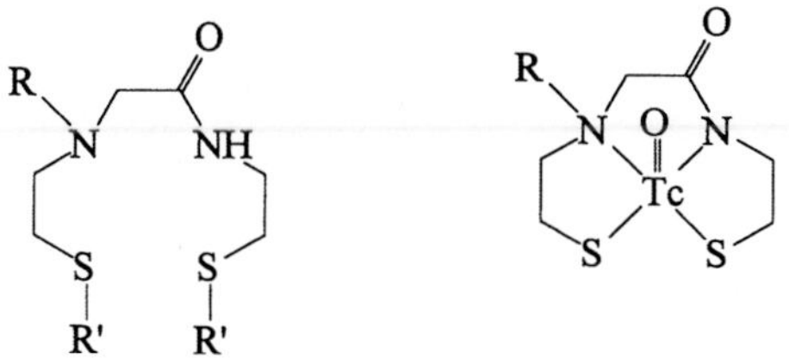

Fig. 24. Monoamine Monoamide Dithiols (MAMA) – a new N_2S_2 ligand system [170, 171]

4.7 Schiff Bases

Schiff bases provide useful mixed donor sets. The carbonyl function of the most frequently used ligands is derived from either 1,3-dicarbonyl compounds or salicylaldehyde. Favourable combinations involve O-, N- and S-donor atoms. A range of technetium and rhenium complexes exist with bi-, tri-, tetra- and pentadentate ligands. The geometry of these complexes depends on the number and type of coordinating atoms as well as on the chain length between the donor atoms in the Schiff-base ligands.

In the reaction with $[TcOCl_4]^-$, the **bidentate ligand** N-phenylsalicylidene imine forms complexes with the formula $[TcOCl_3L]^-$ and $[TcOClL_2]$ [174, 175]. In $[TcOClL_2]$, one Schiff-base ligand occupies two equatorial sites and the other Schiff base ligand bridges equatorial and apical coordination sites. S-alkyldithiocarbazates also act as potentially bidentate Schiff bases, to give cationic bis-ligand complexes $[TcO(L)_2]Cl$ [176].

Tridentate Schiff bases, when coordinated to the MO^{3+} core, require additional ligands in order to produce stable mononuclear complexes. Thus, compounds of the formula [MOLCl] result from the reaction of tridentate ligands with $[TcOCl_4]^-$ [177]. The fourth position may also be occupied by an organic ligand. Interesting combinations of this type lead to mixed-ligand complexes, particularly with thiolates and bidentate Schiff bases or bipyridine as co-ligands they will be discussed in the section on mixed-ligand complexes. The tridentate Schiff-base ligand obtained by condensing 2-aminobenzenethiol with salicylaldehyde reacts in several cases as dianionic tridentate Schiff base ligand, although its stable form is 2-(2-hydrophenyl)benzothiazoline [177, 178, 179, 180].

Tc(V) and Re(V) complexes of **tetradentate Schiff bases** of the O, N, N, O-type, derived from diamines and diketones or hydroxyaldehydes, have been studied extensively. Because of their dianionic nature, they tend to give complexes that involve additional ligands, usually bound in the position *trans* to the M=O bond. Thus, N,N'-ethylenebis(acetylacetone imine) yields the complex $[TcO(ONNO)(H_2O)]^+$, binding a water molecule *trans* to the oxo ligand. This core is intermediate between the monooxo and the *trans*-dioxo cores. A further typical example is the complex with N,N'-ethylenebis(salicylidene iminate). It has a similar coordination geometry but contains the *trans*-oxochlorotechnetium(V) core [181]. Schiff bases containing amino acids have also been used

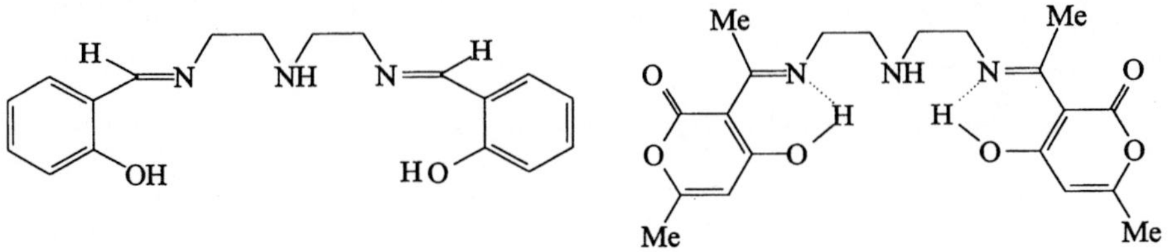

Fig. 25. N, N'-3-azapentane-1,5-diylbis(salicylideneimine) and the H_3apa ligand

as chelating agents for Tc(V) [182]. Tc(V) oxo complexes of unsaturated N_2S_2 ligands exist both as a mononuclear and binuclear species [183].

Potential **pentadentate Schiff base ligands** coordinate in several ways, depending on the nature and position of the fifth donor atom. With ligands of sufficient chain length and flexibility, the additional donor atom in the link between the two bidentate Schiff base units is also expected to participate – occupying the equatorial plane – if it is sufficiently nucleophilic. One donor group of the twisted ligand e.g. OH, will then occupy the position *trans* to the M=O group. This, as well as the alternative tetradentate behaviour, is exemplified by the potential pentadentate N_2O_3 Schiff base ligand N, N'-2-hyroxypropane-1.3-bis(salicylidene imine). On reaction with $[MOCl_4]^-$ it produces both a mononuclear and a binuclear six-coordinated complex, the latter containing the quasi-linear $M_2O_3^{4-}$ bridge [12]. With the N_3O_2 pentadentate Schiff base ligand shown in Fig. 25a, rhenium(V) forms a neutral complex [ReOL]. One oxygen of the ligand is located *trans* to the rhenium oxo bond, whereas the remaining four coordination atoms are in the equatorial plane of the octahedron [184]. This class of ligands is quite versatile and can be modified by substituting aza-nitrogen by oxygen or sulphur. Another pentadentate Schiff base N_3O_2 ligand, H_3apa (Fig. 25b), also produces a highly distorted octahedral Tc(V) oxo complex [185]. All five donor atoms of the ligand are coordinated. Under the influence of the electrophilic $[Tc=O]^{3+}$ core the coordinated central amine-N is deprotonated, resulting in a neutral complex. The ligand wraps around the metal center in such a way that one of the two O-donors is located *trans* to the Tc=O.

5 Mixed-ligand Complexes

From the point of view of radiotracer design, mixed-ligand complexes deserve increasing interest. Mixed-ligand complexes in the special sense are understood to be stable complexes derived from a combination of at least one chelating ligand and a widely variable organic co-ligand. (The relatively unstable complexes with loosely bound organic monodentates in the *trans*-position, although

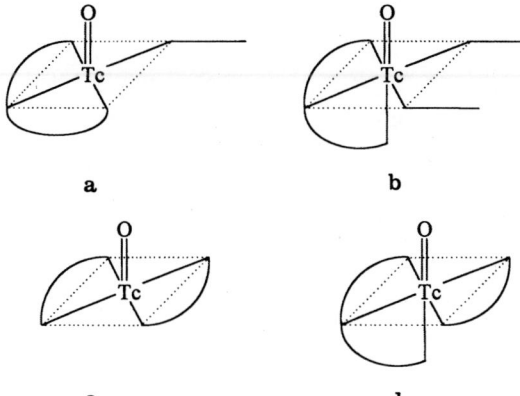

Fig. 26. Types of mixed-ligand oxotechnetium(V) complexes: combination of tridentate ligand with one (a) or two (b) monodentate ligands, combination of a bidentate with a second bidentate (c), or a tridentate (d) ligand

they formally meet the definition, do not come within the scope of compounds relevant for nuclear medicine, and will therefore not be discussed here). The thus defined mixed-ligand complexes are categorized according to type (monodentate or bidentate) and number (one or more) of the respective co-ligands, as shown in Fig. 26. Several authors argue [92, 106, 186, 187] that the flexibility of the co-ligand group, combines versatility in modifying the molecular parameters with the specific advantages of the chelated ligands, may generate a family of potentially useful radiopharmaceuticals. In order to favour formation of mixed-ligand type complexes over the homo ligand complexes, one has to consider the competition of both types of ligands for the metal. Matched reactivity of both types of ligand is required.

5.1 Oligodentate/Monodentate ("n + 1")

"3 + 1" mixed-ligand complexes (Fig. 26a) are made accessible by combining appropriate tridentate and monodentate ligands. Thus, complexes [MO(SXS)-(SR)] have been obtained by the reaction of **tridentate dithiol** ligands $HS-CH_2CH_2-X-CH_2CH_2-SH$ (HSXSH where X is O, S, or NR) and a large variety of monothiols RSH with oxotechnetium or oxorhenium precursors [92, 188, 189, 190]. Owing to the neutral character of the donor group X in the tridentate ligand, the compounds are neutral species. Similarly, **tridentate Schiff bases** react with Tc(V) gluconate or [TcOCl_4]$^-$ in the presence of monothiols to form neutral 3 + 1 oxotechnetium complexes [179, 191]. Apart from their ability to be easily reduced by tertiary phosphines [192], these complexes are a very stable species. For example, they do not undergo exchange reactions with dithiols. The mixed ligand complexes of the 3 + 1 type appear to be quite suitable for designing new Tc and Re compounds of relevance for nuclear medicine. Being neutral, of low molecular weight and stable, they have the potential to mimic biochemical substrates and drugs, mainly by modifying their highly variable ligand RSH. In the bifunctional concept of tracer design, the principal usefulness of [3 + 1] complexes is shown by the preparation of Tc and Re complexes containing simple functional groups, as well as more biologically relevant molecules such as glucose or cholesterol [92, 188, 189, 106]. This is illustrated in Fig. 27. Tridentate ligands containing the SNS donor atom set were used as a novel backbone for the development of technetium brain-imaging agents [193].

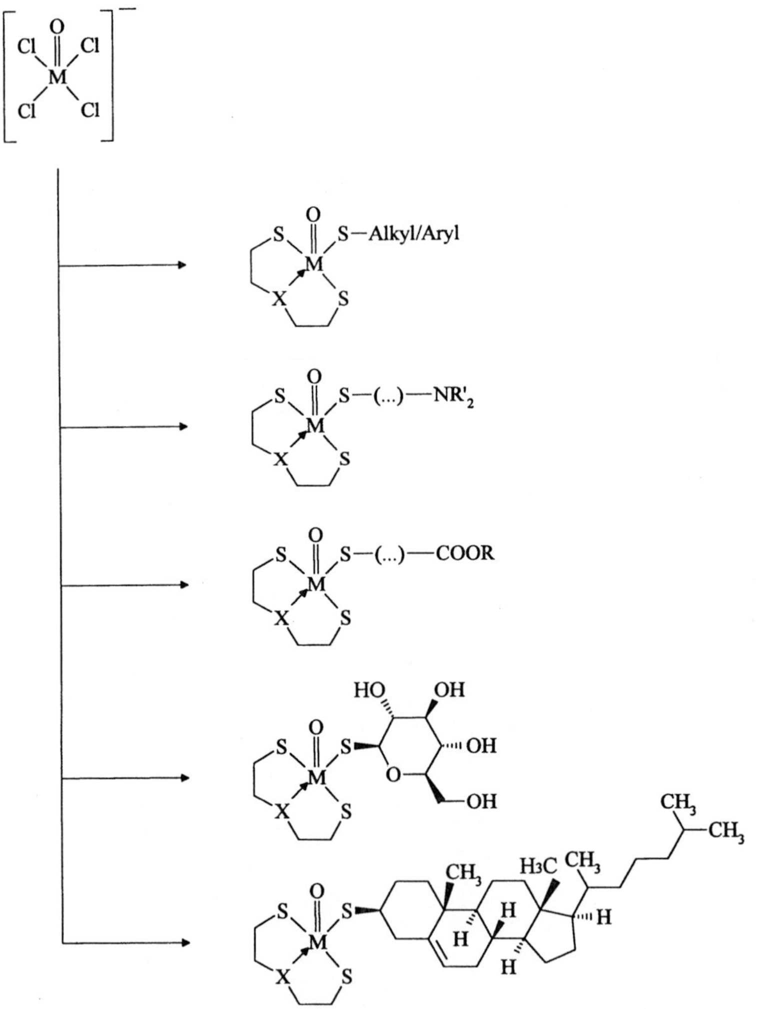

Fig. 27. 3 + 1-complexes [MO(SXS) (SR)] (M = Tc, Re; X = O, S, NR) with different co-ligands RSH

5.2 Oligodentate/Bidentate ("n + 2")

Another type of mixed ligand technetium or rhenium complexes in the above sense contains two or more monodentate ligands (Fig. 26b). The six-coordinated polypyridyl-thiolato complexes of rhenium(V) $[ReO(terpy)(SR)_2]^+$ [186] are an example. Further representatives are a variety of mixed-ligand complexes, combining tetramethylthiourea with dimethyldithiocarbamate $[TcO(tu)_2\text{-}((CH_3)_2NCSS)]^{2+}$ [194].

Heterodimeric "2 + 2" complexes (Fig. 26c) contain two different bidentate ligands. The first selective synthesis of those heterodimeric oxorhenium complexes, which contain two different aminothiols, and their characterization have been reported by Chi and Katzenellenbogen [187]. The mixed-ligand complex is formed in strong preference to the corresponding homodimeric compound (Fig. 28). The compounds $[ReO(SN)(SN)']$ are remarkably stable under ambient conditions and an analogous technetium-99m complex seems to have reasonable stability in vivo. These results offer encouragement to the design of complexes whose size, shape and functional groups allow them to mimic naturally occurring ligands. This was shown for Tc complexes that mimic steroid hormones [195]. The "2 + 2" type is furthermore exemplified by $[TcOCl(NN)(OCH_2CH_2O)]((NN) = bpy, phen)$ [196, 197].

"3 + 2" **type** mixed-ligand complexes (Fig. 26d) derive preferably from tridentate Schiff bases. Tridentate Schiff bases, when coordinated to the MO^{3+} core require additional ligands in order to produce stable mononuclear complexes. Bidentate Schiff bases and bipyridine in particular, lead to mixed-ligand complexes when introduced as co-ligands. Thus, mixed-ligand complexes result either directly from concomitant reaction of the tridentate Schiff base and the additional ligand on an appropriate M^VO precursor, or in a two-step procedure via $[MOLCl]$ (H_2L = tridentate ligand). There are hexa-coordinated neutral compounds, for example, where the co-ligands are bidentate Schiff base ligands [175, 177, 198, 199] or 8-quinolinol [198]. Cationic species result in a similar manner when the tridentate Schiff-base ligand is combined with bidentate aromatic nitrogen-donor ligands of the dipyridine type (2,2'-dipyridine, 1,10-phenanthroline etc. [200]). The "3 + 2" type has furthermore been verified in complexes of the type $[TcO(ONS)(SN)]$, where H_2ONS is S-methyl-β-N-(2-hydroxyphenylethylidene)dithiocarbazate [201].

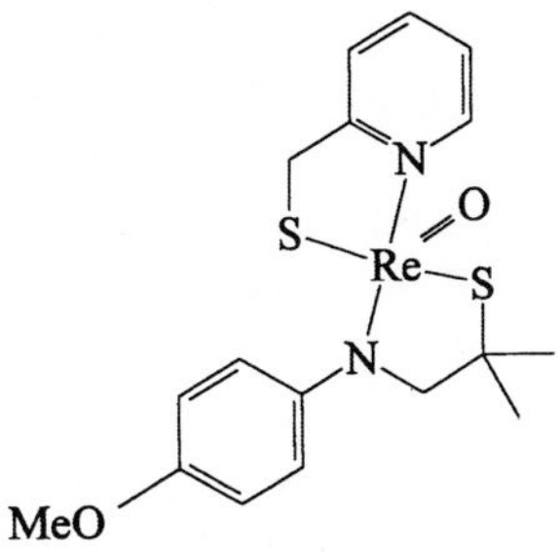

Fig. 28. Heterodimeric complex: $[ReO(SN)(SN)']$

6 Reactivity

The stability of Tc complexes has been considered to be an important criterion in the design of ^{99m}Tc radiopharmaceuticals. This appears to be reasonable as far as in vitro stability over several hours – i.e. radiochemical purity in the radiopharmaceutical formulation – is concerned. Inertness in vivo, however, is contradictory to various specific biodistribution patterns. Formally, Tc(V) complexes are kinetically inert towards ligand exchange reactions, in according to the Taube definition of kinetic inertness; oxotechnetium(V) complexes mainly undergo ligand exchange reactions with half-lives greater than 1 min. The quantitative structure-stability relationships of some oxotechnetium N_2S_2 complexes have been described, and the rank order of the calculated in vitro stability was confirmed by competitive ligand-exchange reactions [202]. In the rational design of ^{99m}Tc radiopharmaceuticals, a desired reactivity – not present in the parent complex – is imposed by introduction of functional groups into the ligand molecule. Metabolizable ester groups, redox-active moieties, or mimics of receptor-binding ligands are the most favoured tools at present.

6.1 Ligand Exchange Reactions and Core-induced Reactivity

In the last decade, there has been increasing interest in ligand exchange reactions, particularly at the Tc(V) oxo core [30]. Already in the early 80s, Tc(V) gluconate proved to be a most convenient starting material for ligand exchange to synthesize e.g. Tc(V) oxo bis(thiolato) compounds under mild conditions [76, 86]. The O-coordinated complex is readily converted into S-coordinated Tc(V) oxo products. Because of their medium position in the rank order of stability, N-coordinated species can also be formed from Tc(V) gluconate, and further transformed into S-coordinated complexes in the presence of competing S-donor ligands [40]. The high reactivity of $[TcOCl_4]^-$ – exploitable in the synthesis of ^{99}Tc complexes – is not considered here, because it is not used in the preparation of ^{99m}Tc radiopharmaceuticals.

Various starting O-donor complexes used to synthesize specific Tc complexes are listed in Table 1. Analogous Tc(V) and Re(V) complexes differ substantially in their relative rates of substitution. The ratio of pyridine exchange rates for the M(V) complexes trans-$[MO_2(py)_4]^+$ (M = Tc, Re) at room temperature in a nonaqueous solvent is ca. 8000 [203]. The substitution process is considered to be dissociative in character; accordingly, the rate is primarily influenced by the nature of the leaving ligand rather than that of the entering ligand. These results can be generalized: The Tc(V) complexes react faster by three to four orders of magnitude than the Re(V) analogues, independently of whether the complexes are cationic or anionic, whether substitution occurs cis or trans to the M=O bond, and whether the reaction was carried out in aqueous or organic media [203, 204]. Ligand exchange reactions are very beneficial in the synthesis of technetium complexes or radiopharmaceuticals, but they do not play an important role, if any, in vivo.

The phenomenon of **inducement of ligand reactivity** by the Tc core has been found in a series of mono-*S*-alkylated derivatives of bis(amine)bis(thiol) ligands. In some instances, when the alkyl substituent was $CH_2CH_2NR_2$ definite anionic dealkylated products were obtained instead of the well-known neutral Tc(V) oxo species. Systematic studies of the dealkylation process revealed nucleophilic attack by amines, halides, or water on the carbon adjacent to a coordinated thioether, and disclosed interesting details about the factors influencing the reactivity of these and some related neutral complexes [105, 170]. An interesting phenomenon of great relevance to nuclear medicine is the isomerism of ligands and the resulting complexes.

DMSA provides an excellent example for study. The ligand itself occurs in the form of dissymmetric *racemic* and *meso* diastereoisomers. Each of them gives rise to isomeric complexes (Fig. 29) with different arrangements of the carboxyl groups relative to the Tc=O group [85, 88, 97, 98]. With the *meso* diastereoisomer ligand, one *anti* and two chemically different *syn* isomers, *exo* and *endo* to the axial Tc=O group respectively, occur. The complex of the *racemic* ligand also exists in three stereoisomeric forms [97]. Isomers of Tc complexes, such as complexes of ECD, HMPAO or NEP-DADT, usually differ in their biodistribution and thus in their suitability as radiopharmaceuticals. As to the tumour affinity of ^{99m}Tc(V)oxo-*meso*-DMSA complexes, it is unknown whether all three isomers in the mixture produced in the preparation of the radiopharmaceutical are equally suitable, or whether the uptake in tumours might improve by application of a pure isomer. Studies of the isomers of the Tc and Re complexes revealed an inter-conversion of the isomers in aqueous solution, that prevents the application of a pure isomer [98, 205]. It is difficult to reconcile the results with any mechanism that does not assume cleavage of the metal–sulphur bond; rupture of at least one bond is required for *anti/syn* conversion.

Another type of complex reactivity was observed with [TcO(*meso*-DMSMe)$_2$]$^-$ [97]. When the complex is exposed to bases, the ligand is converted from the *meso* to the *racemic* form. This process obviously proceeds by

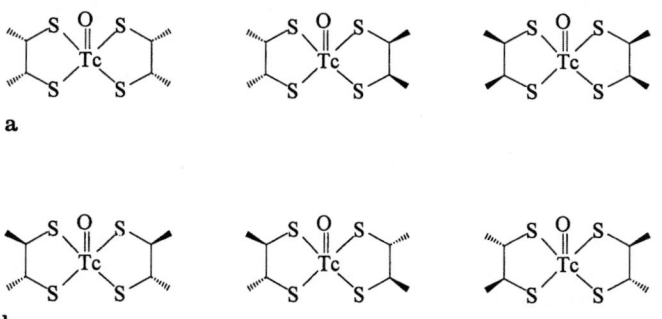

Fig. 29. Isomeric complexes [TcO(DMSA)$_2$]$^-$ with the *meso* (*a*) and *racemic* (*b*) ligand

means of a simple mechanism: formation of a carbanion intermediate, generated by α-proton abstraction in basic media, and subsequent reprotonation from the opposite site. Carbanion formation is favoured by the electron-withdrawing capacity of both the ester and the core; the ligand itself will not racemize under the same conditions.

An unexpected cleavage has also been described for the coordinated Schiff base of 2-aminobenzenethiol in alkaline solution. Instead of neutral complexes of the Schiff bases, anionic 1:2 Tc(V) oxo complexes are formed; the donor groups are now the thiol and generated amino groups [206].

6.2 Reactivity Through Coordinated Ligands

The introduction of metabolizable ester functions may be a very useful means of obtaining retention of the radiotracer within the target, as known for ^{99m}Tc-ECD. It may also help to eliminate nontarget radioactivity more rapidly. Therefore, systematic studies have been performed, in order to understand the enzymatic hydrolysis of oxotechnetium(V) complexes with up to four ester groups in the complex [205]. Fig. 30 shows the partial hydrolysis by pig liver esterase as is was observed for 1:2 complexes of derivatized dimercaptosuccinic acid complexes.

It has been anticipated that biodistribution can be altered through the appropriate introduction of **ether** linkage groups into the ligand molecule. Rapid hepatobiliary clearance of the myocardial agent hexakis(isonitrile)technetium (I) has been promoted by means of ether derivatives. Studies of the effect of ether substitution in phosphine ligands of cationic complexes showed that it induces rapid blood clearance and rapid hepatobiliary clearance without significantly lowering myocardial uptake [123, 207], as shown for tetrofosmin (Fig. 17).

Unlike some Tc(III) complexes, which are reducible in vivo and then play an important role, for example in the design of heart- or brain imaging agents [124, 208], normally Tc(V) or Re(V) complexes need to be derivatized in order to be used as redox-active markers. For this purpose, Tc(III) [209] and various Tc(V) complexes have been derivatized with nitroimidazole (Fig. 31). Nitroimidazoles can be subject to enzymatic reduction of the nitro group in the absence of oxygen and then trapped in hypoxic tissue. With appropriate complexes, the redox potential of the coordinated ligand can apparently be kept close enough to that of misonidazole to allow reasonably fast reduction by nitroreductases, such as xanthine oxidase (XOD), as well as trapping. Xanthine oxidase serves as a model enzyme. The XOD-catalyzed reduction is another example of the recognition of functionalised Tc(V) and Re(V) complexes as substrates for enzymes. Which enzyme system in vivo is responsible for the trapping of the agents is still unknown. Since large differences in the ratio of binding to hypoxic and normoxic cells exist for a series of Tc PnAO complexes – with all complexes showing a redox potential close to that of mis-

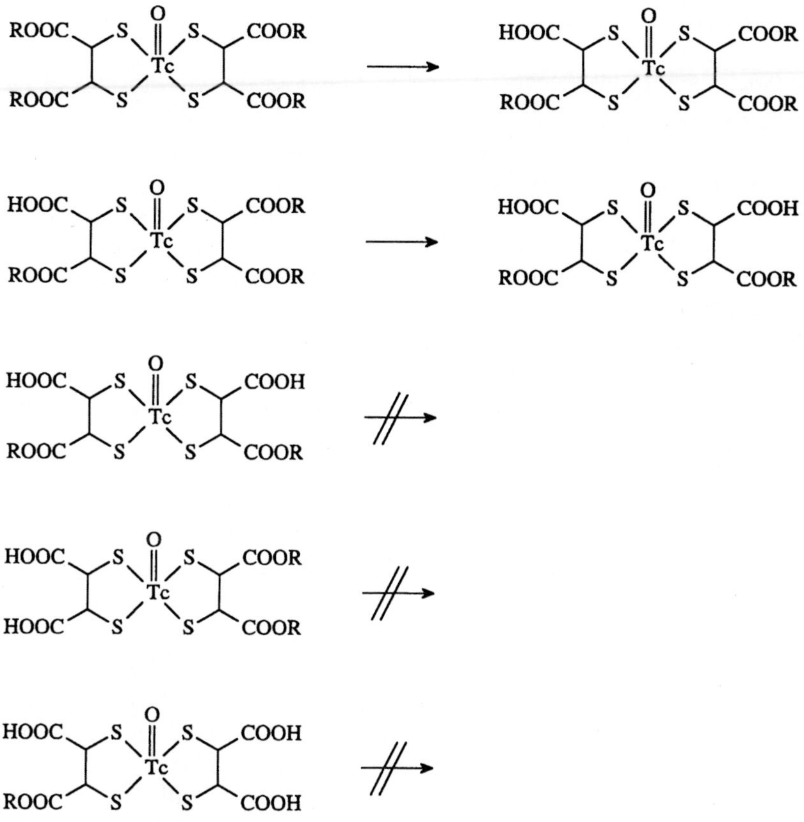

Fig. 30. Pig liver esterase-catalyzed hydrolysis of anionic DMS-ester complexes

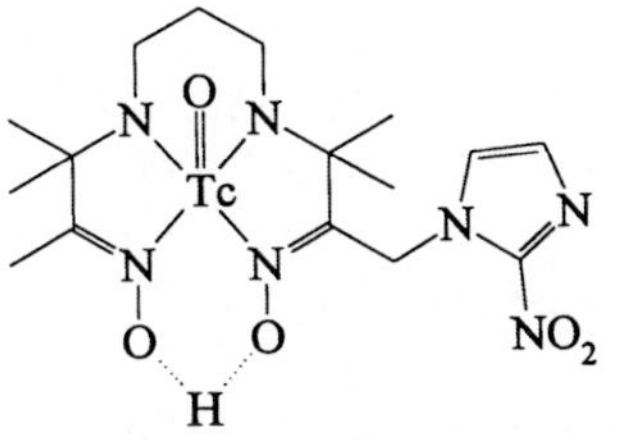

Fig. 31. Tc complex of the PnAO nitroimidazole ligand [63, 210]

omidazole [210, 211] – a favourable redox potential alone does not seem to be decisive. Instead of a nitroimidazole moiety attached to the Tc chelate, the chelating ligand itself can incorporate an aromatic ring with a single nitro group substituent [212]. According to recent studies on complexes that have not yet been defined structurally, Tc complexes – even without the nitro group – seems to provide potential hypoxia agents [213].

6.3 Protein- and Receptor Binding

Apart from various data published on the percentage of protein binding of Tc complexes in plasma or serum protein solution, only a few studies have been specifically concerned with protein binding of Tc complexes [214]. The reaction between Tc complexes and human serum albumin is governed by the law of mass action. In analogy to the conditions prevailing in radioimmunoassays, the binding may be described at very low protein saturation – simulating the conditions used in diagnostic applications – by the equations used in radioimmunoassays. A similar theoretical calculation of bound-to-free ratios has been derived for receptor binding of radiotracers [215]. Exploitation of receptor-ligand interactions – a field of modern nuclear medicine – would be a highly attractive prospect, provided that technetium chemistry permits the design of Tc complexes that mimic antagonists or agonists in their high affinity and selectivity to the specific receptor.

In developing Tc-99m receptor imaging agents, the concept of bifunctional chelating agents seems to be a promising approach, aimed at modified receptor ligands substituted with a chelating moiety, e.g. an N_2S_2 donor set, located far from the pharmacologically essential site of the molecule. If there is no substantial tolerance for such an additional pendant chelated technetium, the Tc complex might serve as an integral replacement for a region of the receptor-binding molecule. As shown by Lever and Wagner [216], replacement of a quite lipophilic portion in quinuclidinyl benzilate (QNB) – a ligand with high affinity for m-acetylcholine receptors – by a Tc=O complex of an N_2S_2 tetradentate ligand, permits the synthesis of QNB analogues of similar overall size. The *syn* and *anti* Tc N_2S_2-QNB analogues are the first technetium complexes to display in vitro affinity for a neuroreceptor. (Binding of the bulky lipophilic BATO-complex to the complete QNB molecule leads, however, to complete loss of receptor binding properties.) With the aim of making new Tc and Re complexes available as potential 5HT receptor imaging agents, the iodine label in *p*-MPPI was replaced by the $[TcO(N_2S_2)]$ unit. Although in vitro binding assays of the resulting complex displayed a moderate affinity to $5HT_{1A}$ receptors, the Tc tracer shows no uptake in the brain of rats. Obviously, the molecule may be too large to pass the blood brain barrier [217]. Furthermore, recent efforts to mimic organic receptor-binding ligands by Tc compounds containing the $[TcO(N_2S_2)]$ unit also refer to the dopamine D_2 receptor binding agent spiperone [218] and progestin [219] as lead structures. Instead of tetradentate chelating ligands, mixed ligands have very recently been conveniently used to synthesize mimics of receptor binding molecules. In the partial structure approach, in which the metal unit is not a conjugated-but an integrated part of the molecule, steroid imitating complexes [195, 219] or serotonin receptor binding molecules [220] have been prepared, see Table 3. Two approaches, bifunctional chelates and direct chelating methods have been applied for a number of receptor-binding peptides [221, 222].

Table 3.

Complex	Receptor	Ref.
	mACh	[216]
	Progestin	[187, 195]
	Dopamine	[218]
 M=Re; Re-p-MPPI	5HT$_{1A}$	[217]
 M = Tc, Re; E = O, S, N(CH$_3$); Y = CH$_2$, N-alkyl; X = H, F	5HT$_2$	[220]
	Somatostatin	[47]

B. Johannsen and H. Spies

Acknowledgements. Our colleagues R. Berger, T. Fietz and H.J. Pietzsch are gratefully acknowledged for their fruitful discussion and valuable help with the manuscript.

7 References

1. Deutsch E, Libson K, Jurisson S, Lindoy LF (1983) Progr Inorg Chem 30: 75
2. Davison A (1983) In: Deutsch E, Nicolini M, Wagner HN Jr (eds) Technetium in chemistry and nuclear medicine, Cortina International, Verona, p 3
3. Davison A, Jones AG (1982) Int J Appl Radiat Isot 33: 875
4. Bandoli G, Mazzi U, Roncari E, Deutsch E (1982) Coord Chem Rev 44: 191
5. Melnik M, Van Lier JE (1987) Coord Chem Rev 77: 275
6. Mazzi U (1989) Polyhedron 8: 1683
7. Franklin KJ, Howard-Lock HE, Lock CJL (1982) Inorg Chem 21: 1941
8. Thomas RW, Davison A, Trop HS, Deutsch E (1980) Inorg Chem 19: 2840
9. Kuzina AF, Oblova AA, Spitsyn VI (1972) Russ J Inorg Chem 17: 1377
10. Fackler PH, Kastner ME, Clarke MJ (1984) Inorg Chem 23: 3968
11. Kastner ME, Fackler PH, Clarke MJ, Deutsch E (1984) Inorg Chem 23: 4683
12. Tisato F, Refosco F, Mazzi U, Bandoli G, Dolmella A (1989) Inorg Chim Acta 164: 127
13. Pietzsch HJ, Spies H, Leibnitz P, Reck G, Beger J, Jacobi R (1993) Polyhedron 12: 187
14. Roodt A, Leipoldt JG, Deutsch EA, Sullivan JC (1992) Inorg Chem 31: 1080
15. Roodt A, Leipoldt JG, Helm L, Merbach AE (1994) Inorg Chem 33: 140
16. Lu J, Clarke MJ (1989) Inorg Chem 28: 2315
17. Cook RL, Woods M, Sullivan J, Applemann EH (1989) Inorg Chem 28: 3349
18. Hugill D, Peacock RK (1966) J Chem Soc A: 1339
19. Davison A, Trop HS, DePamphilis BV, Jones AG, Thomas RW, Jurisson SS (1982) Inorg Synth 21: 160
20. Hübener R, Abram U (1992) Z Anorg Allg Chem 617: 96
21. Baldas J, Colmanet JF (1989) Aust J Chem 42: 1155
22. Peters G, Preetz W (1981) Z Naturforsch 36b: 138
23. Hwang LLY, Ronca N, Solomon NA, Steigman J (1985) Int J Appl Radiat Isot 36: 475
24. Huber G, Anderegg G, May K (1987) Polyhedron 6: 1707
25. Davison A, De Pamphilis BV, Jones AG, Franklin KJ, Lock CJL (1987) Inorg Chim Acta 128: 161
26. Abrams MJ, Larsen SK, Zubieta J (1991) Inorg Chem 30: 2031
27. Rochon FD, Melanson R, Kong PC (1992) Acta Cryst C48: 785
28. Münze R (1980) Radiochem Radioanal Lett 43: 219
29. Münze R, Hoffman I (1981) Radiochem Radioanal Lett 48: 289
30. DePamphilis BV, Jones AG, Davison A (1983) Inorg Chem 22: 2292
31. Johannsen B, Syhre R (1978) Radiochem Radioanal Lett 36: 107
32. deKieviet W (1981) J Label Comp Radiopharm 18: 136
33. Larsen SK, Caldwell G, Higgins III JD, Abrams MJ, Solomon HF (1993) Xth Int Symp Radiopharm Chem, Kyoto, Oct 25–28, p 1 (abstr)
34. Luo H, Rettig SJ, Orvig C (1993) Inorg Chem 32: 4491
35. Hahn EF, Rupprecht S (1991) Chem Ber 124: 487
36. Hahn EF, Rupprecht S (1991) Chem Ber 124: 481
37. Thomas JA, Davison A (1992) Inorg Chem 31: 1976
38. Dyckhoff B, Schulte HJ, Englert U, Spaniol TP (1992) Z Anorg Allg Chem 614: 131
39. Gerber TIA, Kemp HJ, Du Preez GH, Bandoli G, Dollmella A (1992) Inorg Chim Acta 202: 191
40. Johannsen B (1982) Int J Appl Radiat Isot 33: 429
41. Seifert S, Münze R, Johannsen B (1982) Radiochem Radioanal Lett 54: 153
42. Thornback JR, Theobald AE (1981) Int J Appl Radiat Isot 32: 833
43. Kastner ME, Lindsay MJ, Clarke MJ (1982) Inorg Chem 21: 2037

44. Zuckman SA, Freeman GM, Troutner DE, Volkert WA, Holmes RA, Van Derveer DG, Barefield EK (1981) Inorg Chem 20: 2386
45. Simon J, Troutner DE, Volkert WA, Holmes RA (1981) Radiochem Radioanal Lett 47: 111
46. Franz J, Volkert WA, Barefield EK et al. (1987) Nucl Med Biol 14: 569
47. Maina T, Stolz B, Albert R, Nock B, Bruns C, Maecke H (1994) J Nucl Biol Med 38: 452 (abstr)
48. Nock B, Evard F, Paganelli G, Maecke HR (1994) J Nucl Biol Med 38: 469 (abstr)
49. Riche F, Trimcev I, Godart J, Benabed A, Comet M, Duatti A, Pasqualini R, Vidal M (1992) IXth Int Symp Radiopharm Chem, Paris, April 6–10, p 24 (abstr)
50. Riche F, Pasqualini R, Duatti A, Vidal M (1992) Appl Radiat Isot 43: 437
51. John E, Thakur ML, Wilder S, Alauddin MM, Epstein AL (1994) J Nucl Med 35: 876
52. Davison A, Jones AG, Abrams MJ (1981) Inorg Chem 20: 4300
53. Fackler PH, Lindsay MJ, Clarke MJ Kastner (1985) Inorg Chim Acta 109: 39
54. Deutsch E, Elder RC, Lanje BA, Vaal M, Lay DG (1976) Proc Natl Acad Sci USA 73: 4287
55. Troutner DE, Volkert WA, Hoffmann TJ, Holmes RA (1984) Int J Appl Radiat Isot 35: 467
56. Neirinckx RD, Canning LR, Piper IM, Nowotnik DP, Pickett RD, Holmes RA, Volkert WA, Forster AM, Weisner PS, Marriott JA, Chaplin SB (1987) J Nucl Med 28: 191
57. Jurisson S, Schlemper EO, Troutner DE, Canning LR, Nowotnik DP, Neirinckx RD (1986) Inorg Chem 25: 543
58. Jurisson S, Aston K, Fair CK, Schlemper EO, Sharp PR, Troutner DE (1987) Inorg Chem 26: 3576
59. Sharp PF, Smith FW, Gemmell HG, Lyall D, Evans NTS, Gvozdanovic D, Davidson J, Tyrrell DA, Pickett RD, Neirinckx RD (1986) J Nucl Med 27: 171
60. Di Rocco RJ, Kuczynski BL, Pirro JP, Bauer A, Linder KE, Ramalingam K, Cyr JE, Chan YW, Raju N, Narra RK, Nowotnik DP, Nunn AD (1993) J Cereb Blood Flow Metab 13: 755
61. Koch P, Mäcke R (1992) Angew Chem 31: 1507
62. Paganelli G, Magnani P, Zito F, Lucignani G, Sudati F, Truci G, Motti E, Terreni M, Pollo B, Giovanelli M, Canal N, Scotti G, Comi G, Koch P, Maecke HR, Fazio F (1994) Eur J Nucl Med 21: 314
63. Linder KE, Chan YW, Cyr JE, Malley MF, Nowotnik DP, Nunn AD (1994) J Med Chem 37: 9
64. Pillai MRA, Schlemper EO, Volkert WA (1994) Nucl Med Biol 21: 997
65. Suess E, Malessa S, Ungersbock K, Kitz P, Podreka I, Heimberger K, Hornykiewicz O, Deecke L (1991) J Nucl Med 32: 1675
66. Roth CA, Hoffman TJ, Corlija M, Volkert WA, Holmes RA (1992) Nucl Med Biol 19: 783
67. Cyr J, Linder KE, Nanjappan P, Raju N, Ramalingam K, Nowotnik DP, Nunn AD (1992) IXth Int Symp Radiopharm Chem, Paris, April 6–10, p 4 (abstr)
68. Vanbilloen H, Dezutter N, Boonen C, De Roo M, Verbruggen A (1994) J Nucl Biol Med 38: 468 (abstr)
69. Nicholson T, Zubieta J (1988) Polyhedron 7: 171
70. Abrams MJ, Chen Q, Shaikh SN, Zubieta J (1990) Inorg Chim Acta 176: 11
71. Babich JW, Solomon H, Pike MC, Kroon D, Graham W, Abrams MJ, Tompkins RG, Rubin RH, Fischman AJ (1993) J Nucl Med 34: 1964
72. Schwartz DA, Abrams MJ, Hauser MM, Gaul FE, Larsen SK, Rauh D, Zubieta JA (1991) Bioconj Chem 2: 333
73. Abrams MJ, Juweid M, tenKate CI, Schwartz DA, Hauser MM, Gaul FE, Fuccello J, Rubin RH, Strauss HW, Fischman AJ (1990) J Nucl Med 31: 2022
74. DePamphilis BV, Jones AG, Davis MA, Davison A (1978) J Am Chem Soc 100: 5570
75. Smith JE, Byrne EF, Cotton FA, Sekutowsky JC (1978) J Am Chem Soc 100: 5571
76. Spies H, Johannsen B (1981) Inorg Chim Acta 48: 255
77. Hamor TA, Hussain W, Jones CJ, McCleverty A, Rothin AS (1988) Inorg Chim Acta 146: 181
78. Davison A, De Vries N, Dewan J, Jones A (1986) Inorg Chim Acta 120: L15
79. Davison A, Jones AG, DePamphilis BV (1981) Inorg Chem 20: 1617
80. Du Preez JGH, Gerber TIA (1985) Inorg Chim Acta 110: 59
81. Bandoli G, Mazzi U, Abram U, Spies H, Münze R (1987) Polyhedron 6: 1547
82. Colmanet SF, Mackay MF (1987) Aust J Chem 40: 1301
83. Colmanet SF, Mackay MF (1988) Inorg Chim Acta 147: 173
84. Colmanet SF, Mackay MF (1988) Aust J Chem 41: 151
85. Bandoli G, Nicolini M, Mazzi U, Spies H, Münze R (1984) Trans Met Chem 9: 127
86. Johannsen B, Syhre R, Spies H (1980) Nuc Compact 11: 42
87. Ohta H, Endo K, Fujita T, Konishi J, Torizuka K, Horiuchi K, Yokoyama A (1988) Nucl Med Commun 9: 105

88. Blower PJ, Singh J, Clarke SEM (1991) J Nucl Med 32: 845
89. Jones WB, Elgren TE, Morelock MM, Elder RC, Wilcox DE (1994) Inorg Chem 33: 5571
90. Davison A, DePamphilis BV, Faggiani R, Jones AG, Lock CJL, Orvig C (1985) Can J Chem 63: 319
91. Tisato F, Bolzati C, Duatti A, Bandoli G, Refosco F (1993) Inorg Chem 32: 2042
92. Pietzsch HJ, Spies H, Hoffmann S (1989) Inorg Chim Acta 165: 163
93. Colmanet SF, Williams GA, Mackay MF (1987) J Chem Soc, Dalton Trans: 2305
94. Davison A, Orvig C, Trop HS, Sohn M, DePamphilis BV, Jones AG (1980) Inorg Chem 19: 1988
95. Jones AG, Orvig C, Trop HS, Davison A, Davis MA (1980) J Nucl Med 21: 279
96. Byrne EF, Smith JE (1979) Inorg Chem 18: 1832
97. Spies H, Scheller D (1986) Inorg Chim Acta 116: 1
98. Singh J, Powell AK, Clarke SEM, Blower PJ (1991) J Chem Soc, Chem Commun: 1115
99. Schwarz A, Steinsträßer A (1987) J Nucl Med 28: 721 (abstr)
100. Mather SJ, Ellison D (1990) J Nucl Med 31: 692
101. Pimm MV, Rajput RS, Frier M, Gribben SJ (1991) Eur J Nucl Med 18: 973
102. Goedemans WT, Panek KJ, Ensing GJ, de Jong MTM (1990) In: Nicolini M, Bandoli G, Mazzi U (eds) Technetium and rhenium in chemistry and nuclear medicine 3, Cortina International, Verona, Raven Press, New York, p 595
103. Winnard P Jr, Virzi F, Rusckowski M, Hnatowich DJ (1994) J Nucl Med 35: 260P
104. Brenner D, Davison A, Lister-James J, Jones AG (1984) Inorg Chem 23: 3793
105. Bryson N, Dewan JC, Lister-James J, Jones AG, Davison A (1988) Inorg Chem 27: 2154
106. Spies H, Johannsen B (1995) Analyst 120: 775
107. Drouillard S, Alagui A, Apparu M, Mathieu JP, Pasqualini R, Vidal M (1992) Appl Radiat Isot 10: 1227
108. Marchi A, Marvelli L, Rossi R, Magon L, Bertolasi V, Ferretti V, Gilli P (1992) J Chem Soc, Dalton Trans: 1485
109. Truffer S, Ianoz E, Lerch P, Kosinski M (1988) Inorg Chim Acta 149: 217
110. Ianoz E, Mantegazzi D, Lerch P, Nicolo F, Chapuis G (1989) Inorg Chim Acta 156: 235
111. Pietzsch HJ, Spies H, Leibnitz P, Reck G (1993) Polyhedron 12: 2995
112. Pietzsch HJ, Spies H, Leibnitz P, Reck G, Beger J, Jacobi R (1992) Polyhedron 11: 1623
113. White DJ, Küppers HJ, Edwards AJ, Watkin DJ, Cooper SR (1992) Inorg Chem 31: 5351
114. Bandoli G, Clemente DA, Mazzi U, Roncari E (1982) J Chem Soc, Dalton Trans: 1381
115. Abrams MJ, Brenner D, Davison A, Jones AG (1983) Inorg Chim Acta 77: L127
116. Abrams MJ, Davison A, Faggiani R, Jones AG, Lock CJL (1984) Inorg Chem 23: 3284
117. Du Preez JGH, Gerber TIA, Knoesen O (1987) Inorg Chim Acta 130: 9
118. Du Preez JGH, Gerber TIA, Knoesen O (1985) Inorg Chim Acta 109: L17
119. Abram U, Münze R, Hartung J, Beyer L, Kirmse R, Koehler K, Stach J, Behm H, Beurskens PT (1989) Inorg Chem 28: 834
120. Vanderheyden JL, Ketring AR, Libson K, Heeg MJ, Roecker L, Motz P, Whittle R, Elder RC, Deutsch E (1984) Inorg Chem 23: 3184
121. Deutsch E, Ketring AR, Libson K, Vanderheyden JL, Hirth WW (1989) Nucl Med Biol 16: 191
122. Kelly JD, Higley B, Archer CM, Canning LR, Chiu KW, Edwards B, Forster AM, Gill HK, Latham IA, Pickett RD, Edwards PG, Imran A (1992) IXth Int Symp Radiopharm Chem, Paris, April 6–10, p 40 (abstr)
123. Kelly JD, Forster AM, Higley B, Archer CM, Booker FS, Canning LR, Chiu KW, Edwards B, Gill HK, McPartlin M, Nagle KR, Latham IA, Pickett RD, Storey AE, Webbon PM (1993) J Nucl Med 34: 222
124. Forster AM, Storey AE, Archer CM, Nagle KR, Booker FS, Edwards B, Cill HK, Kelly JD, McPartlin M (1992) J Nucl Med 33: 850
125. Johannsen B, Syhre R, Spies H, Münze R (1978) J Nucl Med 19: 816
126. Baldas J, Boas J, Bonnyman J, Mackay MF, Williams GA (1982) Aust J Chem 35: 2413
127. Kirmse R, Stach J, Spies H (1980) Inorg Chim Acta 45: L251
128. Bandoli G, Gerber TIA (1987) Inorg Chim Acta 126: 205
129. Spies H, Pietzsch HJ, Hoffmann I (1989) Inorg Chim Acta 161: 17
130. Cook J, Davis WM, Davison A, Jones AG (1991) Inorg Chem 30: 1773
131. Kung HF, Molnar M, Billings J, Wicks R, Blau MJ (1984) J Nucl Med 25: 326
132. Kung HF, Guo YZ, Yu CC, Billings J, Subramanyam V, Calabrese JC (1989) J Med Chem 32: 433

133. John CS, Francesconi LC, Kung HF, Wehrli S, Graczyk G, Carrol P (1992) Polyhedron 11: 1145
134. Du Preez JGH, Gerber TIA, Fourie PJ, Van Wyk AJ (1984) Inorg Chim Acta 82: 201
135. Lever SZ, Baidoo KE, Mahmood A (1990) Inorg Chim Acta 176: 183
136. Epps LA, Burns HD, Lever SZ, Goldfarb HW, Wagner HN Jr (1987) Appl Radiat Isot, Int J Radiat Appl Instrum Part A 38: 661
137. Lever SZ, Wagner HN Jr (1990) In: Nicolini M, Bandoli G, Mazzi U (eds) Technetium and rhenium in chemistry and nuclear medicine 3, Cortina International, Verona, Raven Press, New York, p 649
138. Bürgi HB, Anderegg G, Bläuenstein P (1981) Inorg Chem 20: 3829
139. Edwards DS, Cheesman EH, Watson MW, Maheu LJ, Nguyen SA, Dimitre L, Nason T, Watson AD, Walovitch R (1990) In: Nicolini M, Bandoli G, Mazzi U (eds) Technetium and rhenium in chemistry and nuclear medicine 3, Cortina International, Verona, Raven Press, New York, p 433
140. Walovitch RC, Cheesman EH, Maheu LJ, Hall KM (1994) J Cereb Blood Flow Metab 14 (Suppl 1): S4
141. Eisenhut M, Mißfeldt M, Lehmann WD, Karas M (1991) J Label Comp Radiopharm 29: 1283
142. Baidoo KE, Lever SZ (1990) Bioconj Chem 1: 132
143. Lever SZ, Baidoo KE, Kramer AV, Burns HD (1988) Tetrahedron Lett 26: 3219
144. Ohmomo Y, Francesconi L, Kung MP, Kung HF (1992) J Med Chem 35: 157
145. Singh PR, Corlija M, Troutner DE, Ketring AR, Volkert WA (1992) Nucl Med Biol 19: 791
146. Marzilli LG, Banaszczyk MG, Hansen L, Kuklenyik Z, Cini R, Taylor A Jr (1994) Inorg Chem 33: 4850
147. Banaszczyk MG, Hansen L, Cini R, Kuklenyik Z, Taylor A Jr, Marzilli LG (1993) J Nucl Med 5: 39P
148. Pirmettis I, Mastrostamatis S, Papadopoulos M, Raptopoulou CP, Terzis A, Chiotellis E (1994) J Nucl Med 35: 263P
149. Hansen L, Folks R, Malveaux EJ, Marzilli LG, Eshima D, Taylor A Jr (1994) J Nucl Med 35: 262P
150. Najafi A, Alauddin MM, Sosa A, Ma GQ, Chen DCP, Epstein AL, Siegel ME (1992) Nucl Med Biol 19: 205
151. Refosco F, Bolzati C, Bandoli G, Moresco A, Nicolini M, Tisato F (1992) IXth Int Symp Radiopharm Chem, Paris, April 6–10, p 7 (abstr)
152. Dilworth JR, Griffiths DV, Hughes JM, Morton S, Archer CM, Kelly JD (1992) Inorg Chim Acta 195: 145
153. Davison A, Jones AG, Orvig C, Sohn M (1981) Inorg Chem 20: 1629
154. Kasina S, Fritzberg AR, Johnson EL, Eshima D (1986) J Med Chem 29: 1933
155. Jones AG, Davison A, LaTegola MR, Brodack JW, Orvig C (1982) J Nucl Med 23: 801
156. Nosco DL, Manning RG, Fritzberg A (1986) J Nucl Med 27: 939 (abstr)
157. Taylor A Jr, Eshima D, Fritzberg AR, Christian PE, Kasina S (1986) J Nucl Med 27: 795
158. Stepniak-Biniakiewicz D, Chen B, Deutsch E (1992) J Med Chem 35: 274
159. Fritzberg AR, Klingensmith III WC, Whitney WP, Kuni CC (1981) J Nucl Med 22: 258
160. Johannsen B, Noll B, Leibnitz P, Reck G, Noll S, Spies H (1993) Radiochim Acta 63: 133
161. Johannsen B, Noll B, Leibnitz P, Reck G, Noll S, Spies H (1993) Inorg Chim Acta 210: 209
162. Bormans GM, Cleynhens BJ, De Roo MJK, Verbruggen AM (1992) Eur J Nucl Med 19: 271
163. Verbruggen A, Bormans G, Cleynhens B, Vandecruys A, Hoogmartens M, De Roo MJK (1989) J Label Comp Radiopharm 26: 436
164. Hansen L, Cini R, Taylor A Jr, Marzilli LG (1992) Inorg Chem 31: 2801
165. Hansen L, Marzilli LG, Eshima D, Malveaux EJ, Folks R, Taylor A Jr (1994) J Nucl Med 35: 1198
166. Verbruggen AM (1990) Eur J Nucl Med 17: 346
167. Schroff RW, Weiden PL, Appelbaum J, Fer MF, Breitz H, Vanderheyden JL, Ratliff BA, Fisher D, Foisie D, Hanelin LG, Morgan AC, Fritzberg AR, Abrams PG (1990) Antibody Immunoconj Radiopharm 3: 99
168. Guhlke S, Diekmann D, Zamora PO, Knapp FF Jr, Biersack HJ (1994) J Nucl Biol Med 38: 444
169. Jeong JM, Kinuya S, Paik CH, Saga T, Sood VK, Carrasquillo JA, Neumann RD, Reynolds JC (1994) Nucl Med Biol 21: 935
170. Rao TN, Gustavson LM, Srinivasan A, Kasina S, Fritzberg AR (1992) Nucl Med Biol 19: 889

171. O'Neil JP, Wilson SR, Katzenellenbogen JA (1994) Inorg Chem 33: 319
172. O'Neil JP, Carlson KE, Anderson CJ, Welch MJ, Katzenellenbogen JA (1994) Bioconj Chem 5: 182
173. Bandoli G, Nicolini M, Mazzi U, Refosco F (1984) J Chem Soc, Dalton Trans: 2505
174. Bandoli G, Mazzi U, Clemente DA, Roncari E (1982) J Chem Soc, Dalton Trans: 2455
175. Bandoli G, Mazzi U, Wilcox BE, Jurisson S, Deutsch E (1984) Inorg Chim Acta 95: 217
176. Du Preez JGH, Gerber TIA, Knoesen O (1987) Inorg Chim Acta 132: 241
177. Tisato F, Refosco F, Mazzi U, Bandoli G, Nicolini M (1987) J Chem Soc, Dalton Trans: 1693
178. Mazzi U, Refosco F, Tisato F, Bandoli G, Nicolini M (1986) J Chem Soc, Dalton Trans: 1623
179. Pietzsch HJ, Spies H, Hoffmann S, Scheller D (1990) Appl Radiat Isot 41: 185
180. Du Preez JGH, Gerber TIA (1984) Inorg Chim Acta 82: 201
181. Jurisson S, Lindoy LF, Dancey KP, McPartlin M, Tasker PA, Uppal DA, Deutsch E (1984) Inorg Chem 23: 227
182. Du Preez JGH, Gerber TIA, Fourie PJ, Van Wyk AJ (1984) J Coord Chem 13: 173
183. Stassinopoulou CI, Mastrostamatis S, Papadopoulos M, Vavouraki H, Terzis A, Hountas A, Chiotellis E (1991) Inorg Chim Acta 189: 219
184. Refosco F, Tisato F, Mazzi U, Bandoli G, Nicolini M (1988) J Chem Soc, Dalton Trans: 611
185. Liu S, Rettig SJ, Orvig C (1991) Inorg Chem 30: 4915
186. Chang L, Rall J, Tisato F, Deutsch E, Heeg MJ (1993) Inorg Chim Acta 205: 35
187. Chi DY, Katzenellenbogen JA (1993) J Am Chem Soc 115: 7045
188. Pietzsch HJ, Spies H, Hoffmann S (1989) Inorg Chim Acta 161: 15
189. Spies H, Pietzsch HJ, Syhre R, Hoffmann S (1990) Isotopenpraxis 26: 159
190. Spies H, Syhre R, Pietzsch HJ (1990) Plzen lek Sborn 62: 85
191. Bandoli G, Mazzi U, Pietzsch HJ, Spies H (1992) Acta Cryst C 48: 1422
192. Pietzsch HJ, Spies H, Hoffmann S (1990) Inorg Chim Acta 168: 7
193. Mastrostamatis SG, Papadopoulos MS, Pirmettis IC, Paschali E, Varvarigou AD, Stassinopoulou CI, Raptopoulou CP, Terzis A, Chiotellis E (1994) J Med Chem 37: 3212
194. Rochon FD, Melanson R, Kong PC (1992) Inorg Chim Acta 194: 43
195. Chi DY, O'Neil JP, Anderson CJ, Welch MJ, Katzenellenbogen JA (1994) J Med Chem 37: 928
196. Pearlstein RM, Davison A (1988) Polyhedron 7: 1981
197. Pearlstein RM, Lock CJL, Faggiani R, Costello CE, Zeng CH, Jones AG, Davison A (1988) Inorg Chem 27: 2409
198. Refosco F, Mazzi U, Deutsch E, Kirchhoff JR, Heinemann WR, Seeber R (1988) Inorg Chem 27: 4121
199. Wilcox BE, Cooper JN, Elder RC, Deutsch E (1988) Inorg Chim Acta 142: 55
200. Du Preetz JGH, Gerber TIA, Kemp HJ (1992) J Coord Chem 26: 177
201. Gerber T, Kemp H, Du Preez J, Bandoli G (1993) Radiochim Acta 63: 129
202. Kung HF, Liu BL, Wei Y, Pan S (1990) Appl Radiat Isot 41: 773
203. Helm L, Deutsch K, Deutsch EA, Merbach AE (1992) Helv Chim Acta 75: 210
204. Johnson DL, Fritzberg AR, Hawkins BL, Kasina S, Eshima D (1984) Inorg Chem 23: 4204
205. Seifert S, Syhre R, Spies H, Johannsen B (1994) J Nucl Biol Med 38: 463 (abstr)
206. Spies H, Pietzsch HJ (1989) Inorg Chim Acta 161: 151
207. Libson K, Messa C, Kwiatkowski M, Zito F, Best T, Colombo F, Matarrese M, Wang X, Fragasso G, Fazio F, Deutsch E (1990) In: Nicolini M, Bandoli G, Mazzi U (eds) Technetium and rhenium in chemistry and nuclear medicine 3, Cortina International, Verona, Raven Press, New York, p 365
208. Deutsch E (1993) Radiochim Acta 63: 195
209. Linder KE, Raju N, Cyr JE, Chan YW, Ramalingam K, Nowotnik DP, Nunn AD (1992) IXth Int Symp Radiopharm Chem, Paris, April 6–10, p 13 (abstr)
210. Kusuoka H, Hashimoto K, Fukuchi K, Nishimura T (1994) J Nucl Med 35: 1371
211. Nowotnik DP, Cyr JE, Chan YW, Ramalingam K, Linder KE, Nunn AD (1993) J Nucl Med 34: 18P
212. Zhang Z, Mannan RH, Wiebe LI, McEwan AJ (1994) J Nucl Biol Med 38: 472 (abstr)
213. Kelly JD, Archer CM, Platts EA, Storey AE, Canning LR, Edwards B, King AC, Burke JF, Duncanson P, Griffiths DV, Hughes JM, Pitman MA (1994) J Nucl Biol Med 38: 472 (abstr)
214. Johannsen B, Berger R, Schomäcker K (1980) Radiochem Radioanal Lett 42: 177
215. Eckelman WC (1994) Nucl Med Biol 21: 759
216. Lever SZ, Baidoo KE, Mahmood A, Matsumura K, Scheffel U, Wagner HN Jr (1994) Nucl Med Biol 21: 157

217. Kung HF, Bradshaw J, Chumpradit S, Zhuang ZP, Kung MP, Mu M, Frederick D (1994) J Nucl Biol Med 38: 449 (abstr)
218. Samnick S, Brandau W, Schober O (1994) J Nucl Biol Med 38: 483 (abstr)
219. DiZio JP, Anderson CJ, Davison A, Erhard GJ, Carlson KE, Welch MJ, Katzenellenbogen JA (1992) J Nucl Med 33: 558
220. Johannsen B, Pietzsch HJ, Scheunemann M, Spies H, Brust P (1995) J Nucl Med 36: 27p
221. Fischman HJ, Babich JW, Strauss HW (1993) J Nucl Med 34: 2253
222. Lister-James J, McBride WJ, Moyer BR, Buttram S, Vallabhajosula S, Bastidas DA, Lippszyc H, Lee H, Dean RT (1994) J Nucl Med 35: 257P

Technetium-99m Chelates as Radiopharmaceuticals

Wynn A. Volkert[1] and Silvia Jurisson[2]

[1] Research Service, HS Truman Memorial VA Hospital and Department of Radiology, 409 Lewis Hall, University of Missouri, Columbia, MO 65211, USA
[2] Department of Chemistry, 123 Chemistry Building, University of Missouri, Columbia, MO 65211, USA

Table of Contents

Topics in Current Chemistry, Vol. 176
© Springer-Verlag Berlin Heidelberg 1996

W.A. Volkert and S. Jurisson

Technetium-99m (^{99m}Tc) is currently, and will be in the foreseeable future, the most widely used radionuclide used in virtually all clinical diagnostic Nuclear Medicine laboratories in the world. Several ^{99m}Tc radiopharmaceuticals that are routinely used have become important factors in decisions on the diagnosis and treatment of patients with a variety of diseases. More sophisticated biological targeting vectors are being developed by the chemical and biomedical communities. The development of technologies and chelating agents for labeling these vectors with ^{99m}Tc, to produce diagnostic radiopharmaceuticals that have optimal in vivo specificities, will be essential for enhancement of the future of Nuclear Medicine in the health care arena worldwide. This chapter is focused primarily on a review of ^{99m}Tc complexes used for the formulation of currently approved radiopharmaceuticals, and highlights several chelation systems that are being used in the design of future ^{99m}Tc drugs. The chemical and structural properties of different ^{99m}Tc complexes and their relationships to their biolocalization and pharmacokinetic properties are described. Emphasis is placed on the importance of developing ^{99m}Tc complexes with a high degree of physicochemical flexibility, in order to optimize the pharmacological performance of future ^{99m}Tc-labeled drugs.

1 Introduction

1.1 Background

Radiopharmaceuticals, drugs containing a radionuclide, are a fundamental and essential component in Nuclear Medicine studies directed toward diagnosis – in some cases the treatment – of disease in patients. While several important radiopharmaceuticals have been developed for positron emission tomography (PET) imaging (including ^{18}F-FDG, ^{82}Rb$^+$, ^{15}O-H$_2$O, etc.), difficulties in approval by regulatory bodies, cost, and lack of uniform reimbursement policies have limited widespread utilization of this technology. In contrast, radiopharmaceuticals for single photon emission computed tomography (SPECT) are widely used for the vast majority of Nuclear Medicine diagnostic studies performed in the world and will continue to be used in the foreseeable future. Several single-photon emitting radioisotopes are used in the formulation of radioactive drugs (incl. ^{123}I, ^{99m}Tc, ^{201}Tl, ^{111}In, etc.), however, ^{99m}Tc remains the most widely used. The favorable physical properties (i.e., 6 hour half life, 140 KeV γ-ray and decay by isomeric transition), the ready availability and low cost of ^{99m}Tc are the reasons for the continued utility of ^{99m}Tc for medical applications.

Two important advances have led to the rapid proliferation of ^{99m}Tc utilization in Nuclear Medicine: 1) the development of a generator system (the ^{99}Mo/^{99m}Tc generator) that can be routinely and reliably eluted with sterile, pyrogen free N. saline [1]; and 2) the development of single vial kits containing a reducing agent (usually stannous ion) and a chelating agent [2]. These "instant kit" preparations normally produce the ^{99m}Tc radiopharmaceuticals in high yields (> 90% radiochemical purity – RCP) in one step upon addition of ^{99m}TcO$_4^-$ [3].

The development of ^{99m}Tc-radiopharmaceuticals offers many challenges, since this metal is not a component of any naturally occurring biological molecule. Thus, in designing ^{99m}Tc-radiopharmaceuticals, two approaches are used for incorporation of ^{99m}Tc into compounds that are produced as final drug products. The ^{99m}Tc-radiopharmaceutical can be considered to be: 1) ^{99m}Tc-essential, whereby the biological distribution is determined totally or influenced significantly by the properties of the coordination compound; or 2) metal tagged, in which case the properties of the carrier molecule (e.g., monoclonal antibody (MAb), to which the bifunctional complex (BFC) containing ^{99m}Tc is conjugated) determine the biological distribution, whereas the [99m]Tc (or ^{99m}Tc-complex) is simply along for the ride. The coordination chemistry of the metal determines the ultimate geometry and stability of the radiopharmaceutical. The stability requirements of ^{99m}Tc-complexes incorporated in radiopharmaceuticals are often much more stringent than for in vitro systems. While ^{99m}Tc-complexes may exhibit high thermodynamic stability in aqueous or organic solution, they are not sufficiently stable when administered in the blood stream of patients, since dilution or excretion of excess uncomplexed ligand, the physiological pH of blood, and the presence of various proteins, metabolites or enzymes in blood may challenge the integrity of the ^{99m}Tc-complex. Thus, it is essential that the kinetic stability of the complex be high (preferably kinetically inert for most applications), as the ^{99m}Tc-radiopharmaceutical must remain intact sufficiently long to reach its target or be cleared from the body.

The chemical properties of technetium are both a blessing and a curse. Its position in the middle of the second-row transition series imparts to it a diverse chemistry. Many complexes of technetium are known, ranging in oxidation state from -1 to $+7$, having a multitude of coordination geometries, and a variety of donor atoms or groups on ligands that fulfill its coordination requirements [4, 5]. These properties provide tremendous latitude when selecting chelation systems for the production of ^{99m}Tc-complexes with the appropriate physico-chemical properties to optimize in vivo targeting of the pharmaceutical.

The characterization of Tc-compounds over the past two decades has provided the basis for rational development of more sophisticated ^{99m}Tc-radiopharmaceuticals. Essentially all of the ^{99m}Tc-radiopharmaceuticals that have been approved over the past decade were chemically and structurally characterized at both the macroscopic ^{99}Tc level and the tracer (^{99m}Tc) level [6]. This is not true of several of the ^{99m}Tc-radiopharmaceuticals (e.g., ^{99m}Tc-MDP) developed earlier, because the products formed at the macroscopic level are not the same as those formed at the tracer level. Even though ^{99m}Tc-complex structures have been identified, the synthetic route for preparation of Tc-complexes in high yield and in high purity at tracer levels (i.e., with ^{99m}Tc) may be quite different from those used to prepare the corresponding ^{99}Tc-complexes. For this reason it is essential to understand the reactions involved in the complexation processes, in order to ensure reliable formulation of ^{99m}Tc-radiopharmaceuticals in high radiochemical purity (RCP).

The purpose of this chapter is to review most of the ^{99m}Tc-radiopharmaceuticals that are currently used routinely in Nuclear Medicine, and several ^{99m}Tc chelation systems or ^{99m}Tc compounds that exhibit properties in animals or in human clinical trials that suggest their potential as future ^{99m}Tc radiopharmaceuticals. Some chemical and structural information of these systems will be presented because of their importance in drug formulation with tracer levels of ^{99m}Tc and to aid in understanding in vivo structure-activity-relationships. Detailed discussions about the chemistry and the structural aspects of most of these, as well as other Tc complexes, will be presented elsewhere in this book.

1.2 ^{99m}Tc-Cold Kit Formulation

^{99m}Tc-radiopharmaceuticals are usually prepared by the reconstitution of cold kits, using sterile N. saline solutions containing ^{99m}TcO$_4^-$ [3]. These kits contain the chelating agent, a reducing agent, and other components that ensure that the reliability of the final ^{99m}Tc drug product (incl. high RCP and acceptable stability) is dependable over a wide range of reconstitution conditions. Several reductants have been used to prepare ^{99m}Tc-complexes from ^{99m}TcO$_4^-$ (incl. sodium dithionite, hydrazine, tin metal and HCl). However, all current commercially available cold kits, without exception, contain a stannous salt as the reductant [3, 4, 6]. The TechneScan® product (Mallinckrodt Medical), which is under development, uses a monophosphine ligand (TMPP) as the reductant as well as a ligand in formation of the final drug product. Additional kit components include antioxidants, buffers, or other compounds in order to maximize the yield of complex, RCP, or stability. Antioxidants, such as gentisic acid and ascorbic acid, have been used to provide greater in vitro stability [8]. The yield or RCP of the ^{99m}Tc-radiopharmaceutical may also be improved by the presence of another ligand. For example, ligands that form weak ^{99m}Tc-complexes (e.g. glucoheptonate, tartrate, citrate, DTPA, etc.) are used for this purpose and provide for ligand exchange reactions [3]. Other components may also play important roles in maximizing yields and reliability. For example, mannitol is used in Cardiolite® and Neurolite®, Cu(I) in Neurolite and Techne-Card® and disodium suphosalicylate in Myoview® [3]. Clearly, continued development of coordination chemistry with ^{99m}Tc at tracer levels must be pursued vigorously in order to design new formulations for clinical applications that provide the high degree of reliability and reproducibility that is required by regulatory agencies.

2 ^{99m} Tc-Radiopharmaceuticals

2.1 Brain Perfusion Imaging Agents

^{99m}Tc complexes developed for brain perfusion (rCBF) imaging must be small (< 500–600 daltons) and neutral, and their log P* value should be in the range: 0.5–2.5 [9]. Even though these general properties are necessary for the complex to penetrate the intact blood-brain-barrier (BBB) efficiently, other structural or electronic features also play important roles. For example, Kung and coworkers [10] and Lever and coworkers [11] showed that the stereochemistry of ^{99m}Tc-diaminodithiol (DADT) derivatives substantially affects the degree of brain uptake and intracellular retention. In addition, the presence of functional groups (e.g., OH's, amides, carbonyls, etc.) on uncharged molecules capable of H-bonding can substantially decrease the log P value and the ability of the molecules to passively diffuse across the BBB [12].

Several different neutral–lipophilic ^{99m}Tc complexes have been evaluated as potential rCBF imaging agents; however, appropriate cerebral uptake properties are exhibited by only a few types of coordinating ligand backbones complexed to ^{99m}Tc. The two ^{99m}Tc rCBF agents most extensively used in humans are derived from either the propyleneamine oxime (PnAO) [13] or the N_2S_2 (DADT) [14, 15] ligand systems. Results of promising studies with ^{99m}Tc complexes of the N-(2(1H pyrolylmethyl)) N'-(4-pentene-3-one-2) ethane-1,2-diamine (or MRP-20) [16] chelation system has also been reported. Interestingly, all three of these systems form Tc complexes with the mono-oxo-Tc(V) core (TcO^{+3}).

N_2S_2 ligands were first shown to form a neutral-lipophilic complex with ^{99m}Tc by Burns et al. [14]. Subsequent work by Kung and co-workers [15] demonstrated that ^{99m}Tc complexes with more lipophilic DADT derivatives exhibit good BBB penetration and high brain extraction efficiencies. After complexation with the TcO^{+3} core, three of the four ionizable protons on the N_2S_2 ligand (two S-H and one N-H) are lost, to neutralize the + 3 charge [4, 11]. Thus, three of the donor groups have sp^3 geometry, while one (the deprotonated N-atom) is sp^2 hybridized [4]. ^{99m}Tc-L,L-ethylenecysteine dimer (^{99m}Tc-L,L-ECD) (Fig. 1), developed for rCBF imaging in humans by DuPont (Neurolite®), is a diester derivative of a neutral ^{99m}Tc-N_2S_2 complex [17]. This small, neutral-lipophilic complex, containing functional groups with minimal H-bonding properties attached to the ligand backbone, is efficiently extracted by the brain. Unlike some ^{99m}Tc-N_2S_2 derivatives [10], the BBB penetration of all optical isomers of ^{99m}Tc-ECD appears to be similar [17].

PnAO was found to form a neutral-lipophilic ^{99m}Tc complex in aqueous solutions and to exhibit a high first-pass cerebral extraction efficiency [13]. All

---

* P = octanol/water partition coefficient

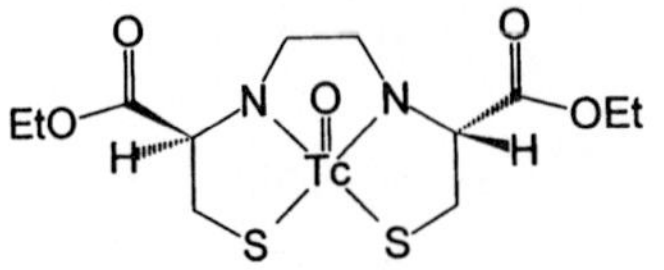

Fig. 1. Tc(V)O-L, L-ECD

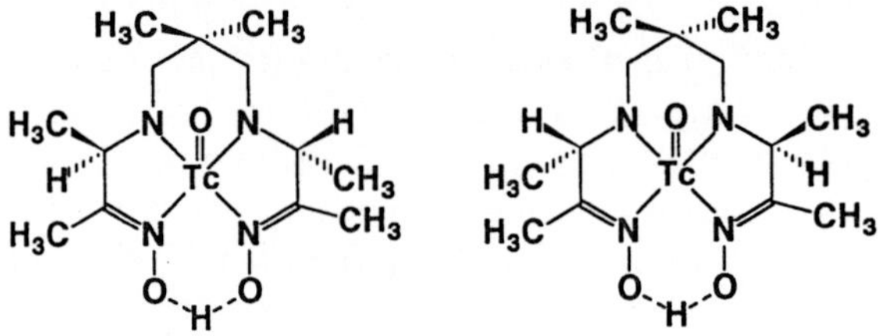

Fig. 2. Tc(V)O-d, 1-HMPAO

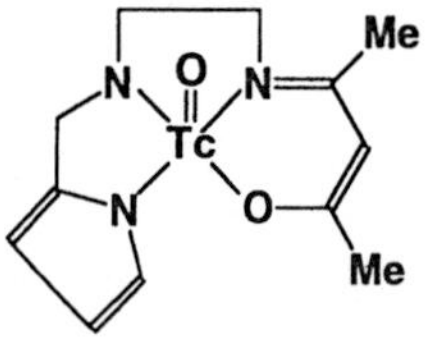

Fig. 3. Tc(V)O(MRP-20)

PnAO ligands and related derivatives, including HMPAO, react with Sn^{+2}-reduced $^{99m}TcO_4^-$ to form a complex with the framework exemplified by the Tc(V)O-d,l-HMPAO structures (Fig. 2). The $+3$ charge of the mono-oxo-Tc(V) core, complexed by all four ligand N-atoms, is neutralized by three negative charges from the ionization of one of the oxime OH groups and the two secondary amines from the PnAO ligand upon complexation [18]. None of the appended groups have significant H-bonding properties. The four N-atoms in this complex are sp^2 hybridized (all bonds lie in one plane), preventing formation of conformational isomers [18]. ^{99m}Tc-d, 1-HMPAO (Figure 2), the tracer developed by Amersham, Int'l (Ceretec®), was the first approved ^{99m}Tc-rCBF imaging agent for human applications [19] and has the same number of pendant methyl groups as PnAO but arranged differently. Its log P value (i.e., ~ 2) is similar to that of ^{99m}Tc-PnAO and it readily penetrates the intact BBB. The various optical isomers of ^{99m}Tc-HMPAO (i.e., the d, 1, and meso stereomers of HMPAO) have similarly high first-pass extraction efficiencies [19].

The brain uptake properties of ^{99m}Tc-complexes with the MRP-20 ligand make it and its derivatives potential candidates for rCBF imaging applications. MRP-20 forms a neutral-lipophilic complex with the $^{99m}TcO^{+3}$ core by loss of three ionizable protons from the ligand backbone (Fig. 3) [16]. ^{99m}Tc-MRP-20 and several of its analogues show high brain uptake and retention and exhibit regional cerebral deposition patterns that are related to rCBF [16]. This complex is in clinical trials in Europe.

2.1.1 Mechanisms of Brain Retention

For conventional SPECT imaging it is beneficial to trap ^{99m}Tc activity for extended periods at the site of the initial local cerebral uptake of the ^{99m}Tc complex. Three general mechanisms are considered to be effective for efficient trapping of ^{99m}Tc in brain: 1) binding of the intact ^{99m}Tc complex to structures, proteins and other molecules in the cells; 2) conversion of the intact neutral complex to a charged, non-diffusible form of ^{99m}Tc in the brain and, 3) intracellular chemical dissociation of the ^{99m}Tc complex to produce other non-diffusible ^{99m}Tc species.

Several N_2S_2 ligands, derivatized with backbone amine groups, were complexed to mono-oxo-Tc(V) in anticipation that the appended amine group would provide avid intracellular binding similar to that operational with ^{123}I-IMP [20]. Even though some degree of retention – which is sensitive to complex stereochemistry – was achieved, trapping was not sufficiently prolonged for most SPECT imaging applications, in comparison to other ^{99m}Tc complexes [10].

Intracellular reactions of ^{99m}Tc-L, L-ECD result in conversion of the neutral complex to a negatively charged species, which is effected by hydrolysis of the complex to the mono or diacid by stereospecific esterases [17, 21]. Using brain homogenates as the source of putative esterases, it was shown that the rate of hydrolysis of ^{99m}Tc-L, L-ECD is much greater than ^{99m}Tc-D, D-ECD; this is consistent with the finding that the D, D isomer undergoes no significant retention in the brain [17]. It is important to note that the ^{99m}Tc-N_2S_2 core complex remains intact during the hydrolysis reactions, reflecting the high inherent in vitro and in vivo stability of the TcO-N_2S_2 complexes. Esterases present in the blood also cause hydrolysis of ^{99m}Tc-L, L-ECD.

Trapping of ^{99m}Tc-d, l-HMPAO in the brain results from rapid dissociation of the complex to form ^{99m}Tc species that cannot backdiffuse across the BBB. No unique ^{99m}Tc product results from the in vivo dissociation of this complex and the chemical reactions responsible for these processes are not fully understood. Neirinckx et al. [22] suggest that intracellular glutathione (GSH) plays a major role in triggering dissociation of ^{99m}Tc-d, l-HMPAO; however, other reactions must also be considered [23]. Regardless of the chemical or biochemical reactions involved, once the non-diffusible ^{99m}Tc products are formed inside of the cells (e.g., small charged ^{99m}Tc species or ^{99m}Tc bound to intracellular proteins or structures), the retention of ^{99m}Tc activity in the brain is prolonged. For example, Andersen et al. [24] reported that ^{99m}Tc-d, l-HMPAO does not redistribute in the brain for at least 8 hr after injection and washout from the brain is insignificant (i.e., approximately 0.4% h^{-1}). The rate of trapping of ^{99m}Tc-HMPAO is stereospecific. Both the d and l-isomers are efficiently trapped in brain while the meso-isomer is not [25]. There is some evidence that the d- and l- enantiomers do not have identical in vivo properties, however, the effects of these differences in brain uptake and retention in humans are not well understood [26, 27].

High stability of the ^{99m}Tc radiopharmaceutical is an important asset. ^{99m}Tc-L,L-ECD has a long shelf–life because of the good stability characteristics of its ^{99m}Tc-N$_2$S$_2$ complex framework. ^{99m}Tc-d, l-HMPAO has limited in vitro stability and must be used within 30 minutes of its formulation [19, 22]. Surprisingly, Tc-d, l-HMPAO is inherently a very stable complex, as evidenced by the observation that minimal dissociation of ^{99}Tc-d, l-HMPAO – if any – occurs at carrier levels (i.e. 10^{-4}–10^{-5} M Tc) for over six hours in aqueous solution [28]. Thus, stability limitations of ^{99m}Tc-d, l-HMPAO are observed only at tracer levels. It has been shown that self-radiolysis of ^{99m}Tc-d, l-HMPAO contributes significantly to complex dissociation in patient preparations, but it is not the only factor [29]. The addition of antioxidants or free-radical scavengers will substantially stabilize ^{99m}Tc-d, l-HMPAO [30], providing a viable approach for increasing its shelf-life.

2.2 Myocardial Perfusion Imaging Agents

Since Deutsch and co-workers [30] first described the potential for ^{99m}Tc lipophilic cations for use as imaging agents to assess regional myocardial perfusion, extensive studies with a wide variety of these types of ^{99m}Tc complexes have been reported [31–34]. All of the ^{99m}Tc-complexes currently approved or are in late stages of their clinical trials have a + 1 formal charge (incl., ^{99m}Tc-sestamibi, Cardiolite®, DuPont-Merck; ^{99m}Tc-tetrofosmin, Myoview®; Amersham, Int'l; and ^{99m}Tc-furifosmin, TechneScan®, Mallinckrodt Medical) with the exception of ^{99m}Tc-teboroxime, Cardiotec®, Bracco, which has a formal charge of zero. Even though these complexes exhibit myocardial uptake and pharmocokinetic properties in humans that make them effective as myocardial imaging agents, a multitude of other complexes with similar structural and physicochemical characteristics are not suitable. Thus, it is important to recognize that a variety of properties must be properly integrated for these ^{99m}Tc-essential complexes to fulfill the requirements for development of clinically useful myocardial perfusion imaging agents.

^{99m}Tc-complexes that are useful as regional myocardial blood flow (rMBF) imaging agents must demonstrate high first-pass extraction, in order to prevent further extraction of the tracer with recirculation [32, 35], and guarantee rapid clearance from the blood and lung. While first-pass extraction should ideally be 100%, so that the initial regional uptake of the tracer would be independent of the flow rate [36], tracers that do not strictly adhere to this criterion are effective imaging agents, provided that their myocardial distribution is directly related to rMBF. With SPECT instrumentation that permits more rapid data acquisition, ^{99m}Tc-agents (e.g., Cardiotec®) that have limited residence times in myocardial tissues can be used [32].

The first report of a heart-imaging agent in animals involved a ^{99m}Tc(III) complex with O-phenylenebis(dimethyl arsine) [30]. Several other bidentate π-acceptor ligands were investigated and early studies focused on a prototype bidentate phosphine ligand, DMPE (1,2-bis(dimethylphosphino)ethane) [31]. Studies with DMPE established that stable, cationic (+ 1), lipophilic complexes with ^{99m}Tc(I), ^{99m}Tc(III) and ^{99m}Tc(V) (vis [Tc(I) (DMPE)$_3$]$^+$, [Tc(III)-Cl$_2$(DMPE)$_2$]$^+$ and [Tc(V)O$_2$(DMPE)$_2$]$^+$, respectively) can be readily prepared [31, 32, 34]. The latter two complexes exhibited good myocardial uptake and retention in animals; however, uptake and/or retention in humans was poor [31]. Several ^{99m}Tc(I) complexes showed promise in animals (incl. ^{99m}Tc (DMPE) $_3^+$) but failed in humans, because of poor extraction from human plasma and the persistence of high blood levels [31]. On the other hand, a compelling, but indirect series of chemical and animal experiments indicated that ^{99m}Tc(III)-diphosphine complexes undergo in vivo reduction of the ^{99m}Tc(III) to Tc(II) [31, 37]. This has serious consequences, since the formal charge on Tc(III)Cl$_2$(DMPE)$_2^+$ changes from + 1 to 0 when the central metal undergoes reduction to form Tc(II)Cl$_2$(DMPE)$_2$. Since mono-cationic ^{99m}Tc complexes are retained in the myocardium whereas neutral complexes are not, premature washout of ^{99m}Tc from the heart muscle can occur [31].

Despite the synthesis of a multitude of monocationic ^{99m}Tc-complexes formed with diphosphine ligands and their in vivo evaluation, only relatively recently has a clinically useful ^{99m}Tc-diphosphine complex been identified. Kelly and coworkers [34] developed a product in which bis[bis(2-ethyoxyethyl)phosphino]ethane (is chelated to the ^{99m}TcO$_2^+$ core (i.e., ^{99m}TcO$_2$(tetrofosmin)$_2^+$) (Fig. 4A). This complex exhibits high myocardial uptake and retention and demonstrates efficient clearance from non-target tissues, especially blood, lung, and liver in both animals and humans [34, 38]. ^{99m}Tc-tetrofosmin exhibits long-term retention in the myocardium and there is no evidence for in vivo reduction of ^{99m}Tc(V) in this complex to lower oxidation states [34].

^{99m}Tc(I) methoxyisobutylisonitrile$_6^+$ (^{99m}Tc(MIBI)$_6^+$) or Cardiolite® (Fig. 4B) was the first ^{99m}Tc-myocardial perfusion agent to receive worldwide approval and is the most widely used [32, 33, 39]. The initial series of hexabis-(alkylisonitrile) Tc(I) complexes showed the existence of a parabolic relationship between molecular weight/lipophilicity and heart uptake, with the t-butylisonitrile (TBI) complex at the maximum of the curve [33]. Since the liver, blood and

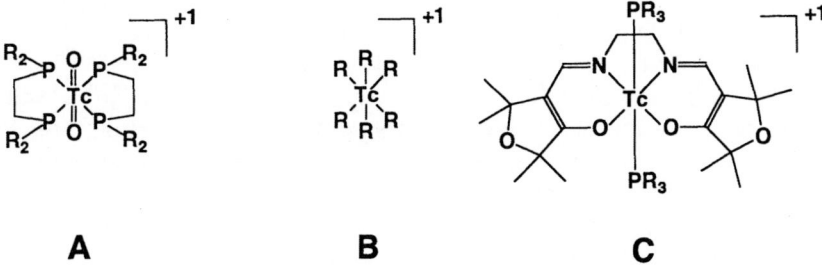

A   **B**   **C**

Fig. 4. Lipophilic-cationic(+1) myocardial perfusion imaging agents; A) ^{99m}Tc-tetrofosmin; R = –CH$_2$CH$_2$OCH$_2$CH$_3$; B) ^{99m}Tc = sestamibi; R = C≡NCH$_2$C (OCH$_3$) (CH$_3$)$_2$; C) ^{99m}Tc-furifosmin; R = –CH$_2$CH$_2$OCH$_2$CH$_3$

lung clearance of $^{99m}Tc(TBI)_6^+$ was unacceptably slow, several isonitrile analogues were designed to improve clearance from these organs [39] $^{99m}Tc(MIBI)_6^+$ was selected from this series as the preferred complex because of its superior rates of clearance from non-target tissues. The presence of methoxy (ether) groups on the periphery of the complex provides for optimal lipophilicity while allowing for in vivo hydrolysis of the ethers to the respective alcohols, and thus to increase the clearance rates from blood and liver. It is important to note that both Myoview® and Technescan® also have multiple peripheral ether groups that accelerate clearance of ^{99m}Tc from non-target tissues.

^{99m}Tc-furifosmin is a mixed ligand $^{99m}Tc(III)$ complex (Figure 4C), in which a tetradentate Schiff-base ligand, trans(1,2-bis(dehydro-2,2,5,5-tetramethyl-3-furanone-4-methylene-amino) ethane) occupies the equatorial Tc(III) coordination sites and tris(3-methoxy-1-propyl)phosphine (TMPP) is coordinated in the two axial positions [7]. ^{99m}Tc-furifosmin is formed with high RCP from a lyophilized kit containing five components: Schiff-base ligand, TMPP, ascorbate, carbonate and γ-cyclodextrin [7]. Interplay among these five components creates an unusually complex and intricate kit chemistry, in which TMPP acts both as the reducing agent for ^{99m}Tc and as a ligand for chelation to the $^{99m}Tc(III)$ core [7, 32]. $^{99m}Tc(III)$ furifosmin is retained in the myocardium. This can be attributed to the greater stability of Tc(III) toward reduction to Tc(II); the smaller the number of phosphine atoms bound to Tc(III) the greater the Tc(III/II) redox potential [31, 40].

The mechanism of uptake and retention of the mono-cationic ^{99m}Tc complexes in the myocardium – or other tissues – has not been fully resolved. Most of the mechanistic studies have been conducted with $^{99m}Tc (MIBI)_6^+$ and ^{99m}Tc-diphosphine complexes. It has been shown that these mono-cationic complexes are not taken up in myocytes via the Na^+/K^+-ATPase pump as is $^{201}Tl^+$ [41]. Instead, these cationic tracers are localized and retained in cellular membranes, including mitochondrial membranes [41].

^{99m}Tc-teboroxime (CardioTec®) is a seven-coordinate neutral-lipophilic complex in which ^{99m}Tc is coordinated to tris-cyclohexanedionedioxime-methylboron (CDO-MeB) and a chloride ion [6, 32, 42]. This analogue was selected as the best ^{99m}Tc complex for myocardial perfusion imaging from a series of BATOs (Boronic Acid Adducts of Technetium diOximes) [32, 43]. It is particularly notable that these complexes are constructed by "template synthesis" during the labeling process [32]. In these kits, the complex is constructed from three vicinal cyclohexanedionedioxime ligands bound to Tc(III) by their six N-atoms, and a boronic acid, which caps the complex by attachment to three oxime oxygens [42]. A chloride ion occupies the seventh coordination site and prevents formation of a bis-capped complex [42]. ^{99m}Tc-teboroxime has high myocardial uptake, but washes out rapidly and thus requires rapid SPECT imaging techniques.

2.3 Hepatobiliary Imaging Agents

^{99m}Tc-complexes that have proven to be effective hepatobiliary imaging agents were developed over a decade ago [44]. While some work on these types of agents continue, no new radiopharmaceuticals for human use have been introduced for over 5 years. Two different classes of ^{99m}Tc complexes for hepatobiliary studies have been approved for human use; 1) derivatives of iminodiacetic acid (IDA) (Fig. 5A) and 2) pyridoxalamino acid derivatives (e.g., Fig. 5B). Extensive structure-activity-relationship studies with these types of complexes have led to the development of specific complexes with optimal properties for hepatobiliary imaging studies [44–46]. Currently, three ^{99m}Tc-IDA derivatives are approved in the U.S.: ^{99m}Tc-Mebrofenin (Choletec®, Bracco), ^{99m}Tc-Disofenin (Hepatolite®, DuPont-Merck) and ^{99m}Tc-Lidofenin® (Mallinckrodt Medical, Inc.) ^{99m}Tc-N-pyridoxyl-5-methyl-tryptophan (^{99m}Tc-PMT; Nihon-Mediphysics), is approved in Japan and other countries outside of the U.S.

All of the ^{99m}Tc radiopharmaceuticals for hepatobiliary imaging exhibit similar pharmacokinetic properties in animals and man. Basically, they are efficiently extracted from the blood by the liver and excreted into bile. As such, they assess disease that affect hepatocyte function, the patency of intrahepatic ducts, the functional status of the cystic duct and gall bladder, and the patencies of downstream portions of the biliary tract [44]. These radiopharmaceuticals have high specificity for hepatocyte uptake, rapid hepatocyte transit, efficient extraction from the blood, and the ability to compete against endogenous compounds (e.g., bilirubin) for binding to cellular transport proteins [44]. The lipophilic properties of these complexes provide increased plasma protein binding and decreased renal clearance (*by GFR*). Important determinants in the binding of these radiopharmaceuticals to hepatocyte transport proteins, the efficiency of hepatocyte uptake, hepatocyte transit rates and canalicular excretion are directly related to the structural and physicochemical relationships between lipophilic and polar groups on the complexes [44].

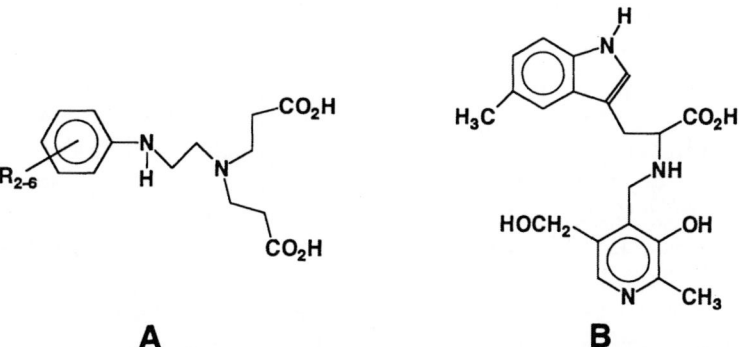

A   **B**

Fig. 5. Chelating agents used to formulate ^{99m}Tc-hepatobiliary imaging agents; A) Mebrofenin©; $R^{2,4,6} = CH_3$, $R^3 = Br$, $R^5 = H$ and Disofenin©; $R^{2,6} = CH(CH_3)_2$ $R^{3-5} = H$; B) N-pyridoxyl-5-methyltryptophan (PMT)

Loberg and Fields [47] demonstrated that the ligand-to-metal ratio in Tc-HIDA (i.e., ^{99m}Tc-Lidofenin) was $2:1$, that the complex did not contain Sn (the reducing agent used in the formulation), and that the complex had an overal-1 charge. FAB mass spectral analysis reported by Costello et al. [48], were consistent with the structure suggested by Loberg and Field [47] for $[Tc(III) (HIDA)_2]^{1-}$. Their results showed a molecular ion at $m/z = 685$ in the positive ion FAB-MS consistent with the formation $H_2 [Tc(HIDA)_2]^{1+}$ and at $m/z = 683$ in the negative ion FAB-MS consistent with $Tc(HIDA)_2^{1-}$. A molecular ion at $m/z = 701/699$ for $TcO(HIDA)_2^{1-}$ may be obscured by the intense peak of the $Sn(HIDA)_2^{2-}$ molecular ion cluster observed in this region. If the ^{99m}Tc-IDA derivatives form Tc(III) complexes as suggested, they are undoubtedly octahedral, bis-anionic IDA complexes, in which the ligand is coordinated to the Tc(III) center through the two imino nitrogens and the four deprotonated carboxylate ligands.

Since Baker et al. [49] first described the potential of ^{99m}Tc-pyridoxal derivatives for hepatobiliary imaging, several studies with these compounds have been performed [50]. Kato-Azuma [51] showed that Tc-complexes of pyridoxylideneamino acids are neutral and have a ligand–to–metal ratio of $2:1$. After evaluation of a series of ligands, Kato-Azuma [52] identified the ^{99m}Tc complex with N-pyridoxal-L-tryptophan (^{99m}Tc-PMT) as having optimal pharmacokinetic properties. ^{99m}Tc-PMT has biliary excretion kinetics, even in the presence of significant competitive molecules (e.g., BSP and bilirubin) [53].

2.4 Renal Imaging Agents

2.4.1 Complexes for Functional Imaging

Two ^{99m}Tc agents are currently marketed for imaging renal function, ^{99m}Tc-DTPA and ^{99m}Tc-MAG$_3$. These agents are used to monitor passage into and through the renal system, and to assess the glomerular filtration and tubular secretion functions.

^{99m}Tc-DTPA, or Technetium Tc-99m Penetate, is prepared by the addition of ^{99m}Tc-pertechnetate to lyophilized kits, that are available from several companies. Depending on the source, each 10 ml lyophilized kit contains between 9.4 and 39.2 μmoles of DTPA and between 0.66 and 2.2 μmoles of SnCl$_2 \cdot$ 2H$_2$O, and has been adjusted so that the pH is between 3.9 and 7.4. This agent is approved for kidney imaging, brain imaging, and the estimation of glomerular filtration function and rate. The identity of ^{99m}Tc-DTPA is not known. The oxidation state of the technetium in ^{99m}Tc-DTPA is postulated to be $+4$ or $+5$ [4, 54]. On the macroscopic Tc-99 level, the principle products from reactions of Tc with aminecarboxylates are dimers containing Tc in the $+3$, $+4$ and/or $+5$ oxidation states [4, 54]. Eckelman reported the oxidation state of Tc in this radiopharmaceutical to be $+4$ when formulated under the reaction conditions present in the ^{99m}Tc-DTPA kits [55]. If this complex is indeed

Tc(IV), the DTPA ligand is undoubtedly coordinated in a hexadentate fashion about the Tc through three nitrogens and three carboxylate oxygens, leaving two uncoordinated carboxylates. Deutsch and Packard [56], however, observed three uncoordinated carboxylates on the complex by titration with base, suggesting that the technetium may be in the $+5$ oxidation state. If Tc(V) is correct, then a 6-coordinate complex would be present with a carboxylate in the position *trans* to the $Tc=O$ group.

^{99m}Tc-MAG_3, Technescan MAG$^{®}_3$ or ^{99m}Tc Mertiatide (Mallinckrodt Medical), is marketed as a renal imaging agent designed to assess the functional integrity of the tubular secretion process in diseased and normal kidneys. Each lyophilized kit contains 1 mg of N- [N-[N[benzoylthio)acetyl]glycyl]glycyl] glycine (MAG_3) (see Fig. 9), 0.02–0.2 mg $SnCl_2 \cdot 2H_2O$, 40 mg sodium tartrate dihydrate, and 20 mg lactose monohydrate. The kits are reconstituted by adding ^{99m}Tc pertechnetate and 2 ml of air, the latter to oxidize any excess stannous ion and prevent the formation of ^{99m}Tc impurities, and to convert undesirable ^{99m}Tc species to the desired product. The preparation requires heating in a bath of boiling water for 5 minutes. $^{99m}Tc(V)O$-MAG_3, first developed by Fritzberg [57], shows good in vivo renal clearance characteristics and, unlike the Tc-DADS compounds [58, 59] which contain tetradentate dimercaptodiamide ligands, no isomer problems were encountered; ^{99m}Tc-MAG_3 has no chiral centers. On coordination to the Tc(V) center, the thiolate sulfur and each of the three amide nitrogens of MAG_3 lose their protons forming an anionic complex [60]. The complex has square-pyramidal geometry, with the oxo group occupying the apical position. The Tc(V) atom is situated above the plane containing the sulfur and the three amide nitrogens. The carboxylate group is not coordinated, even loosely, to the technetium atom [60]. A free carboxylic acid group appears to be required for efficient excretion by the anionic pathway, as has previously been reported for iodohippuran and its analogs [61].

The clearance rate of ^{99m}Tc-MAG_3 ranges from 50–60% of that of ^{131}I-orthoiodohippurate (OIH). However, this rate is proportional to that for OIH (63). ^{99m}Tc-MAG_3 is highly protein-bound and is not retained in the parenchyma of normal kidneys [63, 64]. In spite of the fact that the rate of clearance of ^{99m}Tc-MAG_3 is less than that for ^{131}I-OIH, the renogram curves and the rates at which both tracers appear in the urine are almost identical [63]. Verbruggen et al. [65] reported that the diacid derivative of the brain perfusion agent ^{99m}Tc-ECD has renal clearance properties similar to ^{99m}Tc-MAG_3.

2.4.2 Complexes for Morphological Imaging

Two chelating agents, DMSA (dimercaptosuccinic acid) and GH (glucoheptonate) are used to formulate FDA approved ^{99m}Tc-radiopharmaceuticals for evaluation of kidney morphology. These agents are cleared from the blood stream by GFR, but a significant fraction (i.e., 15–40%) of the injected dose remains fixed in the renal cortex for extended periods [54]. The number of morphological imaging studies performed with these agents has decreased significantly, due to the better resolution and convenience of other imaging modalities (i.e., ultrasound and MRI).

DMSA is present in the ^{99m}Tc-succimer (MediPhysics, Inc.) kit as a 90:10 meso:d, 1 isomer mixture. The structure and identity of the ^{99m}Tc-species produced in this formulation has not been definitively resolved. The complex(es) formed is (are) believed to contain ^{99m}Tc in an oxidation state(s) below five and has been referred to as ^{99m}Tc(III) DMSA [66,67]. Because of the possibility of multiple complexes, it has been impossible to elucidate the mechanism of localization; however, it is reasonable to assume that after GFR, some of the agent is reabsorbed by tubular cells and retained by ligand exchange reactions, in which ^{99m}Tc becomes bound to SH-containing cytoplasmic proteins [67,68].

Ohta and coworkers [69] showed that a basic ^{99m}Tc-DMSA formulation was useful for imaging tumors. It is avidly accumulated in medullary thyroid carcinomas, as well as in some other tumors [67–70]. Preparation of this tumor-imaging ^{99m}Tc-DMSA radiopharmaceutical results from preparation of the no-carrier-added product under alkaline conditions [71]. This preparation is referred to as pentavalent ^{99m}Tc-DMSA or ^{99m}Tc(V)DMSA [67,71]. Its chromatographic characteristics show clearly that ^{99m}Tc(V)DMSA is different than the ^{99m}Tc(III) DMSA product [67,71]. The chromatographic behavior of ^{99m}Tc(V)DMSA, formulated by Yokoyama et al. [71], indicated the formation of a polynuclear technetium complex [72]. In comparison, the pentavalent ^{99m}Tc-DMSA product reported by Blower et al. [67,73] showed a mixture of $[^{99m}TcO(DMSA)_2]^-$ isomeric structures. Once coordinated to Tc(V) as the TcO^{+3} core, *meso*-DMSA can exhibit three possible isomer conformations; *syn-endo*, *anti* or *syn-exo* orientations of the carboxylic acid groups with respect to the Tc=O bond (Fig. 6). The presence of all three isomers has been observed for ^{186}Re-DMSA and has been confirmed by chromatography (HPLC) in comparison to the macroscopically characterized [Bu$_4$N] [ReO(DMSA)$_2$] [73]. Although no crystal structure analysis of the technetium complex has been performed, *syn-endo* [Et$_4$N] [ReO(DMSA)$_2$]·1.5H$_2$O has been structurally characterized; the complex is shown to have square-pyramidal geometry, as expected for five-coordinate monooxo complexes of both Tc(V) and Re(V) [73]. Bandoli et al. [74], reported the crystal structure of (Bu$_4$N) [TcO(1,2-di(carboxymethyl)ethane-1,2-dithiolato)$_2$], a TcO(DMSA)$_2$ analog in which the

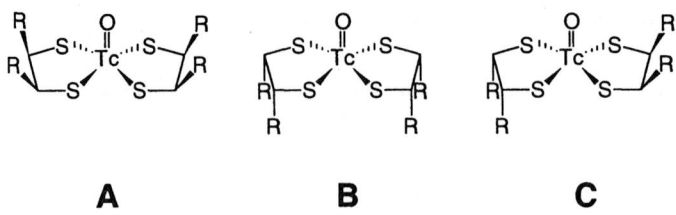

Fig. 6. Tc(V) O-DMSA; A) syn-endo; B) syn-exo; C) anti; R = –COOH

methyl esters of the carboxylic acid groups were observed in the *syn-endo* conformation. But as the authors clearly point out, this stereochemical orientation is the one that happens to crystallize in 21% yield on reaction with *meso*-1,2-dimercaptosuccinic acid dimethyl ester.

[99m]Tc-Gluceptate (Glucoscan, DuPont-Merck; TechneScan Gluceptate, Mallinckrodt) is approved for imaging kidney morphology. This agent is rapidly cleared from the blood by GFR on IV injection, with up to 15% ID retained in the kidneys, with greater retention in the cortex than in the medulla. A great deal of evidence indicates that Tc is present as a Tc(V) core [4.6]. The structure of this compound has not been confirmed by X-ray crystallography, although de Kievet [75] has shown that, in the presence of excess glucoheptonate, more than 1 equiv of Sn(II) per Tc does not yield any more of the Tc-glucoheptonate complex, indicating that the technetium is being reduced from Tc(VII) as (TcO_4^-) to Tc(V). Infrared and Raman spectra of the Tc-glucoheptonate complex are consistent with the formulation of a Tc(V) monooxo core (970 and 975 cm^{-1}, respectively). Titrations with Tc and glucoheptonate (GH) and NMR studies of the complex suggest that two glucoheptonate molecules may be bound to each Tc through the carboxylate oxygen and the *alpha* hydroxyl oxygens. If Tc is complexed by both the carboxylic acid and -OH groups, most α-hydroxycarboxylic acids should form similar complexes. Since lactic and glycolic acids do not form Tc(V) complexes while GH does, the projected structure does not explain the source of greater stability of the longer-chain Tc(V) polyhydroxy carboxylic acid complexes [4]. Thus, the formulation of TcO(glucoheptonate)$_2$ remains to be confirmed.

The ^{99m}Tc-GH complex is a relatively weak, as are, for example, [99m]TcO-(citrate)$^-$ and [99m]TcO(ethylene glycol)$_2$, and requires the presence of excess ligand to stabilize it [76]. As with ^{99m}Tc-citrate, ^{99m}Tc-GH is often used as a donor Tc(V) complex that can be utilized as the synthon in ligand exchange reactions, to prepare other Tc(V) complexes that are more thermodynamically stable but may be kinetically slow to form. ^{99m}Tc-GH offers a practical advantage in this capacity, since it is already available as a lyophilized sterile radiopharmaceutical kit formulation that is approved in most countries.

2.5 Skeletal Imaging Agents

Methylene-diphosphonate (MDP) and MDP derivatives, when complexed with ^{99m}Tc, are the primary approved radiopharmaceuticals used for scintigraphic detection of skeletal abnormalities. Three ^{99m}Tc complexes with MDP derivatives are in routine use: 1) ^{99m}Tc-Medronate (^{99m}Tc-MDP), Osteolite®, DuPont-Merck; TechneScan®, Mallinckrodt; MPI-MDP, Medi-Physics; MDP, Amersham-MDP, Amersham, Int.; AN-MDP®, CIS-US); 2) ^{99m}Tc-oxidronate or ^{99m}Tc-hydroxymethylenediphosphonate (^{99m}Tc-HDP); Osteoscan-HDP®, Mallinckrodt Medical; and 3) ^{99m}Tc-1,3-dicarboxypropanediphosphonate (^{99m}Tc-DPD). These three diphosphonate ligands have different substituents on the α-carbon connecting the two phosphorous atoms. These ^{99m}Tc-radiopharmaceuticals represent second and third generation drugs that were based on the original ^{99m}Tc-polyphosphate kit preparation containing Sn(II), first developed by McAfee and Subramanian [77] for skeletal imaging in patients.

Chromatographic analysis of these products demonstrates that these diphosphonates do not form a single ^{99m}Tc complex. For example, as many as 8–10 different ^{99m}Tc species are resolvable on HPLC analysis [68–78]. Results from these and other studies indicate that ^{99m}Tc complexes with diphosphonate ligands form a mixture of oligomers and polymers that may include Sn [78]. The only Tc–MDP structure, reported by Libson et al. [79], shows Tc-MDP to be a 1:1 polymer in which each Tc atom is coordinated to two diphosphonate ligands and each diphosphonate is coordinated to two Tc centers.

^{99m}Tc–MDP is believed to contain a mixture of ^{99m}Tc complexes that have high affinity for sites of actively growing bone, localizing the areas rich in osteoblastic activity [68, 78]. The mechanism of localization of ^{99m}Tc-diphosphonates is presumably bridging of ^{99m}Tc to hydroxyapatite by the phosphate oxygen atoms [79]. This bridging mechanism is important for metals like Tc with acidic oxides, that have no affinity for hydroxyapatite themselves. The presence of the -OH substituent on HDP and of the carboxyl groups on DPD provide additional donor atoms that can be used in binding the ^{99m}Tc complexes to Ca^{+2} in mineralized bone. The presence of these groups increases complex polarity, which may facilitate more efficient GFR and faster blood clearance.

^{99m}Tc-pyrophosphate (TechneScan PYP®, Mallinckrodt Medical; Pyrolite®, DuPont-Merck; PhosphaTec®, Bracco ; AN-Pyrotec, CIS-US) is used to a limited extent in Nuclear Medicine and primarily for imaging myocardial infarcts or in vivo RBC labeling. Originally, ^{99m}Tc complexes with polyphosphate and pyrophosphate (PYP) were used as skeletal imaging agents, however, the in vivo instability of the P–O–P bond to enzyme hydrolysis prompted development of the diphosphonates, which have an enzyme-stable P–C–P linkage. The PYP kits are useful for in vivo RBC labeling, because of the large quantities of Sn(II) present that is needed to ensure high labeling efficiencies and longer shelf lives.

2.6 Imaging Agents for Hypoxic Tissues

Hypoxic tissue is a marker for several disease states, and this tissue may be at risk of permanent damage. Hypoxic regions are those in which the oxygen supply is low. An in vivo hypoxia imaging agent could be useful clinically for several indications: (1) identifying viable tissue in the border zone of ischemic myocardium and thereby determining the most useful treatment regimen, (2) assessing and treating jeopardized tissue in stroke victims, and (3) assessing tumors that may potentially be resistant to radio-and/or chemotherapy because of their hypoxic status [80]. Studies showing the uptake and retention of

[18]F-fluoromisonidazole [81] in hypoxic tissue have led to the development of such an agent based on [99m]Tc. All of these studies to date have involved the incorporation of the 2-nitroimidazole function onto a known stable complex for technetium.

The Bracco group (formerly Bristol-Myers Squibb) have incorporated the 2-nitroimidazole function onto the BATO core [82] and the PnAO core [80]. Derivatization at the 1 position of the PnAO core with 2-nitroimidazole has proven to be the most successful complex for technetium in this regard. The structure of TcO(PnAO-1-2-nitroimidazole) (BMS-181321), is shown in Fig. 7A. The mechanism of retention of the 2-nitroimidazole based agents in hypoxic tissue is believed to involve an enzymatic one electron reduction of the nitroimidazole to a radical anion [84]. The enzymatic reduction is believed to occur in both normal and hypoxic cells, however, under normal oxygen pressure the reduced nitroimidazole is readily oxidized back to nitroimidazole, and thus clears. Under hypoxic conditions, the radical anion may undergo further reduction and bind to cellular components. Cell permeability of potential hypoxia imaging agents must be high, since the enzymes responsible for the reduction and, thus for retention of these agents, are found inside the cell [83, 84]. The cell permeability of the 2-nitroimidazole derivatives of the BATO technetium-99m complex was found to be poor and have anoxic/oxic retention ratios of very close to 1.0 [84]. BMS-181321 showed good permeability and an anoxic/oxic retention ratio of 3.03 was observed [83].

In vitro and in vivo studies with BMS-181321 clearly show its utility as a potential hypoxia imaging agent. Studies using isolated, perfused rat hearts showed that BMS-181321 washed out of normoxic hearts quickly, while it was retained after reperfusion in ischemic hearts when administered before ischemia [85]. SPECT images obtained from dogs that were administered BMS-181321 just prior to a 10 minute occlusion of a small branch of the LAD clearly showed retention of activity in the hypoxic region 100 to 200 minutes post-injection [86]. Although liver uptake of this compound is relatively high, the ischemic regions of the heart were visualized. SPECT images obtained from cats that were administered BMS-181321 one hour after middle cerebral artery occlusion (MCAO) showed selective brain retention in the ischemic territory of the MCA

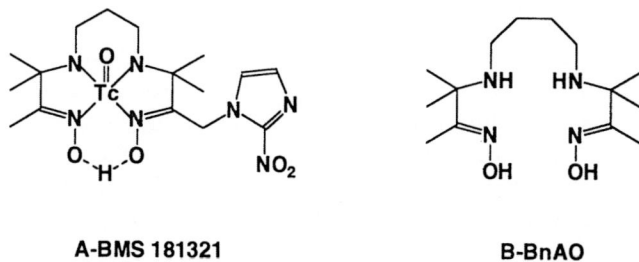

A-BMS 181321   **B-BnAO**

Fig. 7. Hypoxic cell imaging agents; A) TcO (PnAO 1, 2-nitroimidazole); B) Butyleneamineoxime

[87]. In both the cerebral and myocardial SPECT images, the ischemic regions are observed as "hot spots". Imaging with heart agents such as [99mTc]-MIBI or [201Tl] or brain agents such as [99mTc]-HMPAO result in negative images where the defects are observed as "cold spots".

The Amersham group recently reported their efforts in developing a [99mTc] labeled hypoxia imaging agent [88]. They have also derivatized several amine oxime ligands with the 2-nitroimidazole group at the 1 position on the PnAO and n-butyleneamineoxime (BnAO) (Fig. 7B) and penteneamine oxime backbones. The [99mTc] complexes formed with these amine oxime ligands were analyzed in isolated perfused hearts for their retention in hypoxic vs. normoxic tissue. Interestingly, the complex showing the greatest retention in hypoxic tissue contained no 2-nitroimidazole function. [99mTc]-BnAO without a nitroimidazole side-arm showed the greatest hypoxia selectivity, with an anoxic/oxic retention ratio of 55 and the best SPECT images in dogs with 75–90% occlusions of the left circumflex artery [88]. At this time, neither the structure of [99mTc]-BnAO nor the mechanism for its retention in hypoxic tissue is known, although bioaccessible redox processes are probably involved in the retention mechanism.

3 Chelating Moieties for Preparation of High Specific Activity ^{99m}Tc-Radiopharmaceuticals

3.1 Site-directed Biomolecules

The availability of sophisticated molecular probes for the design of new [99mTc]-radiopharmaceuticals will promote future advances in diagnosis of diseases [89]. The rapid pace of development of many new site-directed synthetic derivatives (e.g., immunologically derived molecules, receptor-avid compounds, etc.) will provide a multitude of exciting opportunities for further technological advances in this area. Labeling biomolecules with [99mTc] to produce effective in vivo tracers presents many challenges. It is essential to produce [99mTc] labeled drugs that have high in vitro and in vivo stabilities. Several different types of ligand frameworks have been developed that form [99mTc] complexes exhibiting minimal in vitro or in vivo decomposition – if any – within the time frame of the desired study. The formation of [99mTc] products in high yield and high radiochemical purity, however, usually requires the presence of large quantities of excess ligand during the drug formulation processes.

Preparation of a final drug product of high specific activity is an essential consideration when developing radiopharmaceuticals that are designed to target low capacity biochemical processes (e.g., labeled receptor avid molecules or immunologically derived agents) [89]. Three general approaches for labeling proteins or small biomolecules that have being employed include: a) direct labeling; b) preformed complexes or c) post-conjugation labeling. The direct labeling method has been successfully used by several investigators primarily for preparation of high specific activity ^{99m}Tc labeled proteins (incl. MAbs and fragments) [90]. While this is an effective approach for labeling large bio-molecules, it has limited utility for smaller receptor-avid molecule applications.

High-specific-activity ^{99m}Tc-agents can be prepared both by using the preformed ^{99m}Tc bifunctional chelate complexes (BFCs) and by post-conjugate chelation with ^{99m}Tc of the ligand that is already appended to the biomolecular targeting agent. Clearly, maximization of the specific activity of either the ^{99m}Tc-BFC or the final ^{99m}Tc-labeled drug is achieved by separation of the ^{99m}Tc-labeled molecules from those molecules containing the uncomplexed ligand. If the ^{99m}Tc-labeled product is separated from unlabeled molecules, there is considerable latitude in selecting ligand frameworks for complexing [99m]Tc. Theoretically, even chelation systems that require large quantities of chelator to be present during formulation prior to separation can be used, as long as the yield, radiochemical purity (RCP), and stability of the final product is high. Practically, however, it is more desirable to employ chelation systems that require small quantities of the chelator to form stable ^{99m}Tc-complexes in high yields.

The use of small amounts of chelators for ^{99m}Tc is generally more amenable to radiopharmaceutical applications and effective and rapid separation techniques. In addition, the cost of many site-directed biomolecular targeting agents is high, and recovery of minute fractions of targeting molecules labeled with ^{99m}Tc from the large excess of unlabeled molecules could be prohibitively expensive. Finally, in the formulation of products for ultimate use as approved ^{99m}Tc radiopharmaceuticals for routine patient-care applications, it is most desirable to keep to a minimum the number of steps in the formation of the drug product, ideally to one step, as is the case in most "instant kits". Thus, ligand systems that require only minimal quantities of the chelator to form well defined, stable complexes with ^{99m}Tc in high yields would be the most attractive candidates for use in the formulation of new, site-directed drug carrier radiopharmaceuticals.

Unfortunately, relatively few ligand systems have been shown to be effective for the preparation of high-yield ^{99m}Tc complexes using small quantities of the chelator. Most of these ligands are tetradentate containing either one or two thiol groups, the remainder being N-donors – usually in the form of amine or amido nitrogens. Diaminedithiol (DADT) based BFCAs (Fig. 8) have been used to specifically attach ^{99m}Tc to biomolecules using relatively small quantities of the ligand or the preconjugated biomolecule [91, 92]. For example, Baidoo et al. [92], report labeling of DADT thiolactone BFCAs conjugated to amino groups on B72.3 MAbs. In this study, 0.6 mg of the MAbs, with an average of 3 DADT chelators attached per MAb molecule, was labeled by transchelation, using only 2 mCi ^{99m}Tc-GH at pH 7, to produce specific activities estimated to be approximately 0.5 Ci/µmole (18.5 GBq/µmole). Since ^{99m}Tc-GH does not pro-duce measurable labeling of unconjugated MAbs, it can be assumed that specific

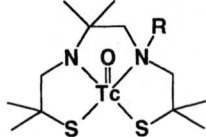

Fig. 8. Example of [99m]TcO-DADT bifunctional chelating agents R = sidechain for attachment to biomolecule

transchelation into DADT chelators occurred, however, the potential for generation of free -SH groups by exposure of the MAb to the -SH group on the BFCA must be considered.

The diaminedithiol ligand backbone has also been used for specific labeling of smaller, receptor-avid targeting agents, in which the final ^{99m}Tc complex-conjugated moiety is neutral and lipophilic. For example, DiZio et al. [93] synthesized a progestin conjugate, in which the DADT backbone is coupled to a steroid molecule that exhibited high receptor-binding affinity after complexation with [99m]Tc. The lipophilicity of the appended ^{99m}Tc-DADT complex added to that of the steroid molecule itself produced a product with increased non-specific binding [93]. The use of neutral-lipophilic ^{99m}Tc complexes (like ^{99m}Tc-DADT or PnAO) are generally considered to be essential for agents developed to target intracellular receptors, since the entire ^{99m}Tc-labeled molecule should be neutral and lipophilic for efficient transport across cell membranes. Interestingly, this type of drug design may make it feasible to use relatively low-specific-activity formulations, as a neutral-lipophilic ^{99m}Tc product is able to diffuse through cell membranes whereas a product containing a charged-polar conjugated, uncomplexed, ligand moiety may be unable to penetrate intracellularly and clear from non-target tissues. The effectiveness of peptide-based receptor-avid pharmaceuticals may also be enhanced – or decreased – by the presence of a lipophilic ^{99m}Tc-BFC conjugated at the appropriate site. Many peptide receptor agents have been synthesized in which the presence of a lipophilic group or molecule at specific sites on the peptide chain can produce major enhancements in binding affinity.

Ligands with one or two thiol groups and one to three amido groups are effective in forming stable ^{99m}Tc complexes at low concentrations. One of the first BFCAs of this type was based on the diamidodithiol (DADS) backbone (Fig. 9A) ligand [94]. Since then a variety of N_2S_2 and N_3S amido-thiol frameworks have been used to synthesize BFCAs. In addition to preparing amido-thiol ligand BFCAs for conjugation to molecules, some peptide targeting agents have been synthesized to include thiol groups strategically placed (or fused) on the peptide chain to create amido-thiol analogues within or at the end of the peptide chain. For example, incorporation of the Cys–Gly–Cys [95] or Cys–Thr–Cys [96] sequences produce the DADS core which will complex [99m]Tc. In comparison to the diaminedithiol ligands (e.g., DADT), the amidothiols form complexes with ^{99m}Tc that are more polar. In order to reduce lipophilicity, O'Neil et al. [97] used an amido-thiol chelator, to form a ^{99m}Tc-labeled progestin agent with lower non-specific binding than when the corresponding ^{99m}Tc-DADT chelator was used [93].

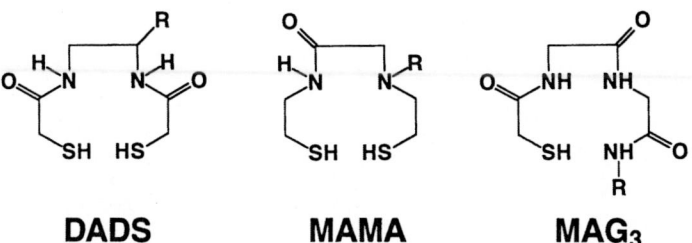

DADS   **MAMA**   **MAG₃**

Fig. 9. Examples of amido-thio bifunctional chelating agents; R = side chains for conjugation [⁹⁹ᵐTc(V) O]⁺ to biomolecule

The presence of the thiol group on ligand cores promotes high stability and rapid rates of formation of complexes with ^{99m}Tc(V). These properties permit formation of ^{99m}Tc-amido-thiol complexes using low concentrations of the uncomplexed ligand by either direct labeling (e.g., Sn^{+2}) or transchelation. For example, Rao and co-workers [98] reported the use of N$_2$S$_2$-mono-aminemonoamide (MAMA) BFCA for specific ^{99m}Tc chelation. With some MAMA derivatives, only 1–10 μg are needed to produce high ^{99m}Tc complex yields. Studies on compounds with MAMA cores (Fig. 9) indicate that complex formation with these ligands by transchelation from ^{99m}Tc-GH is faster than ^{99m}Tc complexation with the corresponding DADS ligands [98].

MAG$_3$ BFCAs (Fig. 9C), an N$_3$S ligand, has been used for preparation of high specific activity ^{99m}Tc-labeled IL2 [99]. Preparation of a "preformed" ^{99m}Tc-BFC, using exchange-labeling from ^{99m}Tc-GH to approximately 16 μg of the BFCA, produced specific activities of approximately 3.3 Ci/μmole (111 GBq/μmole) [99].

Propylene amine oxime (PnAO) has also been used to form high-specific-activity ^{99m}Tc-labeled bioconjugates. PnAO forms a well-defined neutral-lipophilic complex with ^{99m}Tc(V) [13, 18]. Koch and Mäcke [100–102] reported preparation of a BFCA with a PnAO core which, when conjugated to a somatostatin analogue and biotin (for pre-targeting applications, used in combination with avidin conjugated MAbs), were capable of chelating ^{99m}Tc in > 99% yield. For these studies, up to 40 mCi (1480 MBq) of ^{99m}Tc was used in the initial labeling, producing specific activities that can be estimated to be approximately 0.3 Ci/μmole (11.1 GBq/μmole) [100, 102]. Because ^{99m}Tc-PnAO is lipophilic, the ^{99m}Tc-PnAO-conjugated drugs and their respective catabolic products resulted in high liver accumulation and slow blood clearance of ^{99m}Tc, Mäcke and coworkers [104, 105] synthesized a tetraamine BFCA (Fig. 10) that would form the more hydrophilic [TcO$_2$tetraamine]$^+$ complex [103]. They prepared both a tetraamine conjugated with biotin [104] and D-Phe-octreotide [105]; labeling with ^{99m}Tc was performed under basic conditions to produce high specific activity products [104, 105].

The hydrazino nicotinamide (HYNIC) system (Fig. 11) developed by Abrams and co-workers [106] has been successfully used for preparation of

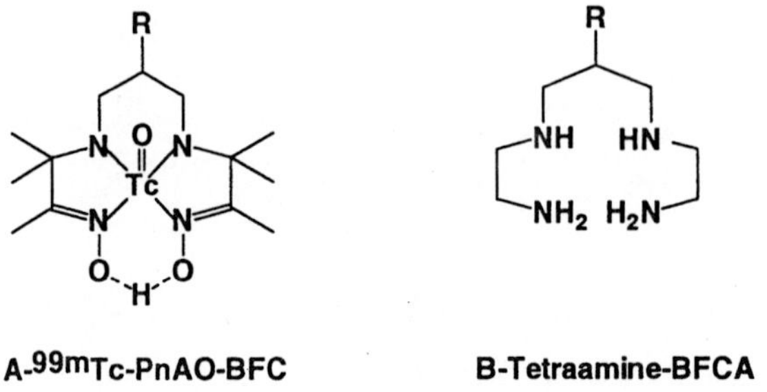

A-99mTc-PnAO-BFC

B-Tetraamine-BFCA

Fig. 10. N_4 ligand systems for conjugation ^{99m}Tc to biomolecules; R = in vivo targeting biomolecules (e.g., biotin [100, 101], somatostatin receptor-avid peptide [103, 104])

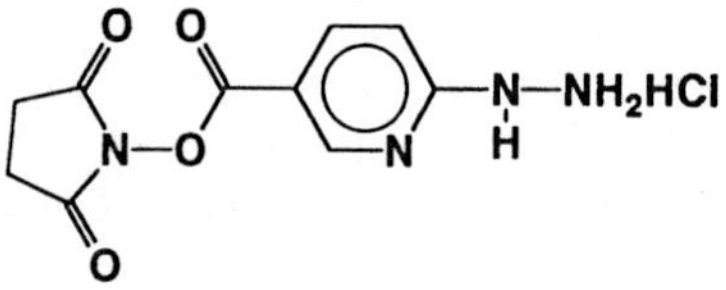

Fig. 11. Succinimidyl-6-hydrazinopyridine-3-carboxylate HCl (i.e., HYNIC-BFCA)

specifically labeled proteins and peptides with high specific activities. Labeling is accomplished by transchelation from ^{99m}Tc-GH or other donor complexes to produce ^{99m}Tc-products that have yet to be characterized [107]. Very high specific activities have been reported with HYNIC-conjugated peptides and proteins. For example, Babich et al. [108], reported that specific activities of > 3 Ci/μmole (111 GBq/μmole) were produced with HYNIC-conjugated chemotactic peptides, without purification of the ^{99m}Tc-labeled agent from unlabeled materials. Results from these studies suggest that HYNIC may be uniquely Tc-selective, providing a means for preparation of high-specific-activity ^{99m}Tc-labeling in the presence of other metals [109]. Better structural and chemical characterization of the ^{99m}Tc-HYNIC complex or complexes, and the dependence on the identity of the ^{99m}Tc-donor complex will be instrumental in broadening this application. A crystal structure of a model hydrazino compound indicates that a Tc=N bond is formed [110].

3.2 Tc-nitrido Complexes

A great deal of progress has been made in the preparation of ^{99m}Tc-nitrido based complexes. Many ^{99m}Tc radiopharmaceuticals contain Tc(V), present as either the mono-*oxo* $[^{99m}$TcO$]^{3+}$ or the *dioxo* $[^{99m}$TcO$_2]^{3+}$ core [4, 10]. The $[^{99m}$Tc$\equiv$N$]^{2+}$ moiety constitutes another Tc(V) core that could be used to form complexes with structures complementary to those containing the ^{99m}Tc(V)-oxo cores. The potential for the design of $[^{99m}$Tc$\equiv$N$]^{2+}$-based radiopharmaceuticals has been made feasible by the development of a reliable and efficient method for preparing Tc$\equiv$N multiple-bond synthon(s) directly from ^{99m}TcO$_4^-$ under sterile and apyrogenic conditions [111]. This method is based on the reaction of ^{99m}TcO$_4^-$ with N-methyl, S-methyl dithiocarbazate [H$_2$N-N(CH$_3$)

$C(=S)SCH_3]$ in the presence of reducing agents, such as Sn^{+2} and tris(m-sulfophenyl)phosphine [111]. This provides a more convenient route to formation of these types of complexes than that used by Baldas and Bonnyman [112] in their pioneering work that first described the synthesis of complexes with the $Tc\equiv N^{+2}$ core. The dithiocarbazate reactant serves as a donor of nitride nitrogen atoms (N^{-3}) to yield the $(TcN)^{+2}$ group efficiently at the tracer level [111]. This approach might provide a multitude of ^{99m}Tc complexes that may be used for selected development of new ^{99m}Tc-radiopharmaceuticals.

4 Conclusions

The past decade has witnessed the synthesis of a variety of well-defined ^{99m}Tc-complexes that are – or will become – approved radiopharmaceuticals. As our understanding of diseased states and ethical drug design progresses, more sophisticated ^{99m}Tc radiopharmaceuticals must be developed. For this reason, further work on new ^{99m}Tc-complex development has to be done. This must include complete structural characterization of all complexes, since the complementary of shape and functionality that characterizes the interaction of ^{99m}Tc-labeled biomolecules with targeted sites has yet to be elucidated. Optimization of the pharmacologic performance of ^{99m}Tc-labeled drugs will rely heavily on the development of new or modified ^{99m}Tc-chelation systems that have a high degree of physiochemical flexibility. Strategies adopted for the design of relevant chelation systems must recognize the importance of convenience and simplicity in the formulation of ^{99m}Tc-labeled drugs intended for use in humans as approved radiopharmaceuticals.

5 References

1. Richards P, Tucker WD, Srivastava SC (1982) Int J Appl Radiat Isot 33: 793
2. Eckelman WC, Richards P (1970) J Nucl Med 11: 761
3. (a) Nowotnik DP (1994) in: Simpson CB (ed) Textbook of radiopharmacy, Gordon and Breach Publishers, Reading, UK, p 29
4. Steigman J, Eckelman WC (1992) The chemistry of technetium in medicine, National Academy Press, Washington, D.C.
5. (a) Bandoli G, Mazzi U, Roncari E (1982) Coord Chem Rev 44: 191; (b) Mazzi U, Nicolini M, Bandoli G, Refosco F, Tisato F, Moresco A, Duatti A (1990) in: Nicolini M, Bandoli (eds) Technetium and rhenium in chemistry and nuclear medicine 3, Corinal International, p 39
6. Jurisson S, Berning D, Jia W, Ma D (1993) Chem Rev 93: 1137
7. (a) Bugaj JE, DeRosch MA, Marmion ME, Quint RH, Deutsch KF, Deutsch E (1994) J Nucl Med 35: 139P; (b) Marmion ME, DeRosch M, Bushman M, Wolfangel R, Webb E, Deutsch K, Deutsch E (1994) J Nucl Biol Med 38: 455; (c) DeRosch MA, Brodack JW, Grunmon GD, Marmion M, Nosco DL, Deutsch KF, Deutsch E (1992) J Nucl Med 33: 850

W.A. Volkert and S. Jurisson

8. (a) Tofe A, Francis M (1976) J Nucl Med 16: 414
9. Dischino DD, Welch MJ, Kilbourn MR, Raichle ME (1983) J Nucl Med 24: 1030
10. (a) Kung HF, Guo Y-Z, Yu C-C, et al. (1989) J Med Chem 32: 433; (b) Kung HF (1990) Semin Nucl Med 20: 150
11. Lever SZ, Burns HD, Kervitsy TM, et al. (1985) J Nucl Med 26: 1287
12. (a) Partridge WM, Mietus LJ (1981) Endocrinology 1138; (b) Stein WD (1967) The movement of molecules across cell membranes, Academic Press, New York
13. (a) Volkert WA, Hoffman TJ, Seger RM, et al. (1984) Eur J Nucl Med 9: 511; (b) Troutner DE, Volkert WA, Hoffman TJ, et al. (1984) J Appl Radiat Isot 35: 467
14. Burns HD, Manspeaker H, Miller R, et al. (1979) J Nucl Med 280: 326
15. Kung HF, Molnar M, Billings J, et al. (1990) J Nucl Med 25: 326
16. (a) Morgan GF, Deblaton M, Thornback J, et al. (1991) J Nucl Med 32: 500; (b) Bossnyt A, Morgan GF, Deblaton M, Pirotte R, Chirico A, Clemns P, Vandenbroeck P, Thornback JR (1991) J Nucl Med 32: 399
17. Walovitch RC, Hill TC, Garrity ST, et al. (1989) J Nucl Med 30: 1892
18. (a) Fair CK, Troutner DE, Schlemper EO (1984) Acta Cryst C40: 1544; (b) Jurisson S, Schlemper EO, Troutner DE, et al. (1986) Inorg Chem 25: 543
19. Neirinckx RD, Canning L, Piper IM, et al. (1987) J Nucl Med 28: 191
20. Winchel HS, Baldwin RM, Lin T-H (1990) J Nucl Med 21: 940
21. Holman BL, Hellman RS, Goldsmith SJ, et al. (1989) J Nucl Med 30: 1018
22. Neirinckx RD, Burke JF, Harrison RC (1988) J Cereb Blood Flow Metab 8: S4
23. Roth CA, Hoffman TJ, Corlija M, Volkert WA, Holmes RA (1992) Nucl Med Biol 19: 783
24. Andersen AR (1989) Cereb Brain Metab Rev 1: 288
25. Sharp PF, Smith FW, Germmell HG, et al. J Nucl Med 27: 171
26. Andersen AR, Friberg HH (1988) J Cereb Blood Flow Metab 8: S23
27. Hoffman TJ, Corlija M, Reitz R, Volkert WA, Holmes RA (1992) J Nucl Med 33: 1032
28. Tubergen K, Corlija M, Rammamoorthy N, et al. (1991) J Labelled Compd Radiopharm 30: 50
29. Tubergen K, Corlija M, Volkert WA, et al. (1991) J Nucl Med 32 : 111
30. (a) Deutsch E, Glavan KA, Sodd VJ (1981) J Nucl Med 22: 897; (b) Deutsch E, Bushong W, Glavan KA, Elder RC, Sodd VJ, Scholz KL, Fortman DL, Lukes SJ (1981) Science 214: 85
31. (a) Deutsch EA, Ketring AR, Libson K, Vanderheyden J-L, Hirth WW (1989) Nucl Med Biol 16: 191; (b) Deutsch E (1993) Radiochem Acta 63: 195; (c) Deutsch E, Hirth W (1987) J Nucl Med 28: 1497
32. (a) Nowotnik DP, Nunn AD (1992) Drugs, News and Persp 5: 183 (b) Nunn AD (1991) Semin Nucl Med 20: 111
33. (a) Abrams MJ, Davison A, Jones AG, Costello CE, Pang H (1983) Inorg Chem 22: 2798; (b) Jones AG, Abrams MJ, et al. (1984) Nucl Med Biol 22: 225
34. Kelly JD, Forster AM, Higley B, Archer CM, Booker FS, Canning LR, Chiu FW, Edwards B, Gill HK, McPartlin M, Nagle KR, Latham IA, Pickett RD, Storey AE, Webbon PM (1993) J Nucl Med 34: 222–227
35. Sapirstein L (1958) Am J Physiol 193: 161
36. Mousa SA (1991) Am J Physiol Imag 6: 16
37 .(a) Libson K, Woods J, Sullivan JL, Watkins JW, Elder RC, Deutsch E (1988) Inorg Chem 29: 999; (b) Vanderheyden J-L, Heeg MJ, Deutsch E (1985) Inorg Chem 24: 1666
38. Higley B, Smith FW, Smith T, Gemmell HG, Gupta PD, Gvozdanovic DV, Graham D, Hinge D, Davidson J, Lahiri A (1993) J Nucl Med 34: 30
39. (a) Holman BL, Jones AG, Lister-James J, Davison A, Abrams MJ, Kirshenbaum JM, Tumeh SS, English RJ (1984) J Nucl Med 25: 1350; (b) Holman BL, Sporn V, Jones AG, Sia STB (1987) J Nucl Med 28: 13; (c) Sands H, Delano ML, Gallagher BM (1986) J Nucl Med 27: 404
40. (a) Jurisson SS, Dancy K, McPartlin M, Tasker PA, Deutsch E (1984) Inorg Chem 23: 4743; (b) Ichimura A, Heinemen WR, Deutsch E (1985) Inorg Chem 24: 3134
41. (a) Sands H, Delano ML, Camin LL, Gallagher BM (1985) Biochim Biophys Acta 812: 665; (b) Piwnica-Worms D, Kronauge JF, Holman BL, Lister-James J, Davison A, Jones AG (1988) J Nucl Med 29: 55
42. (a) Treher EH, Francesconi LF, Malley MF, Gougoutas JZ, Nunn AD (1989) Inorg Chem 28: 341; (b) Linder KE, Malley MF, Gougoutas JZ, Unger SE, Nunn AD (1990) Inorg Chem 29: 2428; (c) Jurisson S, Francesconi LF, Linder KE, Treher E, Malley MF, Gougoutas JZ, Nunn AD (1991) Inorg Chem 30: 1820

43. (a) Narra RK, Eckelman WC, Kuczynski BL, Silva D, Feld T, Nunn AD (1989) J Label Cpmpd Radiopharm 26: 491; (b) Narra RK, Nunn AD, Kuczynski BL, Feld T, Wedeking P, Eckelman WC (1989) J Nucl Med 30: 1830
44. (a) Fritzberg AR (1986) In: Simpson, CB (ed) Radiopharmaceuticals: progress and clinical perspectives, CRC Press, Inc, Boca Raton, FL p 61; (b) Nunn AD, Loberg MD, Connely RA (1983) J Nucl Med 24: 423
45. (a) Loberg MD, Cooper M, Harey E, Callery PS, Farth W (1976) J Nucl Med 17: 633; (b) Loberg MD, Fields AT (1977) Int J Appl Radiat Isot 28: 687
46. (a) Jansholt A-L, Scheeke PO, Vera DR, Krohn KA, Stodalnik RC (1981) J Label Cpd Radiopharm 18: 198; (b) Wistow BW, Subramanian G, VanHeertum RL, Henderson RW, Gagne CM, Hall C, McAfee JE (1977) J Nucl Med 18: 455
47. Loberg MD, Fields AT (1978) Int J Appl Radiat Isot 29: 167
48. Costello CE, Brodack JW, Jones AG, Davison A, Johnson DL, Kasina S, Fritzberg AR (1983) J Nucl Med 24: 253
49. Baker RJ, Bellen JC, Ronai PM (1975) J Nucl Med 16: 720
50. (a) Baker RJ, Bellen JC (1978) Int J Appl Radiat Isot 29: 167; (b) Kato-Azuma M, Hazue M (1978) J Nucl Med 19: 397
51. Kato-Azuma M (1981) Int J Appl Radiat Isot 32: 187
52. Kato-Azuma M (1982) J Nucl Med 23: 517
53. (a) Fritzberg AR, Bloedov DC, Eshima D, Johnson DL, Klingensmith WC (1983) J Nucl Med 24: P10; (b) Sakahara H, Yamamoto K, Tamaki N, Hayashi N, Fujita T, Moreta R, Torizuka K (1983) Jpn J Nucl Med 20: 568; (c) Hazegawa Y, Nakano S, Ibuka K, Hashizume T, Sasaki Y, Imaoka S, Ishiguro S, Kasugai H, Obano Y, Tanaka S (1984) J Nucl Med 25: 1122
54. Chilton HM, Thrall JH (1990) in: Swanson D, Chilton H, Thrall J (eds) Pharmaceuticals in medical imaging. Macmillian Pub, New York, p 305
55. Eckelman WC, Meinken G, Richards P (1972) J Nucl Med 13: 577
56. Deutsch E, Packard A, private communication
57. Fritzberg AR, Kasina S, Eshima D, Johnson DL (1986) J Nucl Med 27: 111
58. (a) Fritzberg AR, Klingensmith WC, Whitney WP, Kuni CC (1981) J Nucl Med 22: 258; (b) Kasina S, Fritzberg AR, Johnson DL, Eshima D (1986) J Med Chem 29: 1933
59. Rao TN, Adhikesavalu D, Camerman A, Fritzberg AR (1990) J Am Chem Soc 112: 5798
60. Nosco DL, Tofe AJ, Dunn TJ, Lyle LR, Wolfangel RG, Bushman MJ, Grummon GD, Helling DE, Marmion ME, Miller KM, Pipes DW, Strubel TW, Wester DW (1990) in: Nicolini M, Bandoli G, Mazzi U (eds) Technetium and rhenium in chemistry and nuclear medicine 3, Cortina International, Verona p 381
61. Kung HF (1990) Semin Nucl Med 20: 150
62. Despopoulos A (1965) J Theor Biol 8: 163
63. (a) Taylor A (1991) Crit Rev Diag Imag 32: 1; (b) Eshima D, Fritzberg AR, Taylor A (1990) Semin Nucl Med 20: 28
64. (a) Taylor A, Ziffer JA, Steves A (1989) Radiology 170: 721; (b) Russell CD, Thorstad B, Dubovsky EV (1988) J Nucl Med 29: 2053
65. (a) Verbruggen A, Nosco D, Van Nevon C, Bormuns G, Adriaens P, DeRoo M (1990) Eur J Nucl Med 16: 429; (b) Verbruggen AM (1992) Eur J Nucl Med 17: 346
66. Ikeda I, Inoue O, Kuratu K (1976) Int J Appl Radiat Isot 27: 681
67. Blower PJ, Singh J, Clarke SEM (1991) J Nucl Med 32: 845
68. Dewanjee MK (1990) Semin Nucl Med 20: 5
69. Ohta H, Yamamoto K, Endo K, Mori T, Hamanaka D, Shimazu A, et al. (1984) J Nucl Med 25: 323
70. (a) Watkins JC, Lazarus CR, Mistry R, Shahein OH, Maisey MN, Clarke SEM (1989) J Nucl Med 30: 174; (b) Ohta H, Endo K, Fujita T, et al. (1985) Clin Nucl Med 10: 855
71. (a) Yokoyoma A, Hata N, Horiuchi K, et al. (1988) Int J Nucl Med Biol 12: 273; (b) Sampson C (1987) Nucl Med Comm 8: 184
72. Horiuchi K, Yomoda I, Yokoyama A, et al. (1986) Technetium in chemistry and nuclear medicine 2. Cortina International, Verona, p 155
73. (a) Singh J, Powell AK, Clarke SEM, Blower PJ (1991) Chem Soc Chem Comm 1115; (b) Bisunandan MM, Blower PJ, Clarke SEM, Singh J, Went M (1991) J Appl Radiat Isot 42: 167
74. Bandoli G, Nicolini M, Mazzi U, Spies H, Münze R (1984) Transition Met Chem 9: 127
75. deKieviet W (1981) J Nucl Med 22: 703
76. Clarke MJ, Podbielski L (1987) Coord Chem Rev 78: 253

77. Subramanian G, McAfee JG (1971) Radiology 98: 192
78. (a) Pinkerton TC, Heineman WR, Deutsch E (1980) J Am Chem Soc 102: 2476; (b) Wilson GM, Pinkerton TC (1985) Anal Chem 57: 246
79. Libson K, Deutsch E, Barnett BL (1980) J Am Chem Soc 102: 2476
80. Linder KE, Chan YW, Cyr JE, Malley MF, Nowotnik DP, Nunn AD (1994) J Med Chem 37: 9
81. Martin GV, Gerqueira MD, Caldwell JH, Rasey JE, Embree I, Krohn KA (1990) Circ Res 67: 240
82. Linder KE, Chan YW, Cyr JE, Nowotnik DP, Eckelman WC, Nunn AD (1993) Bioconjugate Chem 4: 326
83. Ramalingam K, Raju N, Nanjappan P, Linder KE, Pirro J, Zeng W, Rumsey W, Nowotnik DP, Nunn Ad (1994) J Med Chem (Accepted)
84. Varghese AJ, Whitmore GF (1980) 40: 2165
85. Kusuoka H, Hashimoto K, Fukuchi T, Nishimura (1994) J Nucl Med 35: 1371
86. DiRocco RJ, Bauer A, Kuczynski BL, Pirro JP, Linder KE, Narra RK, Nunn AD (1992) J Nucl Med 33: 865
87. DiRocco RJ, Kuczynski BL, Pirro JP, Bauer A, Linder KE, Ramalingam K, Cyr JE, Chan YW, Raju N, Narra RK, Nowotnik DP, Nunn AD (1993) 13: 755
88. Archer CM, Edwards B, Kelly JD, King AC, Burke JF, Riley ALM (1994) Technetium labelled agents for imaging tissue hypoxia in vivo, 4th International Symposium on Technetium and Rhenium in Chemistry and Nuclear Medicine, Bressanone, Italy, September 12–14
89. (a) Eckelman WC, Gibson RE (1993) in: Burns HD, Gibson RE, Dannals RF, Siegl PKS (eds) Nuclear imaging in drug discovery, development and approval. Birkhäuser, p 113; (b) Fischman AJ, Babich JW, Strauss HW (1993) J Nucl Med 34: 2253
90. (a) Rhodes BA, Torvestaad DA, Burchiel SW, Anstring RK (1980) J Nucl Med 21: 54; (b) Thakur ML, DeFulvio J, Richard ME, et al. (1991) Nucl Med Biol 18: 223; (c) Paik CH, Reba RC, Eckelman WC, et al (1985) Nucl Med Biol 12: 3
91. (a) Baidoo KE, Scheffel U, Lever SZ (1990) Cancer Res (Suppl) 50: 79s; (b) Baidoo KE, Lever SZ, Scheffel U (1994) Bioconj Chem 5: 114; (c) John CS, John EK, Li J, et al. (1994) J Label Cpd Radiopharm 35: 65 (d) Del Rosario RB, Jung YW, Baidoo KE, et al. (1994) Nucl Med Biol 21: 197
92. Baidoo KE, Lever SZ (1990) Bioconj Chem 1: 132
93. DiZio JP, Fiaschi R, Davison A, et al. (1991) Bioconjugate Chem 2: 353
94. Fritzberg AR, Abrams PG, Beaumier PL, et al. (1988) 85: 4025
95. (a) Lister-James J, McBride WJ, et al. (1994)) J Nucl Med 35: 257; (b) Lister-James J (BR, Moyer, Buttram S, et al. (1994)) J Nucl Med 35: 257P
96. Knight LC, Radcliff R, Maurer AH, et al. (1994) J Nucl Med 35: 282
97. O'Neil JP, Carlson KE, Anderson CJ, et al (1994) Bioconj Chem 5: 182
98. (a) Rao TN, Gustavson LM, Srinivasan A, et al. (1992) Nucl Med Biol 19: 889; (b) Gustavson LM, Rao TN, Jones DS, Fritzberg AR (1991) Tetrahedron Lett 32: 5485
99. Mather SH, Ellison D (1994) J Nucl Biol Med 38: 481
100. Koch P, Mäcke R (1992) Angew Chem Int Ed Engl 31, 1507
101. Maina T, Stolz B, Albert R, et al. (1994) Eur J Nucl Med 21: 314
102. Maina T, Stolz B, Albert R, et al (1994) Eur J Nucl Med 21: 437
103. Zuckman SA, Freeman GW, Troutner DE, Volkert WA, Holmes RA, Van Derveer DG, Barefield EK (1981) Inorg Chem 20: 2386
104. Noch B, Evard F, Paganelli G, Mäcke HR (1994) J Nucl Med Bio Med 38: 460
105. Maina T, Stolz B, Alberto R, Nock B, Bruns C, Mäcke H (1994) J Nucl Biol Med 38: 452
106. (a) Abrams MJ, Juweid M, tenKate CI, et al. (1990) J Nucl Med 31: 2022–2028; (b) Schwartz DA, Abrams MJ, Hauser mm, et al. (1991) Bioconjugate Chem 2: 333–336; (c) Abrams M, Juweid M, et al. (1990) J Nucl Med 31: 2022–2028
107. Larsen SK, Caldwell G, Higgins JD, et al. (1994) J Label Cpd Radiopharm 35: 3–5
108. Babich JW, Graham W, Barrow SA, et al. (1993) J Nucl Med 34: 2176–2181
109. Babich JW, Soloman H, Pike MC, et al. (1993) J Nucl Med 34: 1964–1974
110. Abrams MJ, Larsen SK, Zubieta J (1990) Inorg Chim Acta 173: 133–135
111. (a) Pasqualini R, Veronique C, Bellande E, Duatti A, Marchi A (1992) Appl Radiat Isot 11: 1329; (b) Pasqualini R, Duatti A, Bellande E, Comazzi V, Brucato V, Hoffscher D, Fagset D, Comet M (1994) J Nucl Med 35: 334
112. (a) Baldas J, Bonnyman J (1985) Int J Appl Radiat Isot 36: 133; (b) Baldas J, Bonneyman J (1985) Int J Appl Radiat Isot 36: 9

High- and Low-Valency Organometallic Compounds of Technetium and Rhenium

Roger Alberto

Division of Radiopharmacy, Paul Scherrer Institute, CH-5232 Villigen/Switzerland

Table of Contents

Topics in Current Chemistry, Vol. 176
© Springer-Verlag Berlin Heidelberg 1996

Roger Alberto

Organometallic technetium chemistry is a relatively poorly developed field, although the location of this element in the periodic table implies the possession of a very diverse chemistry. This article aims to review critically the progress that has been made in this field during the past five to ten years. In particular, the focus is put on synthetic strategies that start from a few important synthons. To evaluate the usefulness and potential of these starting materials, subsequent reactions, that have been published will be discussed in detail, particularly in relation to the prospects for further development. For this reason, this article is not divided into sections defined by oxidation state or type of ligand, but by the relevant synthon. Thus, similar compounds may appear in different sections of the article, whereas differing types of compound may be discussed in the same section, because of their common starting material. Where necessary, the chemistry of technetium will be compared with that of its homologs, especially in the case of any observed catalytic activity or other practical applications. Furthermore, structural characterization or another physical measurement, such as i.r. spectroscopy, often affords an insight into the electronic properties of a compound and consequently allows predictions about chemical properties. For most of the compounds discussed, X-ray structure analysis and i.r. spectroscopy has been performed; thus emphasis will be put on these physical measurements.

1 Introduction

This review, which covers the development of organometallic technetium chemistry during the past five to ten years, is particularly intended to summarize the progress of its synthetic aspects. It should help the reader to extrapolate from the syntheses and reactions that were performed during this period, and to initiate new approaches in this relatively poorly developed field of chemistry. For this reason the general structure of this review does not follow the different oxidation states or ligand, types but rather the synthesis and reactions of the most convenient starting materials. In this respect, a convenient starting material or synthon is one that is directly available from pertechnetate or a primary derivative thereof, such as [Tc₂O₇] (**1**). Only a few reviews on the general chemistry of technetium have appeared in the past ten years, ignoring those which were predominantly concerned with the nuclear-medical applications of this element. [1–5]

Well defined organometallic technetium compounds exist in oxidation states ranging from −I up to +VII, with the exception of the +IV and +V valencies. Compared to the organometallic chemistry of the technetium homologs, manganese and rhenium, that of technetium is strikingly underdeveloped; the reason for this surprising fact should be clarified. In contrast, technetium chemistry in its intermediate oxidation states is very well documented, and the reactivity of a wide variety of different compounds has been completely characterized and carefully investigated. Thus, in the valencies +III, +IV and +V, this diversity allows a systematic comparison between the different characteristics of the homologs [6]. The interest in the intermediate oxidation states results unambiguously from the fact that compounds in these valencies are considered to have potential for use in radiopharmaceuticals, and are therefore interesting to the field of nuclear medicine. Nuclear medicine is the only field where the element technetium has found broad application. In particular, the short lived isotope Tc-99m serves as a very favorable diagnostic tool, due to its weak γ-emission in a well detectable energy range. In contrast, the main isotope Tc-99, which is available at least in the kg scale, is unlikely to have any application in industrial chemistry as a catalyst, owing to the radioactivity associated with it. Consequently, from a practical point of view, the interest in developing organometallic technetium compounds was low. In addition, the use of organometallic compounds for nuclear-medical purposes has scarcely been considered, due to the difficulties in synthesizing such compounds. The situation was completely reversed in the case of rhenium. The catalytic activity of this element was recognized very early, and followed by extensive investigation into the catalytic activity of its organometallic compounds. Such investigations are still being pursued. Only in the very recent past have the rhenium isotopes Re-186 and Re-188 come under serious consideration for application in radioimmunotherapy. The convenient decay characteristics of these isotopes make them superior to others, such as the most widely applied I-131.

Beside the lack of a potential application for organometallic technetium compounds, a further reason for the poor development of the field might be the scarcity of useful starting materials such as $[Tc_2(CO)_{10}]$ (**2**). The syntheses of homoleptic carbonyls described in the early literature were low in yield and technetium at that time a very expensive element. Only in recent years as technetium has become rather inexpensive have some high yield procedures been described (see Sect. 3.1). However, in most modern laboratories it is almost impossible to perform high pressure high temperature reactions with radioactive material for reasons of safety. Consequently, **2** is still scarcely available but for other reasons than before. Important starting materials for the development of organometallic technetium chemistry have thus to be synthesized under "normal" conditions. To find a pathway to stable but reactive technetium precursors that allows the facile introduction of the Tc-carbon bond stands as a challenge to the technetium chemistry of the future. For convenience, such chemistry should start directly from commercially available pertechnetate salts.

In the past few years, continuous rapid progress in the development of organometallic re compounds in the high oxidation states could be observed, particularly in the field of mixed oxo-alkyl and oxo-dienyl species, leading to a broad variety of new complexes. While catalysis with such rhenium compounds was the background of the extensive research in this field, similar applications of technetium were not expected; consequently, not very much was done in the analogous field. Again, the most important starting material **1** is not commercially available, is volatile, and tends to be rapidly hydrolysed. Nevertheless, a few compounds have been prepared, and structural as well as catalytic comparisons have been performed.

Comparison of the development of the organometallic chemistry of the two elements reveals furthermore that the chemistry of technetium is often a copy of that of rhenium. Type reactions, starting materials and compounds have been first synthesized, almost without exception, with rhenium. Some of the rare exceptions where the chemistry of rhenium followed that of technetium will be discussed. In addition the scarcity of **2** was the basis for a very different approach to low-valency organotechnetium chemistry, and resulted finally in the fact that almost all compounds originally derived from **2** can be synthesized from **3** in better yield.

Rhenium chemistry will not be discussed explicitly in this contribution. Various excellent reviews on the progress of organometallic rhenium chemistry have been published over recent years. [7] Rhenium is only mentioned when its comparison to new technetium compounds is interesting in the present context and the focus is on significant differences in the behaviour of the two homologs.

1.1 Starting Materials

Tc-99 is usually available in the form of an aqueous $(NH_4)[TcO_4]$ (**3**) solution or as Tc-metal. Most of the syntheses originally started from one of these two materials. The arrangement of the following sections relates to Tc-compounds which can be synthesized in a one step reaction from one of these two available forms and can be considered, due to their chemical reactivity and the numbers of compounds prepared, as synthons. These precursors either already contain a Tc-carbon bond or allow the convenient formation of such a bond.

2 [TcO₄]⁻ as a Starting Material for the Formation of the Tc–C Bond under Low-Pressure Conditions

The most convenient route to organometallic technetium complexes is directly from **3**, under reaction conditions which allow working in a normal laboratory. There are basically three essential compounds that fulfil these conditions and can be prepared in a one step synthesis starting from **3**, and which are convenient precursors for subsequent chemistry due to their reactivity (Scheme 1).

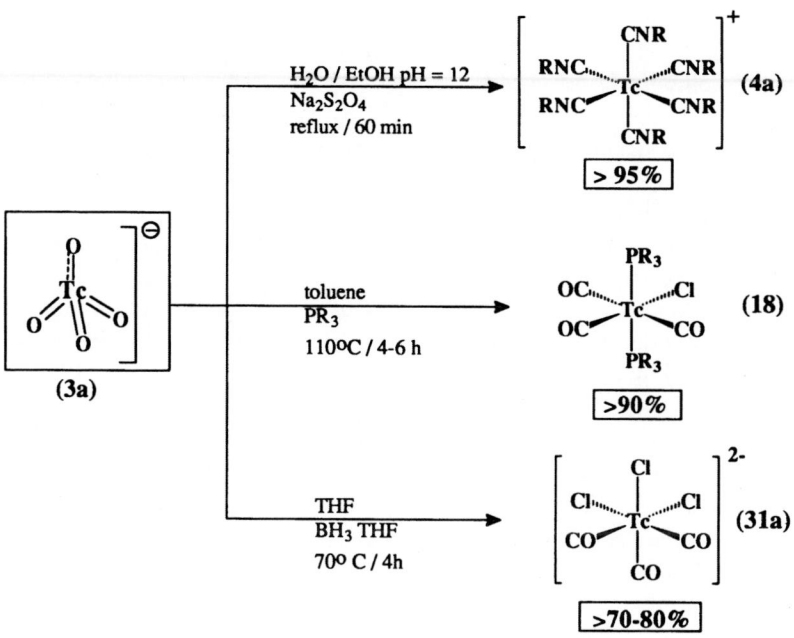

Scheme 1. Pathway to important organotechnetium synthons under low CO pressure conditions

2.1 Synthesis and Chemistry of the Homoleptic Isocyanide Complexes [Tc(CN–R)₆]⁺

The first compound to be described is the homoleptic Tc(I)-isocyanide complex [Tc(CN–R)₆]⁺ (**4a**), where R stands for a wide variety of organic groups. **4a** can be prepared in quantitative yield directly from a refluxing aqueous solution in the presence of the isocyanide, with $[S_2O_4]^{2-}$ as a reducing agent [8]. This 18e⁻ complex is completely water- and air stable and can also be prepared in tracer amounts with metastable short-lived Tc-99m. These chemical properties and the facile synthesis allowed the introduction of the first class of organometallic technetium compounds into the field of nuclear medicine, where technetium has found world-wide application as very important myocardial imaging agents. In contrast, it was not possible to prepare the analogous rhenium compounds with [ReO₄]⁻ (**5**) as starting material, as **5** is a much weaker oxidant than **3**. Only the use of the well known Re(V) complex [ReOCl₃(PPh₃)₂] (**6**) enabled the authors to obtain [Re(CN–R)₆]⁺ (**4b**) in moderate yield. A comparison with the isoelectronic Ru(II) complex [Ru(CN–R)₆]²⁺ was carried out . The latter reacts with excess methylamine to form diaminocarbene compounds under mild conditions [9], but neither rhenium nor technetium could be forced to form carbene complexes under these conditions. It seems that the higher positive charge of the ruthenium center is essential for the formation of carbenes. As mentioned, the compounds (**4b**) are exceptionally stable. The origin of this stability is kinetic as well as thermodynamic. The d^6 low-spin electronic configuration in an almost ideal octahedral ligand field is the reason for the kinetic stability. Infrared spectroscopic studies of these compounds revealed that the strong $\nu_{C\equiv N}$ absorptions appear between 2130 and 2040 cm^{-1} and are red-shifted by up to 80 cm^{-1} compared to the corresponding frequencies in the free isocyanide compounds. This lowering of wave numbers indicates extensive π-backbonding from the metal center to the isocyanide carbon, with concomitant loss of triple-bond character in the $C\equiv N$ bond. This results in a strong Tc–C bond and thus in increased thermodynamic stability against ligand exchange.

Despite this high stability, or even as a result of it, **4a** has been the subject of many subsequent studies of its chemical behaviour and reactivity. A reason for the interest in this compound as a starting material was, of course, the fact that it found extensive application in nuclear medicine. In order to improve myocardial uptake for better imaging, the compound was synthesized with a wide variety of isocyanides and these have been partially substituted by different types of ligands, resulting in Tc(I) complexes containing various isocyanide ligands. Such studies are still being pursued, and examples of some of the more recent results are given in Scheme 2.

Special attention has been paid to oxidation reactions with the halogens Br$_2$ and Cl$_2$. These oxidants react with **4a** under ambient conditions to form the seven-coordinated compounds [TcX(CN–R)$_6$]$^{2+}$ quantitatively, these have been fully characterized for technetium (**7a**) and for rhenium (**7b**) [10]. Only a very limited number of seven coordinated Tc(III) compounds have been described in the technetium literature, although these 18e$^-$ systems are supposed to form a class of very stable compounds. The first and most prominent example described in the early literature was the seven coordinated Tc(III) complex [TcCl$_3$(CO)(PR$_3$)$_3$] (**8**) with C$_{3V}$ geometry [11]. Complexes **7a** and **7b** with different organic groups have been described to be completely water- and air stable. In the i.r. spectra, along with several weak $\nu_{C\equiv N}$ bands, there is, a strong $\nu_{C\equiv N}$ absorption near 2250 cm^{-1}, almost 100 cm^{-1} higher than in non-coordinated isocyanide. This indicates impressively that π-backbonding in these Tc(III) complexes is not at all significant. Although the complexes **7a** and **7b** are water-soluble and easily prepared, only a very limited number of reactions have started with one of these compounds. In analogy to the corresponding reactions of isoelectronic Mo(II) and W(II) complexes, an attempt was made to perform a reductive coupling of two isocyanides with metallic Zn as a reductant [12, 13]. In contrast to the reductive coupling of the isocyanide ligands of the isoelectronic molybdenum compound with formation of [Mo(CN–R)$_4$(L*)]$^{2+}$ (L*=RHNC$\equiv$CNHR) (**13c**), [MoX(CN–R)$_6$]$^+$ only reduction of the Tc(III) center occurred in the case of **7a** and the well known compound **4a** was isolated. Attempts to substitute two isocyanide ligands by aromatic diamines such as bipyridyl resulted in dealkylation, seven-coordinated compounds of the composition [Tc(CN)X(CN–R)$_5$]$^+$ (**9**) were isolated. The presence of base was not even necessary to initialize dealkylation, as this could be achieved by simple heating in acetonitrile. Further reactions with these interesting Tc(III) starting materials have not yet been performed.

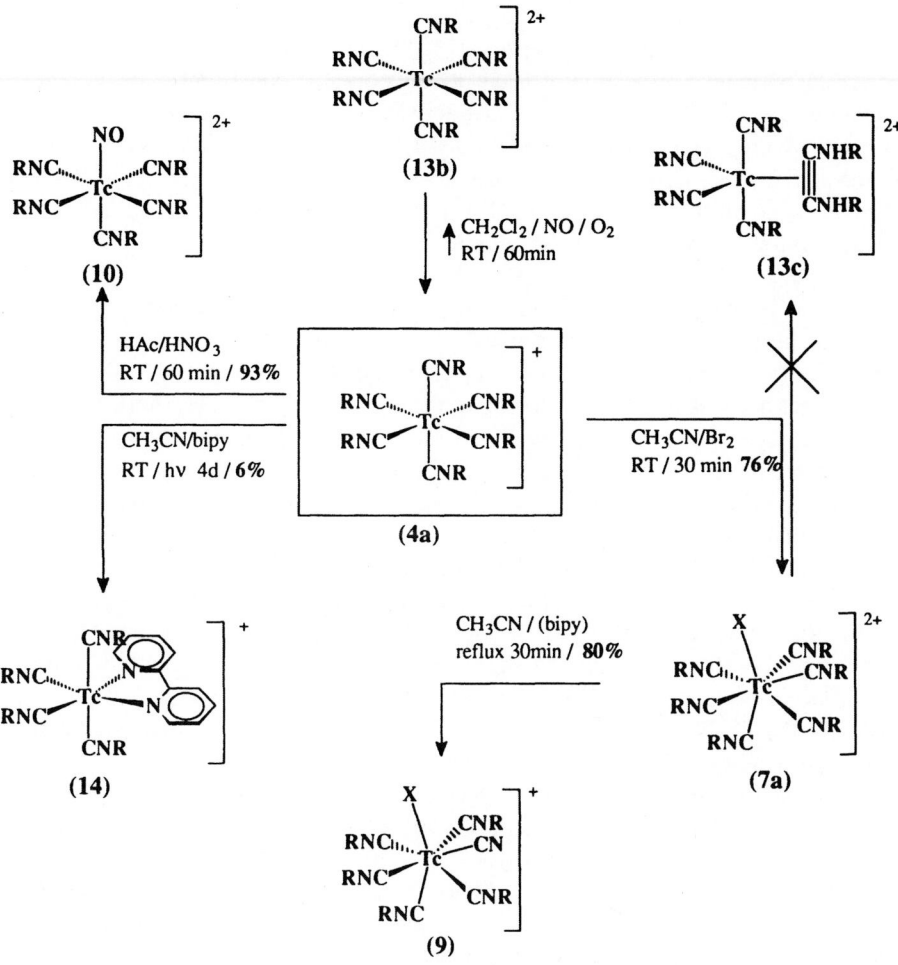

Scheme 2. Important reactions of [Tc(CN–R)$_6$]$^+$ (* see text)

Attempts to prepare mixed isocyanide/nitrosyl complexes starting from **4a** have been described. [14] Applying classical nitrosylation conditions (HNO$_3$/HAc) leads, in a clean reaction, to the substitution of one isocyanide by an NO$^+$ ligand and the Tc(I) complex [Tc(CN–R)$_5$(NO)]$^{2+}$ (**10**) was formed. Using NOPF$_6$ instead of HNO$_3$/HAc resulted in the same compound. Attempts to prepare this compound starting from the Tc(II) complex [Tc(NO)Br$_4$]$^-$ (**11**) resulted only in the reductive substitution of two bromides; the complex [TcBr$_2$(NO)(CN–R)$_3$] (**12**) was isolated and its structure elucidated. A significant difference was found between these nitrosylations and those with the Mn analog [Mn(CN–R)$_6$]$^+$ (**4c**). Under all reaction conditions no introduction of the NO$^+$ ligand took place and it was only possible to obtain the Mn(II) complex [Mn(CN–R)$_6$]$^{2+}$ (**13**) [15, 16]. Reacting the Tc-complex with NO/O$_2$ also yielded a strongly oxidizing paramagnetic complex, which was identified as $[Tc(CN-R)_6]^{2+}$ (**13b**). The $\nu_{C\equiv N}$ absorption in this compound appeared at 2200 cm^{-1}, significantly higher than in the Tc(I) complex, indicating weaker π-backbonding consistent with a higher oxidation state. In all compounds, $\nu_{N\equiv O}$ appeared between 1800 and 1900 cm^{-1}, consistent with a NO$^+$ ligand rather than NO$^-$. This observation was confirmed by the X-ray structure of **12** where the Tc–N–O angle is approximately 176°. In addition, $\nu_{C\equiv N}$ in the latter compound was observed at 2230 and 2160 cm^{-1} and the absorption at the lower wave number was assigned to the C≡N stretching trans to the NO$^+$ ligand.

A very limited number of substitution reactions, starting directly from **4a**, have also been performed. In particular, various aromatic diamines (AA) such as phenanthrolin or bipyridyl have been investigated in this respect [17]. In a photolytic reaction, it was possible to isolate the Tc(I) complex $[Tc(AA)(CN-R)_4]^+$ (**14**) in low yield after several days of irradiation; most of the starting material could be recovered. These compounds could be prepared in slightly better yields directly from **3** in the presence of the ligands and $[S_2O_4]^{2-}$ as the reductant. The good σ-donating properties of the aromatic diamine ligands induced increased π-backbonding to the isocyanide ligands, as is evident from i.r. spectroscopic investigations: all of the $\nu_{C\equiv N}$ were found at lower wave numbers than in the starting material. X-ray analysis of **14** confirmed these spectral characteristics, in that the Tc–C bonds trans to the aromatic ligands are significantly shorter (av. 1.95(2)Å) than those trans to the remaining isocyanide ligands (2.01(2)Å). Similarly the C≡N bonds trans to the bipy ligands are longer than the others. ^{99}Tc-NMR studies on these compounds showed resonances, the chemical shifts of which are in the range where Tc(III) compounds are usually found, proving that backbonding must be very strong and that the Tc(I) center thus mimics a Tc(III) core. The preparation of aminocarbynes has also been attempted by the classical protonation of coordinated isocyanides [18], but the results were ambiguous. Furthermore, it was not possible to substitute more than two isocyanide ligands even under more vigorous conditions. It seems that, due to this backbonding, the four remaining isocyanide ligands have high thermodynamic stability.

An attempt to prepare Tc(I) complexes with isocyanide ligands containing different organic groups has been described [19]. Applying the original reaction conditions in the presence of a mixture of isocyanides produced the complexes with statistically distributed isocyanides in the coordination sphere in good yield. On the other hand, it was not possible to exchange isocyanide ligands with additional isocyanide in solution, demonstrating clearly the high kinetic stability of **4a**.

Complexes with mixed phosphine/isocyanide coordination have been described. Starting with **3**, the original reaction conditions were applied; on varying the phosphine/isocyanide ratio, the compounds $[Tc(PPh_3)(CN-R)_5]^+$ (**15**) and $[Tc(PPh_3)_2(CN-R)_4]^+$ (**16**) were isolated. [20] It was concluded on the basis of spectral investigations that the phosphines in the latter compound are trans to each other, although the cis configuration or fluxional behaviour could not be ruled out. In contrast to the bipyridyl complexes described above, the Tc-center exhibits a nuclear resonance in the typical Tc(I) region, and the internal oxidation observed in the case of, diamines does not occur with phosphines.

In a recent paper interesting results were presented about the hydrolysis of ester groups present as functional groups on the isocyanide ligands. The starting material was the complex $[Tc(CNC(CH_3)_2COOCH_3)_6]^+$ (17) which is of potential interest for nuclear-medical applications. It was possible to hydrolyse the ester group stepwise and to characterize all of the various intermediates $[Tc(CNC(CH_3)_2COOH)_n(CNC(CH_3)_2COOCH_3)_{6-n}]^{(1-n)}$, where $n = 1$ to 6, by means of HPLC and mass spectrometry [21].

No isocyanide complexes have been described for lower oxidation states, probably because the σ-donating properties of this kind of ligand are too strong or the π-accepting properties too weak.

2.1.1 Potential of $[Tc(CN-R)_6]^+$ as a Synthon

Most of the substitution reactions with the homoleptic Tc(I) isocyanide complexes presented in the preceding section had to be performed at elevated temperatures and were often characterized by low yield. The reason for this behaviour is the exceptionally high kinetic and thermodynamic stability of this class of compounds. From this point of view, 4a are not very convenient or flexible starting materials, although they are prepared directly from 3a in quantitative yield. The exceptionally high kinetic and thermodynamic stability is mirrored by the fact that it was not possible to substitute more than two isocyanides under any conditions. On the other hand, oxidation to seven-coordinated Tc(III) complexes occurs very readily. Technetium compounds of this type, which are not expected to be very inert, could open up a wide variety of new compounds, but this particular field has not been investigated very thoroughly. A more convenient pathway to mixed isocyanide complexes that starts with carbonyl complexes of technetium will be described in Sects. 2.3 and 3.2.

2.2 Synthesis and Chemistry of $[TcCl(CO)_3(PR_3)_2]$ and $[TcCl(CO)_2(PR_3)_3]$

Apart from 4a and 4b, no homoleptic organometallic compounds have been described which can be prepared directly from 3, without applying high pressure conditions (see Sect. 3.1). In a recent paper, the high yield preparation of $TcCl(CO)_3(PPh_3)_2$ (18) starting from $(NBu_4)[TcO_4]$ (3b) $(NBu_4)[TcOCl_4]$ (19) has been described. Compound 18 has been investigated in connection with substitution reactions with tridentate facial coordinating ligands [22]. Different Tc(I) and Tc(III) complexes with mixed CO and phosphine ligands had already been described in the earlier literature, but their synthesis started from $[TcCl_4(PR_3)_2]$ (20) or $[TcCl_3(PR_3)_3]$ (21), which itself had to be prepared from 3 [23, 24, 25]. The yields relative to pertechnetate are thus low. An overview of the synthesis of mixed carbonyl phosphine complexes under low pressure conditions is given in Scheme 3.

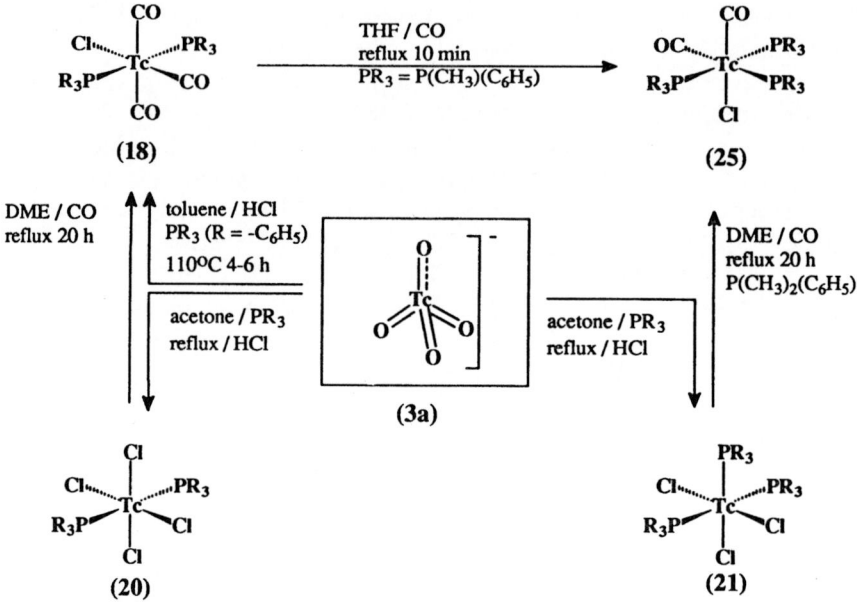

Scheme 3. Pathway to the synthons **18** and **25**

Starting from **3b** or **19** complex **18** has been prepared in a one pot synthesis in almost quantitative yield. Some characteristic intermediates had been expected during the $6e^-$ or $4e^-$ reduction steps, particularly the known compounds $[TcCl_3(NCCH_3)(PPh_3)_2]$ (**22**) or $[TcCl_3(CO)(PPh_3)_2]$ (**23**) [26, 27]. Under the reaction conditions applied the latter two compounds are reasonable intermediates, but neither of them has been detected. They are present if at all, in very low steady-state conditions. The structure of **18** was elucidated and showed the expected *trans* configuration of the two phosphines. In addition the Tc–C bond lengths of the two trans-standing CO ligands are significantly longer than the Tc–C bond *trans* to the Cl ligand (av. 1.982(2)Å vs. 1.887(2)Å), indicating the competing π-backbonding of the two *trans*-situated CO ligands. The synthesis of the corresponding Re(I) complex could be performed with similar yields, but longer reaction times had to be applied. This is in agreement with the general observation that elements of the third transition row react more slowly than those of the second row. The structure of the Re complex has been elucidated and it proved to be isomorphous with its Tc congener. Another important difference was described. While the Tc solution turned black instantaneously after the addition of phosphine, the $(NBu_4)[ReOCl_4]$ (**19b**) solution turned deep green and the complex $(NBu_4)[ReOCl_4(PPh_3)]$ (**24**) was isolated from it and its structure elucidated. As expected, it was shown to be the phosphine ligand *cis*- and a chloride ligand *trans* to the terminal *oxo* group. [28]

18 or $[TcCl(CO)_2(PR_3)_3]$ (**25**) have proved to be convenient starting materials for the studies of the substitution behaviour of Tc(I) compounds. Most of the studies have been performed with N-ligands or ligands of the "Kläui" type. Scheme 4 presents a concise overview.

Reactions with pseudoallyl ligands of the formamidinato or acetamidinato type cleanly replaced the halide and a CO ligand in **18** or a halide and a phosphine ligand in **25**, resulting in neutral Tc(I) complexes of the general composition [Tc(N^N)(CO)$_2$(PR$_3$)$_2$] (**26a**) [29]. Two of these compounds have been characterized by X-ray structure analysis, which showed the two phosphines to be *trans* to each other. The i.r. spectra exhibit $v_{C\equiv O}$ stretching at relatively low frequencies (1920 and 1845 cm^{-1}), indicating good π back-donation to the CO ligands. The Tc–C bonds are correspondingly short (av. 1.875(3)Å). A report has very recently appeared on a compound of the composition [Tc(N^N)(CO)$_2$(PR$_3$)$_2$], where N^N stands for a benzodiazepine derivative. This compound has been synthesized for nuclear-medical purposes and its structure elucidated [30]. A further reaction has been performed with a Schiff-base ligand; the complex [Tc(N^O)(CO)$_2$(PPh$_3$)$_2$] (**26b**) was isolated and structurally characterized [31]. In these substitution reactions with bidentate ligands, the weak σ- or π-donating halide is always substituted first. Subsequently a ligand *trans* to a labilizing CO is replaced, thus resulting in compounds with the same coordination sphere: "TcN$_2$C$_2$P$_2$" or "Tc(N^O)C$_2$P$_2$" respectively.

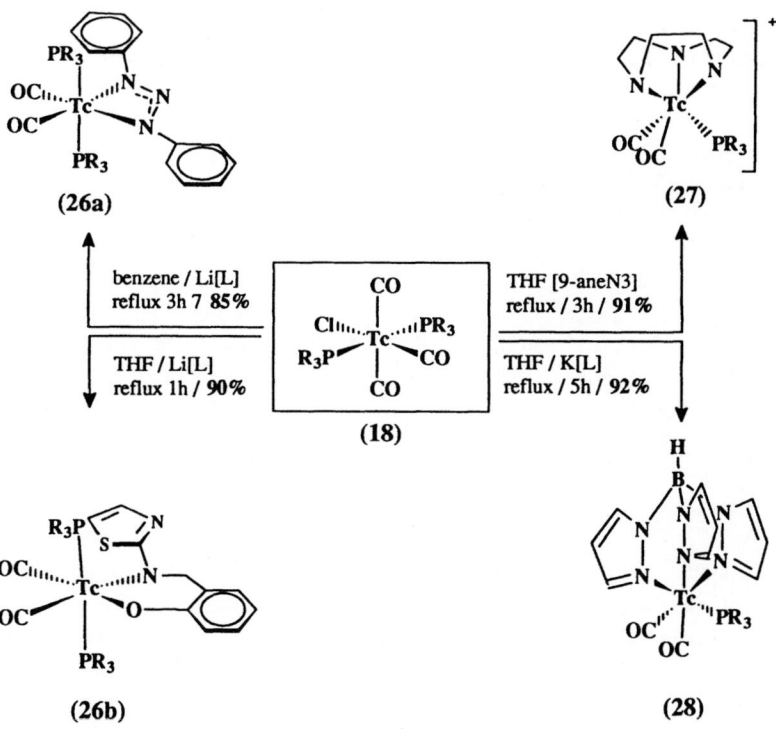

Scheme 4. Important reactions with [TcCl(CO)$_3$(P(C$_6$H$_5$)$_3$)$_2$]

Substitution of **18** with the tridentate facial-coordinating N-ligands [9-aneN3] ((9-aneN3=1, 4, 7 triazacyclononane) and [HB(pyz)$_3$]$^-$ (pyz=pyrazolyl) respectively formed the compounds [Tc(9-aneN3)(CO$_2$(PPh$_3$)]$^+$ (**27**) and [Tc(HB(pyz)$_3$) (CO)$_2$(PPh$_3$)] (**28**) in good yields in THF under moderate conditions. Considering the halide to be the most weakly bound and most strongly labilized ligand, the only compound formed retained two CO and one phosphine. The structures of both compounds have been elucidated. In both cases it was found that the Tc–N bond lengths *trans* to CO are significantly longer than those *trans* to the phosphine ligand (i.e. av. 2.244(4) Å vs. 2.192(3) Å in **27**). ^{99}Tc-NMR exhibits resonances similar to those found in complexes [Tc(NN)(CN–R)$_4$]$^+$ (see section 2.1) (−934 ppm for **27** and −1198 ppm for **28** vs. [TcO$_4$]$^-$=0 ppm), outside of the range characteristic for Tc(I) complexes [32]. Internal oxidation is assumed to be responsible for this shift.

Compounds of the composition [Tc(9-aneN3)(CO)$_3$]$^+$ (**29**) and [Tc(HB(pyz)$_3$) (CO)$_3$] (**30**) have been described as being directly available from the reaction of **19** in toluene under CO, using exactly one equivalent of phosphine as a reductant. In the case of [HB(pyz)$_3$]$^-$, the long-known compound [TcOCl$_2$(NNN)] may be the first intermediate that is subsequently reduced by the phosphine [33].

2.2.1 Potential of [TcCl(CO)$_3$(PR$_3$)$_2$] and [TcCl(CO)$_2$(PR$_3$)$_3$] as Synthons

Both compounds can be synthesized in good yield in a one-step procedure starting from **3b** or **19**. Therefore, when considering availability, they are excellent starting materials for subsequent chemistry. However, kinetic stabilization in the same order of magnitude as in case of the **4a** has to be considered. All of the reactions described in the literature and presented in the previous section have to be performed at elevated temperature, and a maximum of three ligands could be substituted. Probably due to the good σ-donating capacity of the investigated N or mixed N,O ligands, the Tc–C and Tc–P bonds are thermodynamically very stable. In any case, compounds **18** and **25** are more versatile educts than the homoleptic isocyanides, because at least two or three CO ligands can be easily removed at equilibrium. It would be interesting to investigate the oxidation chemistry or these compounds, but – to our knowledge – no investigations in this direction have been published.

2.3 Synthesis and Chemistry of [NEt$_4$]$_2$[TcCl$_3$(CO)$_3$]

Probably the most recent promising Tc(I) starting material is the mixed halide/ carbonyl compound (NEt$_4$)$_2$[TcCl$_3$(CO)$_3$] (**31a**). An analogous Re(I) complex had been described in the earlier literature as a byproduct of the formation of dinuclear Re-carbonyls [34]. The authors started with [ReBr(CO)$_5$] in refluxing diglyme and in the presence of [NEt$_4$]Br, and obtained small amounts of (NEt$_4$)$_2$[ReBr$_3$(CO)$_3$] (**31b**). The strong *trans* effect of the carbonyls seems to be responsible for the release of 2 molecules of CO, and the only complex to be formed is **31b**. This complex could be isolated only thanks to its insolubility, even in refluxing diglyme. Surprisingly, this interesting starting material has never been investigated very extensively in relation to its substitution chemistry. The structure of this compound could be elucidated [35]. Although the bromides are only very weak σ- or π-donating ligands, the CO-stretching frequencies ($v_{C\equiv O}$) are low (1870 and 2001 cm^{-1}) and the pattern of the i.r. spectra is typical for a complex anion of the C_{3v} point group. It indicates strong backbonding and therefore very stably bound CO ligands. Two of the Re–C bond lengths are indeed found in the shorter range (av. 1.90(1)Å) and one Re–C bond in a longer distance (2.01(1)Å). In contrast to other Re(I) compounds that exhibit C_{3v} symmetry but have only monodentate ligands, (i.e. [Re(NCCH$_3$)$_3$(CO)$_3$]$^+$ [36]) two bonds are short and one is long, whereas in the case of all known cationic Re(I) compounds two are long and one is short.

To transfer the original synthesis to technetium was not very convenient, as it would have to start from TcBr(CO)$_5$ (**32a**). A preparation has recently been reported which starts from **3a** or **19a** in refluxing THF and used BH$_3$ as the reducing agent [36]. The complex **31a** could be isolated in 60–70% yield, based on Tc, and has been characterized by i.r. specroscopic methods as well as Tc and Cl analysis, and compared to its rhenium congener (Scheme 5).

The compounds **31a** and **31b** are described as being quite soluble in polar coordinating solvents. I.r. spectroscopic investigations in acetonitrile and in water indicate that the halide ligands are completely exchanged against solvent molecules. In particular, aqueous solutions of **31a** and **31b** exhibit a C_{3v} pattern which is shifted upwards by about 50 cm^{-1} compared to the KBr spectra. Precipitation of the halides with AgPF$_6$ left the solution spectra completely unchanged, so it was concluded that the halides had already been replaced and an "aqua-ion" of the composition [Tc(OH$_2$)$_3$(CO)$_3$]$^+$ (**33a**) formed [37]. **33a** and **33b** are completely water stable even under aerobic conditions. The same behaviour was

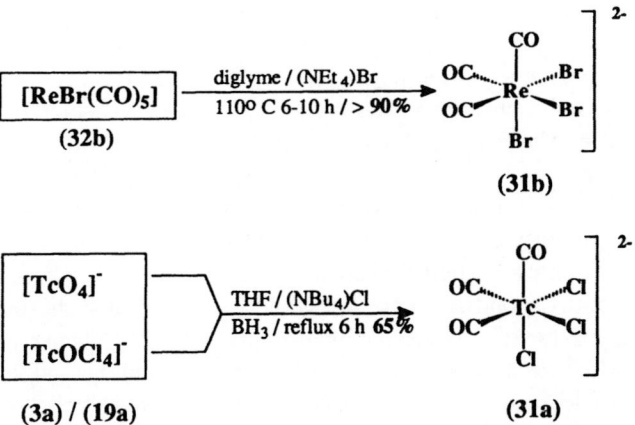

Scheme 5. Pathway to the synthons **31a** and **31b**

found in acetonitrile and thus the complex is $[Tc(NCCH_3)_3(CO)_3]^+$ (**34**). This complex had already been prepared starting from **2** [38, 39] and has proved to be a very useful starting material (see section 3.2) and a convenient source for the "fac-$Tc(CO)_3$" fragment.

This behaviour of **31a** or **31b** at room temperature with coordinating solvent molecules is an impressive illustration of the labilization induced by the CO ligands. In addition, the stability of the pure σ- or π-donating halides seems to be very weak. Although the d^6 metal center undergoes a certain amount of electron depletion by the CO ligands, its electron deficiency is not great enough to affect the electrons of the halides very strongly. This is different from the case of **32b** where five CO ligands backbond electrons from the metal center. In this case the halide is tightly bound and can only be removed by precipitation with Ag- or Tl-salts in the corresponding coordinating solvent [40].

In the case of the rhenium aqua-ion $[Re(OH_2)_3(CO)_3]^+$ (**33b**) the question has been posed whether complex-anion can be considered to be a Brønsted acid. Titrations with hydroxide in water yielded a pK_a value of 7.55 which is exceptionally low for a "+1" cation. After the deprotonation of one coordinated water molecule, polymer formation over (μ-OH) bridges was initiated and the two compounds $[Re_3(\mu_3\text{-}OH)(\mu\text{-}OH)_3(CO)_9]^-$ (**35**) and $[Re_2(\mu\text{-}OH)_3(CO)_6]^-$ were (**36**) isolated and structurally characterized (Scheme 6).

In addition, the long-known cubane-type Re(I)-cluster $[Re(\mu_3\text{-}OH)_3(CO)_{12}]$ (**37**) was also isolated as an intermediate and its structure elucidated [41]. The thermodynamic formation constants have been calculated from titration curves at an ionic strength of 0.1 M KNO_3. Whether technetium behaves similarly has not yet been investigated; however similar behaviour can be expected, since the two homologs behave rather similarly in the low oxidation states.

The complexes **31a** and **31b** proved to be very useful starting materials and can be regarded as the most convenient source of the "fac-$M(CO)_3$" fragment that is known up to now. Some examples for M=Re and Tc are given in Scheme 7.

Compared to **32** (see Sect. 3.2), these compounds have the advantage that they react quantitatively under ambient conditions, in water as well as in polar organic solvents. The behaviour towards the ligands depicted in Scheme 7 proved to be identical for rhenium and technetium. The compounds listed in Scheme 7 have been fully characterized. The heterogenous reaction in THF in the presence of isocyanides yielded quantitatively the neutral complex $[TcCl(CN\text{–}R)_2(CO)_3]$ (**38**), which had already been described, but with **32** as its precursor [42]. I.r. spectra exhibit $v_{C\equiv N}$ absorptions at 2208 and 2190 cm^{-1} and $v_{C\equiv O}$ absorptions at 2049, 1977 and 1909 cm^{-1} with a pattern typical for an asymmetrically substituted "$M(CO)_3$" unit. The Tc–C bond $trans$ to the halide is significantly shorter (1.914(7)Å) than those $trans$ to the isocyanide ligands (av. 1.944(6)Å) due to competing backbonding. Homogenous reaction in water yielded rapidly and in good yield the $tris$-isocyanide compound $[Tc(CN\text{–}R)_3(CO)_3]^+$ (**39**) which has been characterized by $i.r.$ and elemental analysis. In comparison to the i.r. spectrum of **38**, that of **39** exhibits two distinct patterns for symmetric "fac-ML_3"

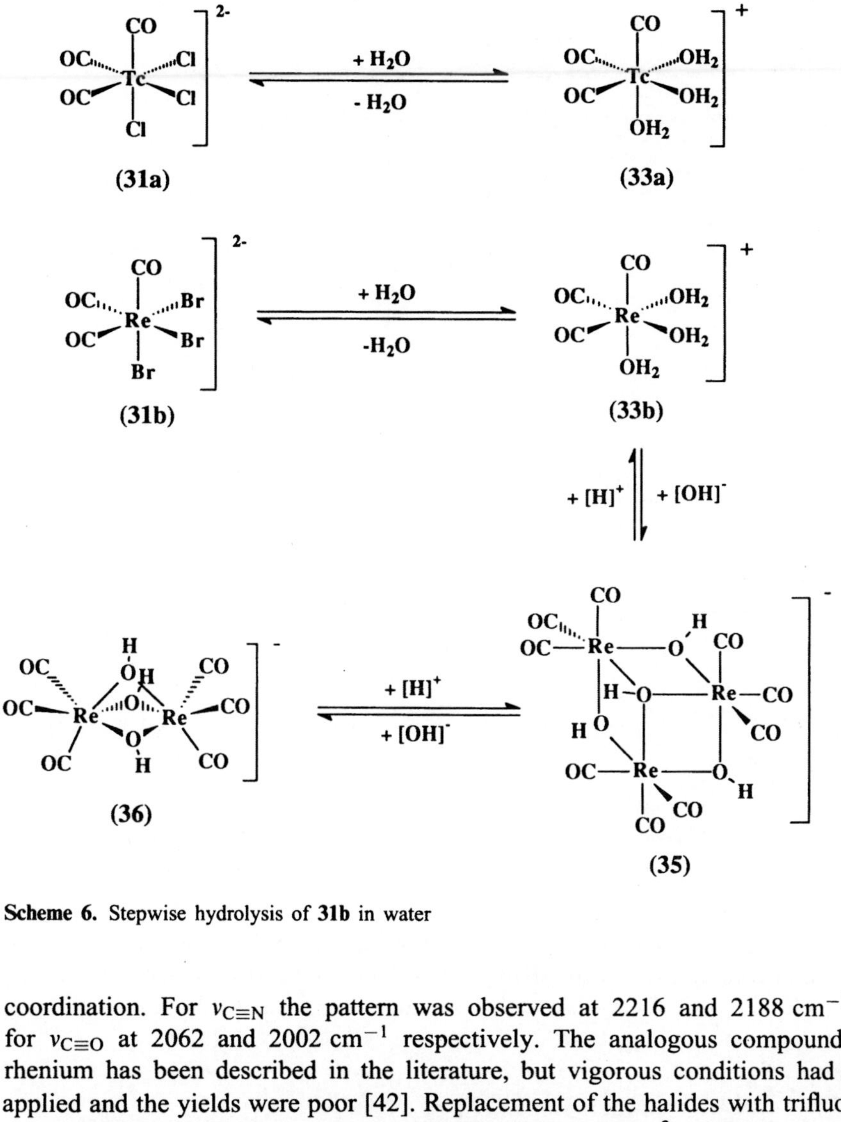

Scheme 6. Stepwise hydrolysis of **31b** in water

coordination. For $v_{C\equiv N}$ the pattern was observed at 2216 and 2188 cm^{-1} and for $v_{C\equiv O}$ at 2062 and 2002 cm^{-1} respectively. The analogous compound with rhenium has been described in the literature, but vigorous conditions had to be applied and the yields were poor [42]. Replacement of the halides with trifluoroacetate yielded the dianionic complex $[M(OOCCF_3)_3(CO)_3]^{2-}$ (**40**), the structure of which had been elucidated for the rhenium complex. Surprisingly, the trifluoroacetate ligands only bind in a monodentate fashion, and the C=O bondlengths for coordinated and not coordinated O-atoms are very similar (av. 1.23 (7)Å vs av. 1.25(6)Å). The Re–O bond distances exhibit the same behaviour as in **31b** in that two bonds are short (av. 2.118(10)Å) and one is long (2.161(10)Å). Access to (η^5-C$_5$H$_5$) chemistry was achieved by exchanging the halides against weakly coordinating anions such as [PF$_6$]$^-$ in THF and subsequent reaction with Tl[C$_5$H$_5$]; thus [(η^5-C$_5$H$_5$)Tc(CO)$_3$] (**41**) was described as having been formed in good yield [36]. A number of other compounds have been prepared directly

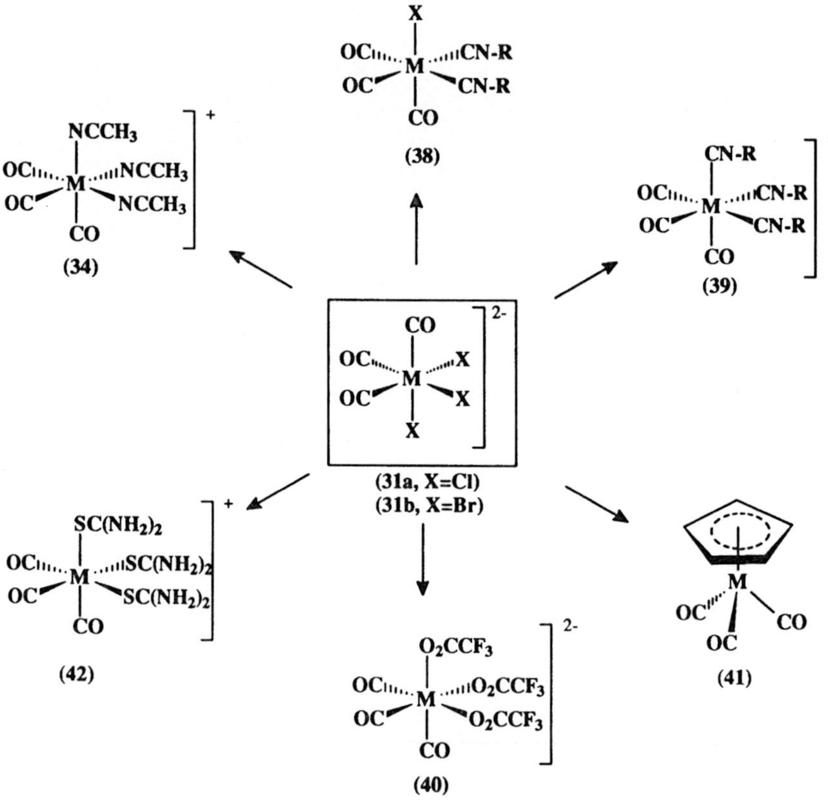

Roger Alberto

X
OC,,,.. M ....ıICN-R
OC CN-R
CO
(38)

NCCH₃ ⁺
OC,,,.. M ....ıINCCH₃
OC NCCH₃
CO
(34)

CN-R ⁺
OC,,,.. M ....ıICN-R
OC CN-R
CO
(39)

CO 2-
OC,,,.. M ....ıIX
OC X
X
(31a, X=Cl)
(31b, X=Br)

SC(NH₂)₂ ⁺
OC,,,.. M ....ıISC(NH₂)₂
OC SC(NH₂)₂
CO
(42)

O₂CCF₃ 2-
OC,,,.. M ....ıIO₂CCF₃
OC O₂CCF₃
CO
(40)

M
OC CO
OC
(41)

Scheme 7. Important reactions with (NEt₄)₂ [MX₃(CO)₃] [M = Re, X = Br and M = Tc, X = Cl]

from aqueous solution, namely [Re(tu)₃(CO)₃]⁺ (tu= thiourea, SC(NH₂)₂) **(42)** and **29**. The Re–C bond lengths in **42** showed the same characteristics as in other cationic complexes in that two Tc–C bonds are short (1.905(6)Å) and one is long (1.922(7)Å). This observation has also been made for the Tc–S bond lengths (two are short av. 2.525(2)Å and one is long 2.536(2)Å).

2.3.1 Potential of (NEt₄)₂[TcCl₃(CO)₃] as a Synthon

In comparison to the two other described educt compounds **4a** and **18**, the anions of **31a** and **31b** can also be prepared in a one step synthesis in satisfying yields, without the application of high pressure conditions starting from **3a** or **19a**. This compound should open intensive exploration of Tc(I) and Re(I) chemistry. In contrast to the other two educts, no competing ligands are present. The halides bind only very weakly and the carbonyls are easily withdrawn from equilibrium by volatility. As shown in the previous section, a number of examples illustrate the versatility of this educt. It can be applied not only for substitutions in organic solvents but also in water, and therefore allows reactions with ligands that are completely insoluble in any organic solvent. The field of organometallic chemistry thus can be merged to that of typical aqueous chemistry. In contrast to **4a**, reactions can be performed in a very short time period at room temperature, and the yields are – at least for the cases described – almost quantitative.

3 [TcO$_4$]$^-$ as a Starting Material for the Formation of the Tc–C Bond under High-Pressure Conditions

3.1 Synthesis and Chemistry of Tc$_2$(CO)$_{10}$

The availability of homoleptic carbonyls has been the decisive starting point, and some interesting reviews have appeared dealing with this kind of compound [44, 45]. During this development, the binary technetium carbonyl **2** was synthesized for the first time in moderate yield and its properties described [46]. In the last five years, several new syntheses have been described for this key-compound and the yields improved. Starting from solid **3a** under very high pressure and temperature conditions gave a yield better than 95% [47]. Pressurizing additional H$_2$ as a reducing agent in toluene produced **2** quantitatively [48]. The yields in both protocols are slightly better than described in a very recent synthesis starting from Na[TcO$_4$] in CH$_3$OH, but the pressure and temperature are much lower in the latter case [49]. However, in all of the cases described, an autoclave is still necessary and the syntheses have to be performed under strict safety precautions. In general, the conditions for the preparation of **2** follows those of the rhenium analog. This was also the case for the procedure described in [49], which has been adapted from a new rhenium protocol for the synthesis of [Re$_2$(CO)$_{10}$] (**2b**) [50]. As will be obvious from the following sections, **2** is not the important starting material and only a very few reactions have been described where oxidation state 0 is retained. Most of the syntheses start with the Tc(I) complexes **32** which can be prepared simply by oxidative substitution of one CO by X$_2$. Several standard protocols have been developed for this synthesis. A very convenient one consists of Br$_2$ oxidation in CS$_2$, providing a 60% yield of the key-compound **32** [51]. The corresponding iodide complex has been prepared in 90% yield by reacting **2** with I$_2$ in THF at 100 °C for 40 h under 70 atm of CO [52].

Since the key-compound **2** is not the actual starting material but is converted in most cases to Tc(I) precursors, it might be useful to develop approaches ending directly with reactive Tc(I) compounds. Two such approaches have been described. The first one, from the early literature, consists of a high pressure reaction directly on (NH$_4$)$_2$[TcBr$_6$] (**43**) in the presence of copper. This protocol produces directly **32** in satisfactory yields [53]. Carbonylation under 50–60 atm of CO for two hours gave the cubane-like cluster [NaTc$_3$(μ_3-OCH$_3$)(μ-OCH$_3$)$_3$(CO)$_9$] (**44**) in good yield. This compound also proved to be a powerful source of the "*fac*-Tc(CO)$_3$" fragment, as will be described in section 3.4 [54]. **44** has been carbonylated furthermore to **2** indicating that the cluster is an intermediate which is formed in a short period of time. From these experiments it was also concluded that the rate determining step is the reduction down to the "0" oxidation state. These reactions are summarized in Scheme 8.

The substitution of CO in **2** by PF$_3$ was recently the subject of an intensive investigation [47], that was followed by a similar study of the Mn-homolog [55]. The authors investigated the substitutions by means of GC-MS, GC-IR and ^{19}F-NMR. Assuming the composition [Tc$_2$(CO)$_x$(PF3)$_{(10-x)}$], there are 77 possible stereoisomers of which 24 could be assigned unambiguously, especially for x= 1, 2, 3. As was expected from the higher π-density in the axial positions, substitution at these two positions was found to be preferred.

Following the classical protocol for the synthesis of (η^5-dienyl) M(I) compounds [56], oxidative addition of H(C$_5$R$_5$) on **2** gave the compounds [(η^5-C$_5$Me$_5$)Tc(CO)$_3$ **(45)**, [(η^5-C$_5$Me$_4$Et)Tc(CO)$_3$] **(46)** and [(η^5-Ind)Tc(CO)$_3$] (ind = indenyle) **(47)** in good yields [57]. The structures of all three compounds as well as those of their homologs have been elucidated and compared. It turned out from this detailed study that the binding parameters are equal, at least within 3σ. Furthermore, the covalent radii of rhenium and technetium are almost the same. In agreement with theoretical considerations [58] the binding mode in **41** can be described as dienyl-like, whereas derivatives of H(C$_5$H$_5$) such as (η^5-C$_5$Me$_5$) fit an allyl-en coordination mode better. I.r. spectra exhibit the typical pattern of neutral compounds of the type [LM(CO)$_3$]. The C $\equiv$ O stretchings occur around 2010+/ − 10 for the A$_1$ mode and at 1920+/−20 cm^{-1} for E. For all the technetium and rhenium compounds described herein, the wave numbers for Tc are shifted upward by 4–6 cm^{-1}. **41** was the subject of an earlier study and has not been discussed in this work. The synthesis of these key-compounds is summarized in Scheme 9.

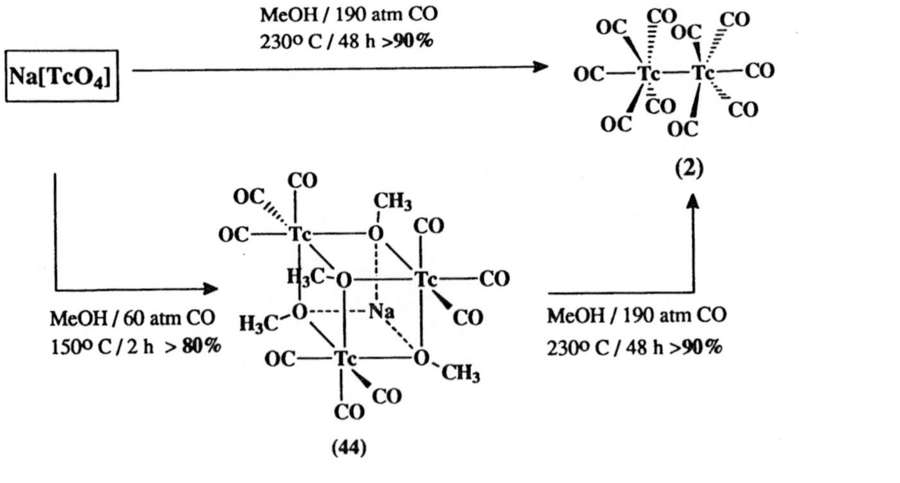

Scheme 8. Pathway to the organotechnetium synthons **2** and **44** under high CO pressure conditions

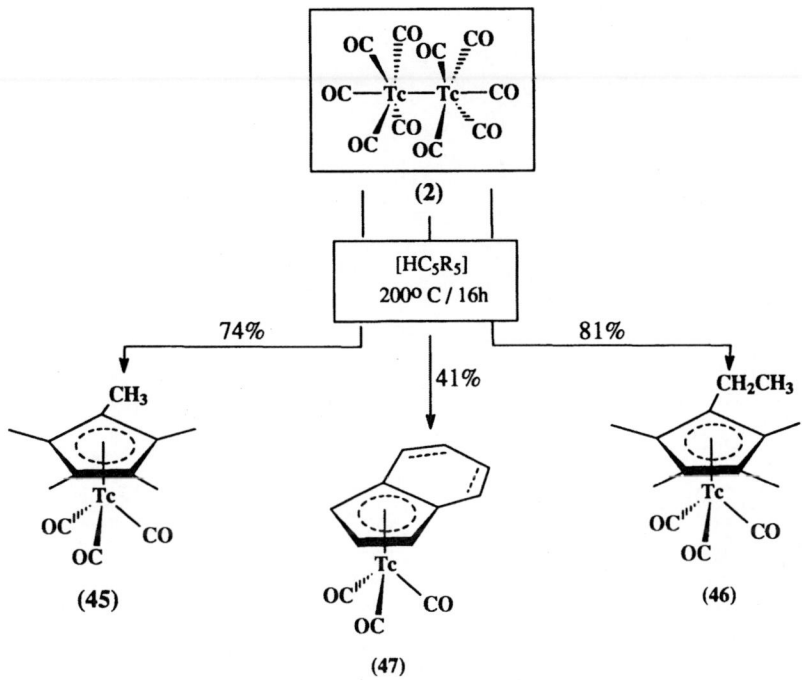

Scheme 9. Oxidative addition of cyclopentadienyl-derivatives to [Tc$_2$(CO)$_{10}$]

45 was subsequently investigated with respect to its photolytic reactions in non-coordinating solvents such as cyclohexane. Similar investigations with its homologs had been performed earlier, in which it was shown that the dinuclear species [(η^5-C$_5$Me$_5$)M(μ-CO)$_3$M(η^5-C$_5$Me$_5$)] (**48a–c**) and [(η^5-C$_5$Me$_5$)M(CO)$_2$(μ-CO)M(CO)$_2$(η^5-C$_5$Me$_5$)] (**49a–c**) were formed (M = Mn, Re). The former contains the first Mn-Mn triple bond described in the literature [59, 60, 61]. Technetium behaves identically; **48b** and **49b** are formed after a 3 h period of irradiation in cyclohexane at room temperature. Chromatographic isolation gave yields of 3% and 28% respectively. The yields for rhenium are significantly higher. **48b** is stable as a solid and in solution, whereas **49b** decomposes over a 12 h period. The structure of **48b** has been elucidated. The Tc–Tc triple bond was found to be 2.413(3)Å, comparable to that of its rhenium analog within 3σ but significantly longer than the Mn–Mn triple bond (2.171Å) [62]. The i.r. spectrum of **48b** exhibits 4 $\nu_{C\equiv O}$ absorptions in the typical region for bridging CO, whereas in solution only one sharp band was found at 1785 cm^{-1}. **49b** exhibits five absorption bands for terminal CO in KBr and four bands in cyclohexane. Furthermore, one band is found in the solid and in solution for the bridging CO, at 1738 and 1756 cm^{-1} respectively. In the case of organotechnetium compounds it has sometimes been found that ^{13}C-NMR does not exhibit all of the expected resonances [57, 63]. For **48b** only the -CH$_3$ signal is found whereas **49b** shows all of the ring carbon- but no CO-resonances. As the corresponding Re-complexes

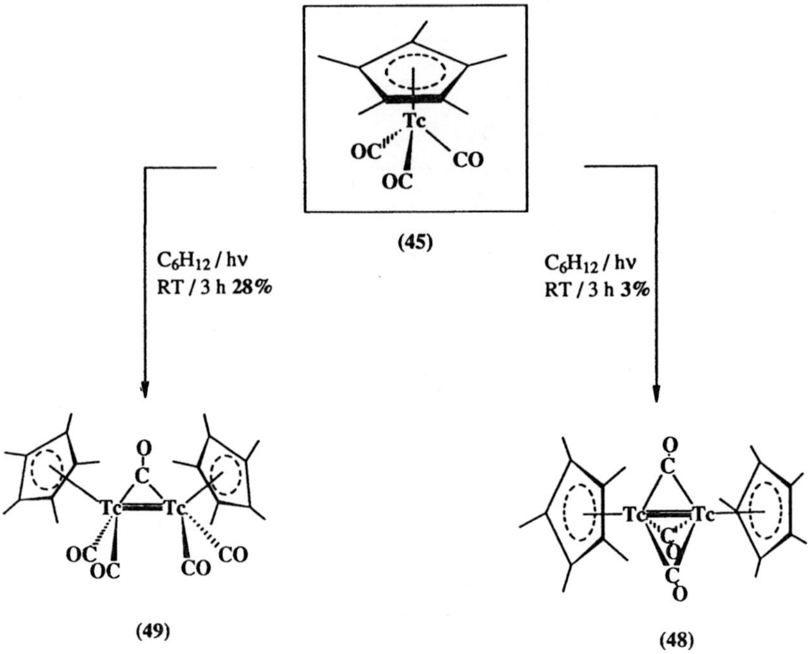

Scheme 10. Photolytic dimerization of **45** in non-coordinating solvents

49c exhibit all of the signals, the explanation of this behaviour must be found in very long relaxation times for the carbon atoms in question, as well as in strong coupling to the quadrupole nucleus Tc-99. The reactions described are shown in Scheme 10.

3.2 Synthesis and Chemistry of [TcX(CO)$_5$]

The synthesis of **32** was described in the previous section. **32** is the most convenient starting material in organotechnetium chemistry. Practically all of the described investigations in this section follow those of its homologs; it is often useful to compare with the respective protocols.

3.2.1 Reactions with Tridentate Facial Coordinating Ligands

An alternative pathway to Tc(I) complexes of the general composition [(η^5-C$_5$R$_5$ Tc(CO)$_3$] has been described. This method follows the salt metathesis procedure of Fischer, in that [TcI(CO)$_5$] is allowed to react with Li(C$_5$R$_5$) or its derivatives, to produce the complexes [(η^5-C$_5$R$_5$)Tc(CO)$_3$] and LiI in good yield, in short reaction time, and at room temperature [52]. Beside the common H(C$_5$R$_5$) ligands, the more complicated system (HC$_5$Me$_4$(CH$_2$)$_3$NMe$_2$)

has also been utilized to prepare the respective tris-carbonyl complex of Tc(I) $[(\eta^5\text{-}C_5Me_4(CH_2)_3NMe_2)Tc(CO)_3]$ (**50**). Subsequently, the tertiary amine group in **50** could be quaternized with CH_3I and the structure of the resulting salt $[(\eta^5\text{-}C_5Me_4(CH_2)_3NMe_3)Tc(CO)_3]I$ (**50a**) resolved. As expected, the coordination of the $(\eta^5\text{-}C_5R_5)$ ring can be described as "dienyl-like" within 3σ. In addition to the synthesis of such piano-stool-type complexes, their reactions with Br_2 in trifluoroacetic acid as well as CO substitutions with $NO[PF_6]$ were investigated. Analogously to the Re-congener **45b** [64], **45a** undergoes oxidative addition of one molecule of Br_2, to yield the *cis*- and *trans*-isomers of $[(\eta^5\text{-}C_5Me_5)TcBr_2(CO)_2]$ (**51**) in 18% and 55% yield respectively. The $\nu_{C\equiv O}$ absorptions are found at high frequencies (2040 and 1972 cm^{-1} for the *cis*- and at 2060 and 1985 cm^{-1} for the *trans* isomer) indicating that π-backbonding is not relevant for these Tc(III) compounds. In the presence of $NO[PF_6]$, one molecule of CO was substituted instantaneously and the compound $[(\eta^5\text{-}C_5Me_5)Tc(CO)_2(NO)]^+$ (**52**) formed in good yield. The $\nu_{N\equiv O}$ absorption occurs at 1745 cm^{-1}, as is consistent with an NO^+ ligand rather than NO^- [52].

Furthermore, the reaction of **32a** and **32b** has been performed with ligands of the Trofimenko type $[HB(pyz)_3]^-$ and $[HB(pyzMe_2)_3]^-$ [65]. Refluxing **32a**

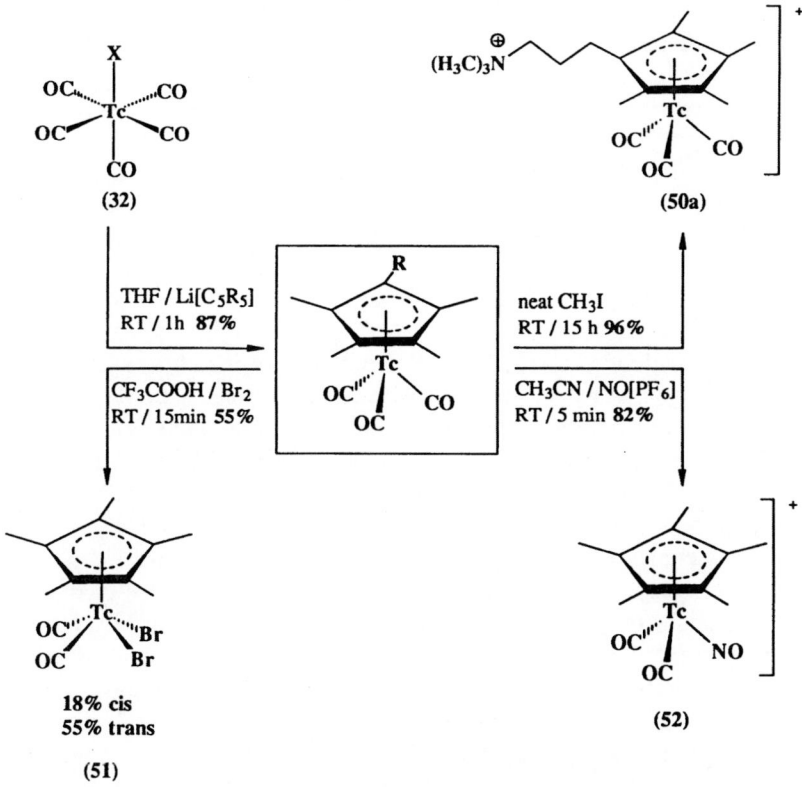

Scheme 11. Important reactions with $[(C_5Me_4R)Tc(CO)_3]$

or **32b** in THF in the presence of the ligand and subsequent extraction with pentane over 14 days gave the neutral complexes **30a**, **30b**, [Tc(HB(pyzMe$_2$)$_3$(CO)$_3$] (**53** and its rhenium analogue. In addition the same compounds with manganese have been prepared. The structures of all six compounds have been elucidated and proven to be isostructural for all three elements bonded to same type of ligands. In addition, all homologs of the ligand [HB(pyzMe$_2$)$_3$]$^-$, crystallize homomorphously in the monoclinic space group P2$_1$/c. I.r. spectroscopy of these compounds revealed the typical "*fac*-M(CO)$_3$" pattern of C$_{3v}$ symmetry. The A$_1$ stretching frequency 2020+/−5 and E is at 1900+/−10 cm^{-1}. Although the CO stretch is, in most cases, not strongly influenced by the metal center, a trend can be discerned in the described series, in that a gap of about 3–4 cm^{-1} was found between Mn, and Tc and between Tc and Re.

Photolytic reactions of the compound (**53**) have been studied[66]. Irradiation for 30 min at room temperature and subsequent evaporation with N$_2$ gave the (μ-N$_2$) bridged dinuclear compound [Tc(HB(pyzMe$_2$)$_3$)(CO)$_2$]$_2$(μ-N$_2$) (**54**) in 15% yield. Its structure was determined and the N–N bond length found to be 1.160(3)Å and Tc–N–N angle to be 174.0(10)°. Only a limited number of (μ-N$_2$) bridged compounds have been described in the literature. The bond length and angle of the N$_2$ bridge fit well into the range of other known compounds of this type. The following interesting difference from Mn and Re was noted. Similar reactions with these elements had already been performed under comparable conditions and only mononuclear compounds [M(HB(pyzMe$_2$)$_3$)(CO)$_2$(N$_2$)] (M = Mn, Re) with "end-on" N$_2$ coordination were found [67, 68]. However, for all three homologs, the initially formed intermediate is the surprisingly stable THF complex [M(HB(pyzMe$_2$)$_3$)(CO)$_2$(THF)] (**55a–c**), which could be isolated and characterized for all homologs. **55a** has also been allowed to react with trifluorodiazoethane, to yield the (μ-N$_2$) bridged Mn-analog of **54** [69].

3.2.2 Reactions with Mono- and Bidentate Ligands

In few other investigations presented in the more recent literature, appropriate mono- or bidentate ligands reacted directly with **32**. Reactions with phosphines, pyridine derivatives and isocyanide ligands in refluxing methanol gave compounds of the general composition [TcBr(CO)$_3$L$_2$] in good yield [70]. While such compounds with L = phosphines are also available via the direct carbonylation pathway described in Sect. 2.2 the latter route is limited, up to now, only to the phosphines. It was concluded from i.r. spectroscopy that the coordination geometry of the three carbonyl ligands is facial. In the case of ^{t}Bu-NC, the elemental analysis of the product revealed that a mixture of di- and tris(isocyanide) complexes must be present. This observation was not investigated in greater detail.

Only a few reactions with bidentate ligands have been described. **32**, reacting with the chiral carbohydrate derivative phenyl-4,6-o-(R)-benzylidene -2,3, 0-bis(diphenyl-phosphino-β-D-glucopyranoside) (Ph-β-glup), gave the compound [TcBr(CO)$_3$(Ph-β-glup)] (**56**) in 53% yield. The complex was characterized by elemental analysis and i.r. spectroscopy [71]. Furthermore, reaction with dithiocarbamate (dtc) produced the neutral compound [Tc(SS)(CO)$_4$] (**57**) [72], which exhibits the characteristic numbers of four intense i.r. absorptions. In the course of this investigation a number of dithio ligands were also reacted with **18** and compounds of the composition [Tc(SS)(CO)$_2$(PPh$_3$)$_2$] were observed. ^{99}Tc-NMR of the latter compounds showed sharp resonances and, in addition, coupling to the phosphorous atoms (J$_{Tc,P} \approx 600$ Hz).

Apart from the direct reactions of ligands on **32**, the compounds [Tc(NCCH$_3$)(CO)$_5$]$^+$ (**58**) and **34a** proved to be very versatile sources for the "Tc(CO)$_3$" fragment. The former has been prepared in good yield (96%) by direct reaction of NO[PF$_6$] on **2** in acetonitrile [39], following the protocols described for Mn and Re [73]. I.r. absorptions were observed at high frequencies 2169(w) $\nu_{C\equiv N}$, 2069(vs) and 2040(s) cm^{-1}, the latter two of which were assigned as $\nu_{C\equiv O}$; the pattern indicates C$_{4v}$ symmetry. It was shown that subsequent reaction with [9-aneN3] type ligands at room temperature yielded the cationic compound **29**. Reaction with the sulfur analog [9-aneS3] was only achieved at elevated temperature. Reaction with bipyridyl-type ligands in refluxing acetonitrile led to the substitution of two CO ligands and formation of [Tc(bipy)(NCCH$_3$)(CO)$_3$]$^+$ (**59**), probably via the initial formation of **34a**. With acetonitrile as solvent, no reaction of **59** occurred with phosphine ligands, whereas in other solvents such as acetone, one CO and the acetonitrile ligand were readily substituted to yield [Tc(bipy)(PR$_3$)$_2$(CO)$_2$]$^+$ (**60**). These reactions are summarized in Scheme 12.

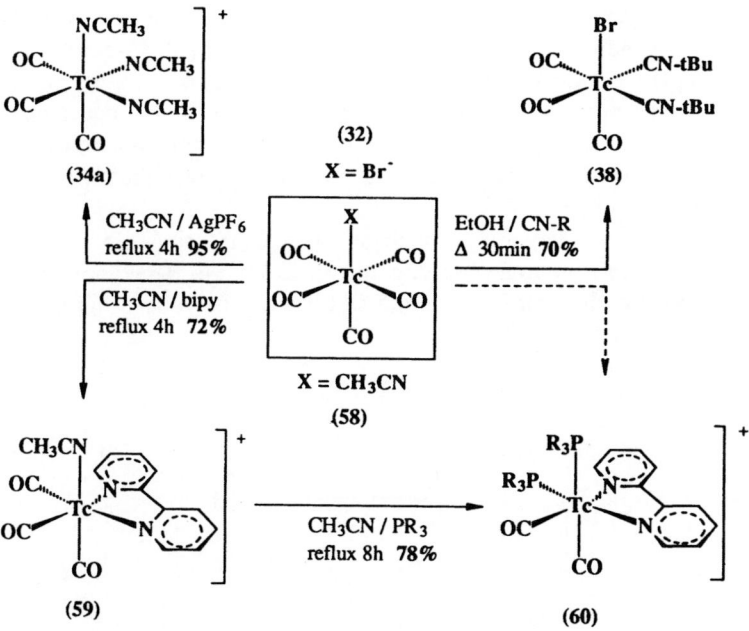

Scheme 12. Important reactions with [TcX(CO)$_5$]

On the basis of i.r. spectroscopic investigations the authors concluded that in compounds of the latter type the carbonyls are *cis* to one another, in agreement with the proposed reaction mechanism. However it could not be shown unambiguously whether the absolute configuration is *cis,cis* or *cis,trans*.

Aside from substitutions of **58** with different ligands, the easy access to reactive **34a** was also an important result of this study, although this compound had already been the subject of an earlier investigation [38], where it had been synthesized by the traditional approach of substituting two COs and one halide of **32** in refluxing acetonitrile. Two of the acetonitrile ligands are readily substituted by two phosphines, confirming the assumption in [39] that the intermediate for the formation of **59** occurs via **34a**.

A series of technetium(I) β-diketonates have been prepared by refluxing **32** in the corresponding neat β-diketone. Due to the instability of the CO ligands *trans* to a σ-donor ligand, one of the COs can be substituted by an additional amine (L), forming compounds of the composition $[Tc(O^\wedge O)(CO)_3L]$ (**61**) [74, 75].

Investigations on the thermal decomposition of **32** revealed the continuous loss of CO with concomitant formation of the di- and tetranuclear Tc(I) clusters: $[TcX(CO)_4]_2$ (**62**) and $[TcX(CO)_3]_4$ (**63**) respectively. The latter is assumed to have cubane structure comparable to $[Tc(OH)(CO)_3]_4$ [76, 77, 78]. The thermal decomposition reactions are depicted in Scheme 13.

3.3 Potential of $Tc_2(CO)_{10}$ and its Primary Derivatives as Synthons

In several investigations, the synthesis of $[Tc_2(CO)_{10}]$ was improved with respect to yield and reaction conditions. The lowest pressure allowing the isolation of $[Tc_2(CO)_{10}]$ in acceptable yield was about 100 atm. However, even under these conditions, the use of an autoclave for high-pressure reactions is still necessary. This offers a clue as to how preparation of larger amounts of this synthon may currently be achieved, thereby providing a basis for further development of the field of organotechnetium chemistry.

$[Tc_2(CO)_{10}]$ and primary derivatives, such as $[TcX(CO)_5$ or $[Tc(NCCH_3)_3(CO)_3]^+$, are thus the most important starting materials for the development of classical organotechnetium chemistry. A number of reactions have been discussed

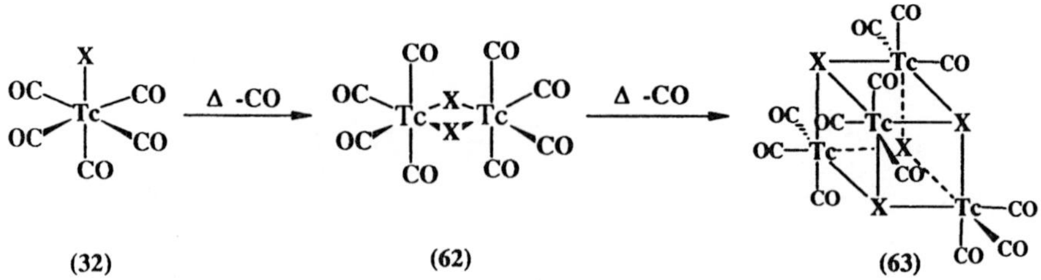

(32)   (62)   (63)

Scheme 13. Stepwise thermolytic release of CO from **32**

in the previous section revealing the versatility of the above mentioned synthons. From a synthetic point of view, practically all approaches followed those of its heavier homolog rhenium. With some minor exceptions the products were the same, although yields were usually lower in case of technetium. Most of the compounds were compared structurally with their rhenium analogs. Many of the compounds were shown to be isostructural within a 3σ limit and they often even crystallized isomorphously. Consequently not much additional structural information could be deduced. The expected close similarity chemical reactivity was confirmed for the low-valency homologs. Rhenium compounds proved to exhibit higher kinetic stability, but – since vigorous reaction conditions often have to be applied – this observation is of minor significance. In conclusion, it should be mentioned that chemistry starting with $[Tc_2(CO)_{10}]$ often follows that of rhenium exactly; consequently, new compounds can hardly been expected with this approach.

3.4 Synthesis and Chemistry of the Cluster $Na[Tc_3(OCH_3)_4(CO)_9]$: A New Source for the "$Tc(CO)_3$" Moiety

Another convenient starting material for Tc(I) organometallics is the recently described cluster compound **44** [79]. Starting from $Na[TcO_4]$ in CH_3OH the compound is synthesized in good yield under reduced CO-pressure (50–60 atm) in a short reaction time (2 h). The structure of this cluster has been elucidated it is best described as an incomplete cubane frame. Three positions are occupied by Re-centers, another three by (μ_2-OCH_3)-, and the last corner by a (μ_3-OCH_3) ligand. Finally, in the solid state, the eighth position is occupied by a sodium ion, to complete the cube. Consequently, the trinuclear cluster acts as a strong anionic crown ether with high affinity for sodium. Even in the presence of crown 2.2.2 it was not possible to release the Na^+ from the cluster coordination. Furthermore, the X-ray structure reveals that the cubane molecules are linked by $Tc–CO–Na^+$ bridges, and thus that infinite chains are formed. The relatively short distance $CO–Na^+$ (2.510 Å) represents one of the rare cases of isocarbonyl coordination, at least in the solid state.

44 seems to be the initial intermediate for the formation of **2**. This assumption was confirmed in that the pure compound **44** could be transformed quantitatively to **2** by increasing pressure and reaction time (see Scheme 8). The reaction also revealed that the +1 oxidation state can be reached under relatively mild conditions and that the subsequent reduction to the "0" valency state is rate determining. This cluster is one of the rare examples where the synthesis of a technetium compound initiated investigations with rhenium rather than the other way round. The carbonylation of **3b** in methanol was systematically: pressure, temperature and reaction time varied over a broad range [80]. It could be shown that even with an initial pressure of 50 atm and at a temperature of 240 °C the analogous rhenium cluster (**44b**) was formed in good yield. Increasing pressure and temperature yielded the dinuclear anionic rhenium compound

$[Re(CO)_3(\mu\text{-}OCH_3)_3Re(CO)_3]^-$ **(64)** whereas decreasing pressure favoured production of rhenium clusters of higher nuclearity with mixed (μ-oxo)- and (μ-methanolato) bridges. The presence of 1.33 equivalent of base per metal atom in the clusters **44** and **44b** facilitates subsequent reactions, that demand base. The reaction of **44** with (HC$_5$Me$_5$), for example, resulted in the quantitative formation of **45** [79]. In the case of the trinuclear rhenium cluster **44b**, reaction with the carbene precursor 1,1'-methylene-bis(3-methyl-1H-imidazolium)-diiodide yielded the corresponding mononuclear Re(I) complex **(65)** with the bidentate carbene ligand (see Scheme 14) [80].

This facile approach to the carbene chemistry of rhenium has not yet been investigated with technetium. Further reactions with the technetium cluster **44a** have been performed in C$_6$H$_6$/HCl to yield the compound $[(C_6H_6)Tc(CO)_3]^+$ **(66)** which previously had only been described for manganese and rhenium [81]. ^{99}Tc-NMR of the latter compound exhibits a resonance at -1983 ppm (relative to [TcO$_4$]$^-$), and it therefore fits very well into the range proposed for Tc(I) complexes.

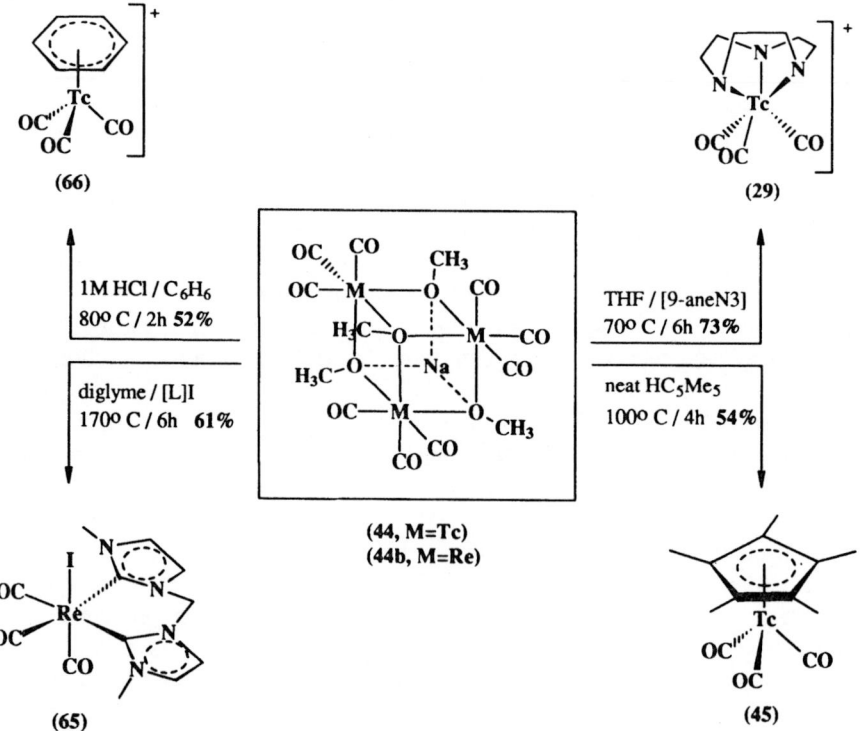

Scheme 14. Important reactions of **44**

4 Formation of the Tc–C Bond from Other Starting Materials

In some recent studies a number of organotechnetium compounds have been prepared from inorganic compounds such as [TcCl$_4$] (67) or coordination compounds not containing the Tc–C bond. Several of these reactions are summarized in Scheme 15.

4.1 Organotechnetium Compounds from [TcCl$_4$]

A very nice example is the reaction of K(C$_5$H$_5$) with 67 in THF [82]. Reduction of 67 and concomitant coordination of (C$_5$H$_5$)$^-$ results in the formation of the diamagnetic 18e$^-$ Tc(III) compound [(η^5-C$_5$H$_5$)$_2$TcCl] (68) in moderate yield as a deep red-brown powder. Crystals could be grown by sublimation and the structure was elucidated. The two (η^5-C$_5$H$_5$) rings form an angle of 143.76° relative to the Tc-center. The ring centers, the Tc and the Cl are coplanar, and the complex thus has a C$_{2v}$ symmetry. The average ring center–Tc distance is 1.877(5) Å, which is rather short and comparable to those in nickelocene (1.850 Å). Reaction with potassium naphthalide in THF led to the dinuclear complex [(η^5-C$_5$H$_5$)$_2$Tc]$_2$ (69) and [(η^5-C$_5$H$_5$)$_2$TcH] (70), both of which have been detected and characterized by mass spectrometry. Reaction of 68 with one equivalent of K(C$_5$H$_5$) in refluxing THF yielded the extremely water- and oxygen sensitive Tc(III) complex [(η^5-C$_5$H$_5$)$_2$Tc(η^1-C$_5$H$_5$)] (71). The structure of the complex was resolved and showed some interesting features [83]. The two ring centers and the technetium form an angle of 169.4°, thus the two (C$_5$H$_5$) rings in 71 are much closer to coplanar coordination than in 68. The mean distance to the ring centers is 1.831 Å, where one distance is long (1.881 Å) and the other short (1.783 Å). The σ-bound (C$_5$H$_5$) ring is twisted and its rotation is hindered, as is evident from ^{1}H-NMR investigations. The distance from the α-H of the σ-bound (C$_5$H$_5$) to Tc is only 1.592 Å. This short distance renders this hydrogen atom agostic and explains the easy formation of 70. In addition, the charge distribution in this complex has been determined as well as its dipole moment.

4.2 Organotechnetium Compounds from [TcH(N$_2$)(dppe)$_2$]

The long-known dinitrogen complex [TcH(N$_2$)(dppe)] (72) [84] has been used as a starting material for the introduction of Tc–C bonds. The labile N$_2$ ligand is a good leaving group and can easily be substituted by π-accepting molecules such as CO or CN–R [85]. Stirring at room temperature in the presence of CO, for example, does not touch the hydride function and a complex of the composition [TcH(CO)(dppe)$_2$] (73) is obtained in 55% yield. The preparation of the same complex with isocyanide instead of CO demanded refluxing conditions for several hours in the presence of excess ligand. The hydride could subsequently

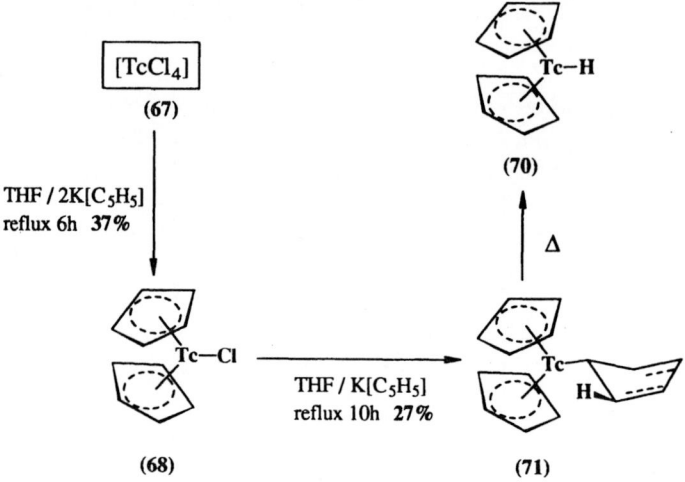

Scheme 15. Typical reactions of **67** with K[C₅H₅]

be replaced by a solvent molecule from refluxing acetonitrile after several hours. The strong trans-influence of the CO or the isocyanide labilizes the strongly bound hydride and it can consequently be exchanged. The i.r. spectrum of **73** shows the $v_{C\equiv O}$ absorption at 1859 cm⁻¹. This low frequency is the result of the strong σ-donating properties of the *trans*-situated hydride ligand that induce strong Tc $\rightarrow$ C back-bonding, thus lowering the C≡O bond order significantly. Reaction of **73** in refluxing methanol in the presence of excess isocyanide yielded cationic compounds of the composition *trans*-[Tc(CN–R)₂)(dppe)₂]⁺ (**74**) directly [86].

4.3 Organotechnetium Compounds from Other Starting Materials

Isocyanide compounds of Tc(III) with the "umbrella" ligand [P(C₆H₄SH)₃] (PS3) have been prepared directly from **3a**. Reduction with [S₂O₄]²⁻ at room temperature gave the neutral Tc(III) complex [Tc(PS3)(CN–R)] (**75**) with a coordination number of five. The structure of **75** could be elucidated. The coordination number in solution can be increased to six if either isocyanide or acetonitrile is present in excess. The sixth ligand is, however only weakly bound and can be removed from the equilibrium mixture by applying a slight vacuum, or simply by evaporating the solvent with a stream of N₂ [87]. Similarly, coordination with small π-accepting molecules has also been found in the case of sterically hindered arenethiolate complexes of Tc(III). Reduction of (NH₄)₂[TcCl₆] with Zn dust in CH₃CN in the presence of [C₆H(CH₃)₄SH] (Htmbt) resulted in the five coordinated species [Tc(tmbt)₃(NCCH₃)₂] (**76**). The acetonitrile ligands could be replaced by CO, CN–R or pyridine ligands to yield the compounds [Tc(tmbt)₃(CO)₂] (**77**), [Tc(tmbt)₃(CO)(CN–R)] (**78**) or [Tc(tmbt)₃(CO)(py)] (**79**) respectively, as well as complexes with permutations of two π-accepting ligands. The structures of [Tc(tmbt)$_3$(NCCH$_3$)$_2$] (**80**), [Tc(tmbt)$_3$(CO)(NCCH$_3$)] (**81**) and **79** have been elucidated [88]. The $\nu_{C\equiv O}$ absorptions are observed between 1960 and 1990 cm^{-1} depending on the fifth substituent. These frequencies, that are low for Tc(III) compounds, mirror the good σ-donating properties of the thiolates that result in an increased electron density at the Tc(III) center, promoting increased π-backbonding.

As described earlier in this review, cationic Tc(I) compounds have received much attention, due to their potential application as myocardial imaging agents. For this purpose, compounds of the type [Tc(η^6-C$_6$R$_6$)$_2$]$^+$ (**82**) have been prepared directly from **3a** in an ultrasonic bath, using Al^0 as reducing agent. The yields are almost quantitative [89]. However, the described procedure allows the synthesis with only trace amounts of the short-lived isotope Tc-99m. The complexes with long-lived Tc-99 could only be obtained in microgram amounts by isotope exchange with the Tc-99m compounds. Such "sandwich" compounds were first prepared a long time ago [90], but the preparation started with **67** and was low in yield. Using this method it was possible to characterize the compounds by mass spectrometry, enabling the unambiguous determination of the retention times by HPLC methods. The compounds have been tested for in vivo application and the myocardial uptake investigated. The results were promising, but the imaging could not compete with that of the homoleptic isocyanide complexes (Scheme 16).

4.4 Status of Low-Valency Organotechnetium Chemistry

Although organotechnetium chemistry has made rapid progress during the past 5–10 years the knowledge about reactivities and fundamental behaviour of such compounds can not be compared at all with that of its homologs. A number of reactions and compounds have been thoroughly investigated and important starting materials were characterized. In the future, the precursors presented herein should allow the broad development of organotechnetium chemistry in both low and medium valencies, since [Tc(CN–R)$_6$]$^+$, [TcCl(CO)$_3$(PPh$_3$)$_2$] or [TcCl(CO)$_2$(PR$_3$)$_2$] and (NEt$_4$)$_2$[TcCl$_3$(CO)$_3$] make most other organometallic

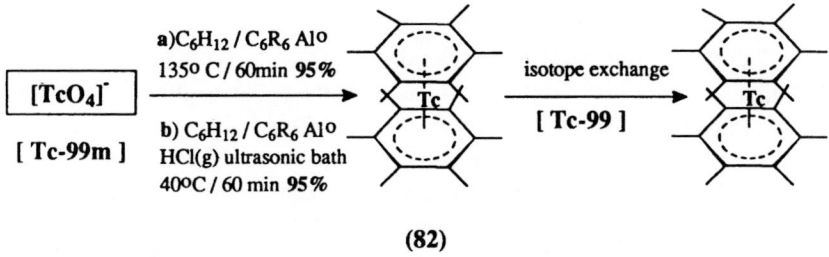

(82)

Scheme 16. Synthesis of cationic bis-arene complexes of Tc(I)

compounds accessible in a relatively simple way, without the necessity of utilizing the high-pressure product $[Tc_2(CO)_{10}]$. A main focus in the future might be the investigations with coordinatively unsaturated complexes, because of the potential catalytic activity of such electron-rich materials.

5 Formation of the Tc–C Bond in High Oxidation States

Organotechnetium chemistry in high oxidation states is a very young field with the first publications appearing only a few years ago. Most of these investigations were performed with compounds containing, additional imido- or terminal and bridging oxo-ligands in addition to the Tc–C bond. Once again, the research was initialiated by extensive studies in the field of high-valency organorhenium compounds. This relatively new class of rhenium compounds has opened a broad field of research. Most of the high-valency organorhenium compounds proved to be stable but reactive, and these characteristics make them interesting for potential catalytic application. A wide variety of reactions have indeed proven to be catalized by such compounds, the oxidation processes of olefins in particular. Therefore, an obvious question to investigate was whether analogous Tc-compounds could be prepared and if their reactivity would differ significantly from that of rhenium. Two Tc(VII) compounds have been widely used as starting materials. The first is the extremely hygroscopic $[Tc_2O_7]$, which is prepared by oxidation of Tc-metal at several hundred degrees centigrade in a stream of O_2 [91]. The second, $[TcO_3(OSiMe_3)]$, is easily prepared from $Ag[TcO_4]$ and $[SiClMe_3]$ [92].

5.1 Synthesis and Chemistry of Mixed Oxo-Alkyl Complexes

The most important high-valency organorhenium complex $[CH_3ReO_3]$ (83b) was originally prepared by the reaction of $[Re_2O_7]$ (1b) with $[SnMe_4]$. This metathesis-like synthesis yielded 50% of 83b and 50% of the stannylester $[ReO_3(OSnMe_3)]$ (84b) [93]. A more recent preparation, starting with mixed perrhenic acid-anhydrides, gave almost quantitative yields of 83b. [94] The synthesis of $[CH_3TcO_3]$ (83a) has been investigated; it turned out that under the same conditions $[TcO_3(OSnMe_3)]$ (84a) was formed but not 83a. Instead, the dinuclear Tc(VI) compound $[(CH_3)_2Tc(O)(\mu\text{-}O)_2Tc(O)(CH_3)_2]$ (85a) was found and structurally characterized [95]. The average Tc–C bond lengths 2.130(1) Å and the Tc=O distance is 1.650(2) Å. The Tc–Tc bond is 2.5617(3) Å, which is a reasonable length for a single bond. However ^{99}Tc- and ^{1}H-NMR investigations showed the very broad lines indicative of paramagnetism, indicating that there is no bonding between the two metal centers. I.r. spectra exhibit two very strong $\nu_{Tc=O}$ absorptions at 1006 and 986 cm^{-1}. The identical Re(VI) compound 85b could only be prepared by reacting the much stronger methylation/reducing agent

$Zn(CH_3)_2$ with **83b**. It can be deduced from this behaviour that **83a** is probably formed as an intermediate which undergoes further reduction/methylation, due to its much higher oxidative power. This assumption has been confirmed by reacting **1a** and $Sn(CH_3)_4$ below 0 °C, which resulted in the formation of **83a** (scheme 18). **83a** has been characterized by ^{99}Tc- and ^1H-NMR, mass spectrometry and i.r. spectroscopy. Two strong $\nu_{Tc=O}$ absorptions were observed at 1002 and 948 cm^{-1}. **83a** is not sensitive to oxygen or water but is very volatile and can be sublimed at room temperature even under 1 atm of N_2.

83b has been shown to be a very powerful catalyst for the epoxidation of olefins with hydrogen peroxide [96]. Reaction of **83a** with olefins revealed different behaviour in that – as in the case of $[OsO_4]$ – oxidative addition of the olefin occurred, affording the formation of a so called technetate-ester. Like $[OsO_4]$, **83a** selectively oxidizes olefins to *cis*-diols. The process runs catalytically with various olefins in the presence of H_2O_2. Due to the thermal lability of **83a**, these investigations have not been pursued further (Scheme 18).

The reactions of olefins with non-organometallic Tc(VII) compounds behaved similarly. In a recent study, $[TcO_3Cl(A^\wedge A)]$ (**86a**) (in which AA stands for aromatic diamine derivatives) was shown to react quantitatively with olefins, and produce the corresponding Tc(V) diolato-complex $[TcOCl(O^\wedge O)(A^\wedge A)]$ (**87a**). The process could not be run catalytically, as Tc(V) complexes tend to undergo disproportionation rather than reoxidation in the presence of water [97]. These alkene-glycol interconversions could not be performed with the analog Re(VII) compound. Rhenium displays completely contrary behaviour, in that alkenes can

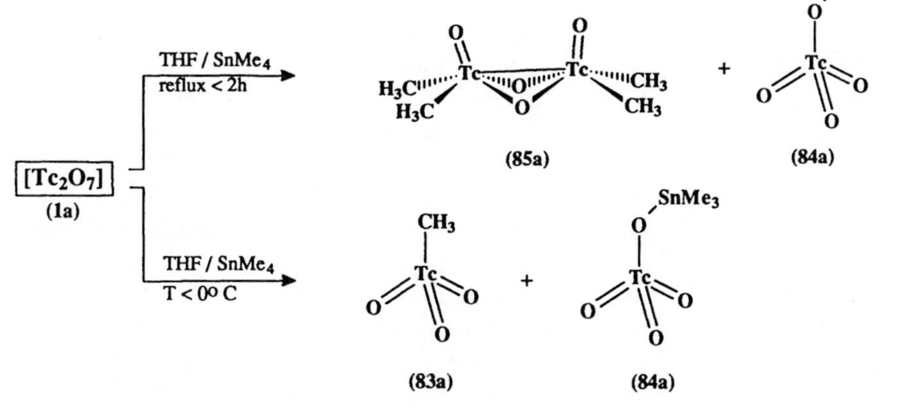

Scheme 17. Synthesis of mixed oxo-alkyl complexes of high valency technetium and rhenium

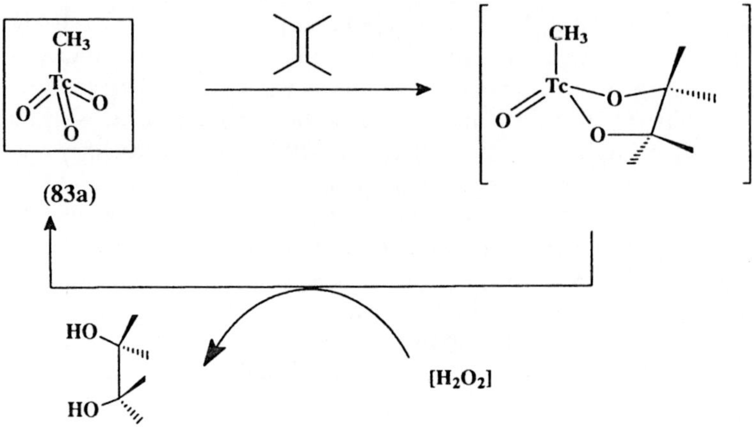

Scheme 18. Proposed oxidation process of olefins catalyzed by **83a**

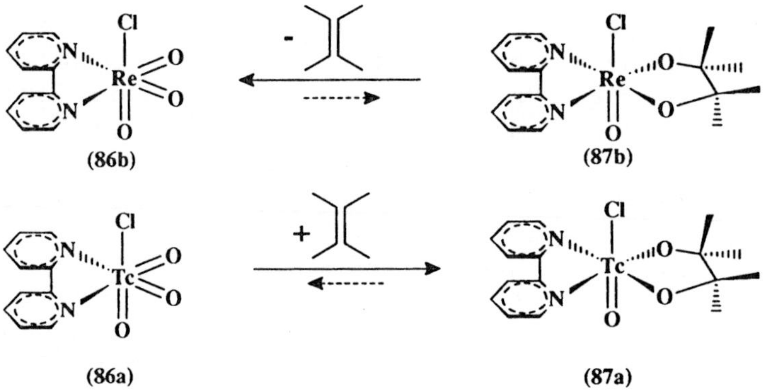

Scheme 19. Fundamental difference in the behaviour of high valency Re- and Tc-complexes with alkenes

be thermally cleaved from the Re(V) diolato complex [ReOCl(O^O)(A^A)] (**87b**) to form the Re(VII)-complex [ReO₃Cl(A^A)] (**86b**) and the corresponding olefin (Scheme 20). In contrast to **83a**, the reaction with **86a** could not be run catalytically due to the water sensitivity of the Tc(V) compound **87a**.

83a and **85a** are the only-high valency oxo-organotechnetium compounds that have been described to date. It is noteworthy that these Tc-compounds behave quite differently from their rhenium analogs. It seems that the significantly higher oxidation power of Tc over Re is responsible for this observation. The chemical behaviour of **83a** is similar to that of [OsO₄], which has so far been the most widely used *cis*-hydroxylation catalyst for olefins. [98] In parallel, it has to be noted that Mn(VII) behaves like Ru(VIII), in that olefins are oxidized completely to ketones or carboxylic acids. Thus the reactivity is not identical within the same transition metal row between neighboring elements, but – in the high oxidation states – group 7 elements of the nth transition row behave identically to isoelectronic (d-shell) group 8 elements of the $(n + 1)$st transition row (Scheme 20).

Various attempts to prepare the Tc-analog **88a** of $[(\eta^5\text{-}C_5Me_5)ReO_3]$ (**88b**) have been described. According to theoretical *ab initio* SCF-calculations, compounds of the general composition $[TcO_3R]$ should be stable [99] under ambient conditions but less so than the rhenium analogues. **45b** is easily oxidized to **88b** with H_2O_2, $[Mn_2O_7]$, or dioxiran derivatives. Similar attempts to oxidize **45a** to **88a** failed, although all of these strongly oxidizing reagents have been used. Especially in the case of oxidation with $[Mn_2O_7]$, catalytic decomposition of $[Mn_2O_7]$ was observed after addition of small amounts of **45a**. Oxidation with H_2O_2 gave tiny amounts of the polymer compound $[Tc(\mu\text{-}O)_3Tc(\eta^5\text{-}C_5Me_5)]_n$ (**89**). [100] The complex was structurally characterized, exhibiting a very short (1.867(4) Å) Tc–Tc distance with a bond order of 2.5. On the basis of the cell constants, which are almost identical to those of **88b**, it has been suggested that compound **89** is in fact $[(\eta^5\text{-}C_5Me_5)TcO_3]$ (**88a**). [95] The latter structure has never been resolved satisfactorily, owing to a high disorder in the crystals.

5.2 Synthesis and Chemistry of Mixed Imido-Alkyl Complexes

While **1** turned out to be the starting material of choice for high valency oxo-organotechnetium compounds, the trimethylsilylester of pertechnetic acid was found to be very convenient for the preparation of imido-group-containing organotechnetium compounds. $[TcO_3(OSiMe_3)]$ could be reacted with different organocyanate derivatives in hexamethyldisiloxan, to form the corresponding tris-imido Tc(VII) complex $[Tc(NR)_3(OSiMe_3)]$ (**90**) [101]. The structure of **90** could be resolved; the coordination geometry around the metal center is approximatively tetrahedral. The strongly π-donating imido-ligands can be expected to be collinear with the nitrogen atom if both pairs of $p-\pi$ electrons can be donated to the metal center. In **90** however, the angles are significantly bent, mirroring a lower $M = N$ bond order than three. The angle at the nitrogen atom is av. 155°, significantly different from the 180° expected for a $M \equiv N$ triple bond. The $Tc = N$ bond lengths are not equal to within one standard deviation; two bonds are long (av. 1.756(6)Å) and one is short (1.749(7)Å). Compound **90** reacted easily with a variety of Grignard reagents or Li alkyls to yield the corresponding tris-imido alkyl (Tc(VII)) complexes $[Tc(NR)_3(R)]$ (**91**), as depicted in Scheme 21 [102].

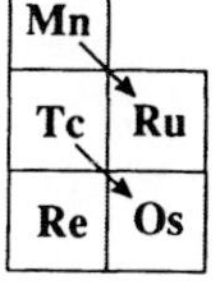

Scheme 20. Chemical relation between group 7 and group 8 transition metals in high oxidation states

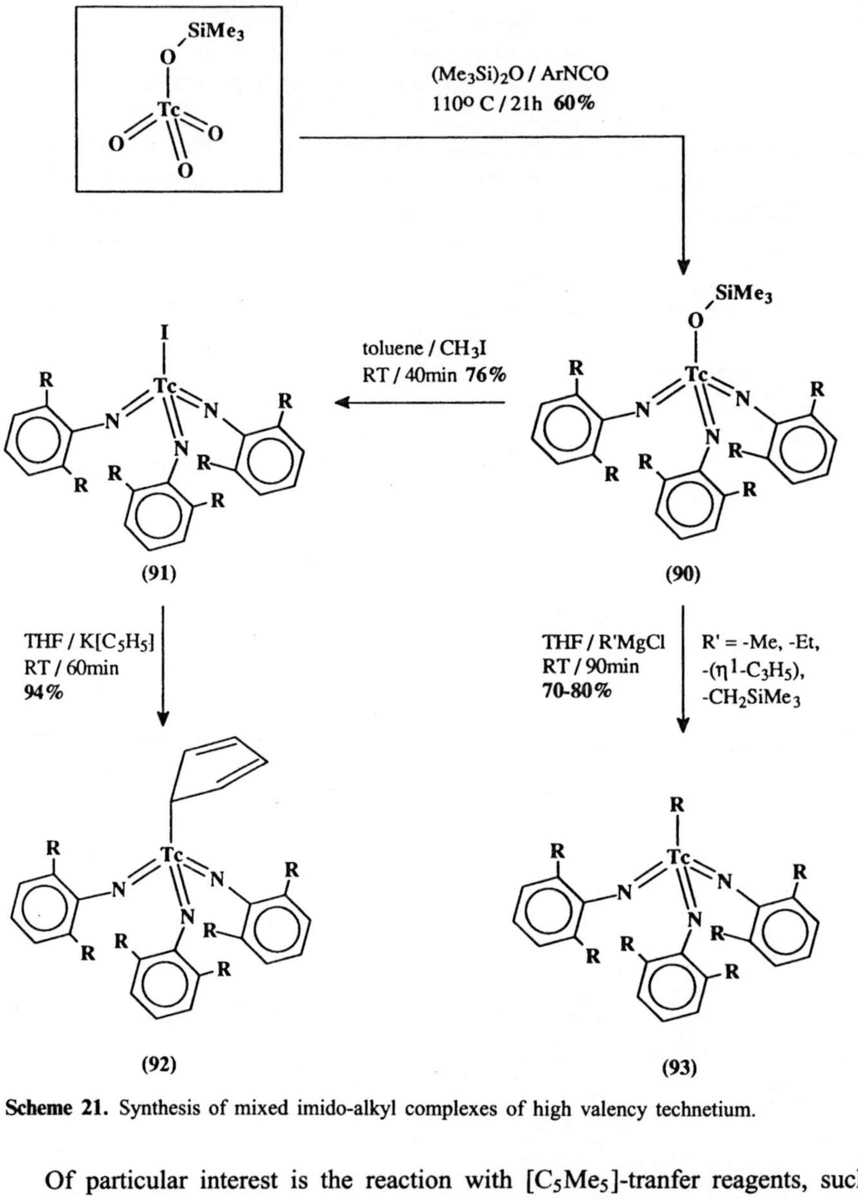

Scheme 21. Synthesis of mixed imido-alkyl complexes of high valency technetium.

Of particular interest is the reaction with [C₅Me₅]-tranfer reagents, such as K[C₅Me₅], Na[C₅Me₅] and others. The preparation of the tris(imido) analogs of **88b** was attempted, but all efforts failed and no reaction taking place in most cases [103]. It has been aruged that the tertiary (butyl-imido) complexes are too sterically hindered and this would inhibit the reaction. Conversion of [Tc(NR)₃(OSiMe₃)] into [Tc(NR)₃I] (**91**) – where R stands for a 2,5-dimethylphenyl group – and subsequent metathesis with K[C₅H₅], afforded the compound [Tc(NR)₃(η¹-C₅H₅)] (**92**) in good yield. The Re-analog could be prepared in the same way. Reaction with K[C₅Me₅] instead of K[C₅H₅] showed immediate color changes, which might be a result of electron transfer reactions rather than of metathesis. The structure of **92** has been elucidated, proving the [C$_5$H$_5$] ring to be σ-bonded. The complex shows only small deviations from tetrahedral geometry and the cyclopentadienyl ring is almost planar (in contrast to the (η^1-C$_5$H$_5$)-ring in [Tc(η^5-C$_5$H$_5$)$_2$(η^1-C$_5$H$_5$)] (see Sect. 4.1). The bond angles at the imido nitrogen atoms are less bent (av. 162°) than in **90** but more bent than in **91** (av. 166°). Comparison within this series reveals very clearly that the donating properties of the fourth ligand influences the Tc = N bond order. Strong donating properties ($-$OSiMe$_3$) decrease the TcN bond order, resulting in a bent angle at the nitrogen atom, while weak donating properties ($-$I) have the contrary effect. It was concluded from the Tc = N bondlength that one imido ligand is significantly more weakly bound than the others two. Further action of excess K[C$_5$H$_5$] on **92** afforded the salt K[Tc(NR)$_3$(η^1-C$_5$H$_5$)$_2$] (**93**), as deduced from elemental analysis, mass spectrometry and NMR experiments [102]. **92** provides first demonstration of the proposed σ-coordination of [C$_5$H$_5$] in tris-(imido) complexes of Tc(VII). It seems that, as a result of the stronger donating properties of imido over oxo ligands, no η^5 coordination is possible.

A very interesting study of the reaction of methyl Grignard reagents with homoleptic imido complexes of Tc(VI) has been performed [104]. Reduction of **91** with Na in THF gave the first homoleptic Tc(VI)-imido complex [Tc$_2$(N–Ar)$_6$] (Ar = 2,6-dimethylphenyl) [105]. Under more vigorous conditions dimerization to the second homoleptic Tc(VI)-imido complex [Tc$_2$(N–Ar)$_4$(μ-N–Ar)$_2$] (**94**)

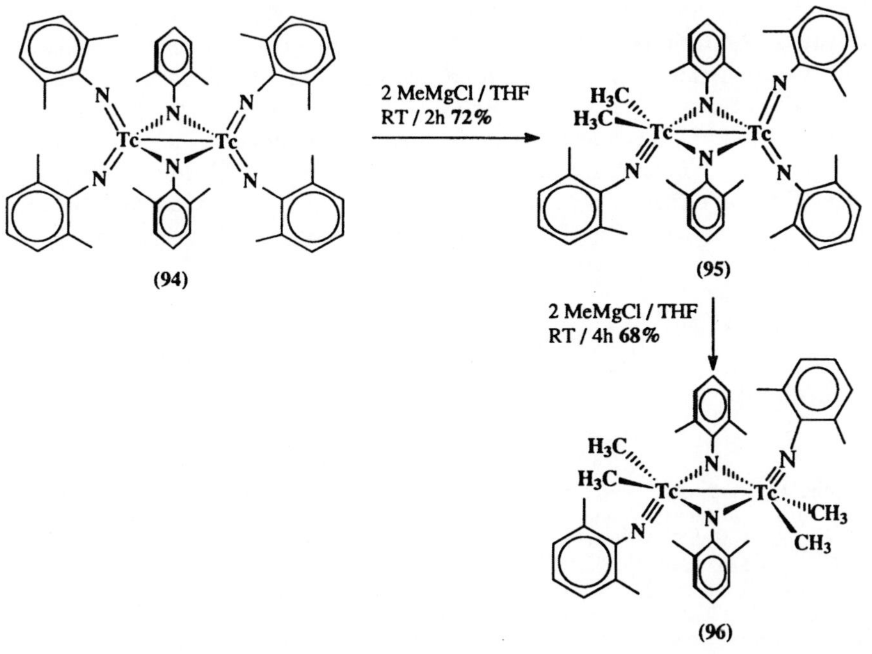

Scheme 22. Reaction of homoleptic dinuclear Tc-imido complex **94** with methyl Grignard reagent

could be performed. [106] Reaction with two equivalents of MeMgCl at room temperature resulted in the unprecedented substitution of one imido group by two methyl groups (**95**). Repeating the procedure resulted in substitution of a second imido group at the other Tc(VI) center (**96**). The structures of both compounds were confirmed by X-ray analysis. As in case of **85a**, the bridging nitrogen atom and the two technetium atoms form a "butterfly", but the angle between the wings is larger (167°). Interestingly, **96** exhibits *trans*-geometry with respect to the remaining terminal imido groups, while the terminal *oxo* groups in **85a** are in the *cis*-configuration.

5.3 Potential of High-Valency Organotechnetium Compounds as Synthons

Organorhenium compounds in high oxidation states can be prepared from easily available starting materials. These compounds proved to be synthons for an unusual diversity of reactive and interesting compounds. Some of these compounds were shown to have very useful catalytic properties. [CH$_3$ReO$_3$], for example, provides catalytic activity and it is known that this complex catalyzes at least three different types of organic reaction. Only a very few studies on analogous technetium compounds have been published. Theoretical studies suggested that such technetium compounds should also be stable for, and the experimental investigations confirmed the calculations. However, it was also found experimentally that the Tc-complexes are less stable than the Re-homologs, a finding that had also been predicted in advance. On the other hand, the decrease in stability is accompanied by a significantly higher reactivity, including – in particular – oxidative power towards organic molecules. These investigations revealed that the differences between Re and Tc in their high oxidation states are therefore not only gradual – as it is in the lower oxidation states – but fundamental. This means that homologous compounds react in different ways and yield different types of products. From this fact, it is to be expected that an enormous synthetic potential emerge from these accessible high valency organotechnetium compounds in the near future.

6 References

1. Nicolini M, Bandoli G, Mazzi U (eds) (1985) In: Technetium in chemistry and nuclear medicine 2, Cortina International, Verona
2. Nicolini M, Bandoli G, Mazzi U (eds) (1990) In: Technetium in chemistry and nuclear medicine 3, Cortina International, Verona
3. Constable EC, Housecroft CE (1994) Coord Chem Rev 131: 153
4. Schwochau K, Pleger U (1993) Radiochim Acta 63: 103
5. Bryan JC, Sattelberger AP (1994) In: Wilkinson G (ed) Comprehensive organometallic chemistry. Pergamon Press, Oxford, in press
6. Mazzi U (1989) Polyhedron 13/14: 1683

7a. Turp JE, Turp N (1987) Coord Chem Rev 80; 173; b) Vites JC, Lynam MM (1994) Coord Chem Rev 131: 127
8. Abrams MJ, Davison A, Jones AG, Costello CE, Pang H (1983) Inorg Chem 22: 2798
9. Doonan DJ, Balch AL (1974) Inorg Chem 13: 921
10. Farr JP, Abrams MJ, Costello AE, Davison A, Lippard SJ (1985) Organometallics 4: 139
11. Bandoli G, Clemente DA, Mazzi U (1978) J Chem Soc Dalton Trans 373
12. Lam CT, Corfield PWR, Lippard SJ (1977) J Am Chem Soc 99: 617
13. Giandomenico CM, Hanau LH, Lippard SJ (1982) Organometallics 1: 142
14. Linder KE, Davison A, Dewan JC, Costello CE, Maleknia S (1986) Inorg Chem 25: 2085
15. Matteson DS, Bailey RA (1969) J Am Chem Soc 91: 1975
16. Treichel PM, Direen GE, Mueh HJ (1972) J Organomet Chem 44: 339
17. O'Connell LA, Dewan J, Jones AG, Davison A (1990) Inorg Chem 29: 3539
18. Chatt J, Pombeiro AJL, Richards RL, Royston GHD, Muir KW, Walker RJ (1975) J Chem Soc Chem Commun 708
19. Abram U, Beyer R, Muenze R, Findeisen M, Lorenz B (1989) Inorg Chim Acta 160: 139
20. O'Connell LA, Davison A (1990) Inorg Chim Acta 176: 7
21. Kronauge JF, Davison A, Roseberry AM, Costello CE, Maleknia S, Jones AG (1991) Inorg Chem 30: 4265
22. Alberto R, Herrmann WA, Kiprof P, Baumgärtner F (1992) Inorg Chem 31: 895
23. Dright D, Ibers JA (1969) Inorg Chem 7: 1099
24. Mazzi U, Bismondo A, Kotsev N, Clemente DA (1977) J Organomet Chem 135: 177
25. Biagini Cingi M, Clemente DA, Magon L, Mazzi U (1975) Inorg Chim Acta 13: 47
26. Trop HS (1979) Thesis Massachusetts Institute of Technology
27. Pearlstein RM, Davis WM, Jones AG, Davison A (1989) Inorg Chem 28: 3332
28. Bryan JC, Alberto R, Sattelberger A, Schubiger PA, unpublished results
29. Marchi A, Rossi R, Duatti A, Magon L, Bertolasi V, Ferretti V, Gilli G (1985) Inorg Chem 24: 4744
30. Mazzi U (1994) In: Nicolini M, Bandoli G, Mazzi U (eds) Technetium in chemistry and nuclear medicine 4, Cortina International, Verona, Vol in press
31. Rossi R, Marchi A, Magon L, Duatti A, Cassellato U, Graziani R (1989) Inorg Chim Acta 160: 23
32. O'Connell LA, Pearlstein RM, Davison A, Thornback JR, Kronauge JR, Jones AG (1989) Inorg Chim Acta 161: 39
33. Abrams MJ, Davison A, Jones AG (1984) Inorg Chim Acta 82: 125
34. Abel EW, Hargreaves GB, Wilkinson G (1958) J Chem Soc 3149
35. Alberto R, Schibli R, Herrmann WA, Abram U, Artus G, Schubiger PA, Kaden TA (1994) J Organomet Chem: in press
36. Alberto R, Schibli R, Egli A, Herrmann WA, Abram U, Schubiger PA (1994) In: Nicolini M, Bandoli G, Mazzi U (eds) Technetium in chemistry and nuclear medicine 4, Cortina International, Verona, in press
37. Alberto R, Egli A, Abram U, Hegetschweiler K, Gramlich G, Schubiger PA (1994) J Chem Soc Dalton Trans 2815
38. Kaden L, Lorenz B, Rummel S, Schmidt K, Wahren M (1988) Inorg Chim Acta 142: 1
39. Knight Castro HH, Hissink CE, Teuben JH, Vaalburg W (1992) Rec Trav Chim Pays-Bas 111: 105
40. Raab K, Beck W (1985) Chem Bar 118: 3830
41. Herberhold M, Süss G (1975) Angew Chem Int Ed Engl 14: 700
42. Lorenz B, Findeiesen M, Olk B, Schmidt K (1988) Z Anorg Allg Chem 566: 160
43. Treichel PM, Williams JP (1977) J Organomet Chem 135: 39
44. Falbe J (1980) In: New synthesis with carbon monoxide, Springer, Berlin Heidelberg New York
45. Cotton FA (1976) Prog Inorg Chem 21: 1
46. Hileman JC (1964) In: Jolly WJ (ed) Preparative inorganic reactions. Interscience, New York, Vol 1, p 77
47. Grimm CC, Clark RJ (1990) Organometallics 9: 1123
48. Calderazzo F, Mazzi U, Pampaloni G, Poli R, Tisato F, Zanazzi PF (1989) Gazz Chem Ital 119: 241
49. Herrmann WA, Alberto R, Bryan JC, Sattelberger AP (1991) Chem Ber 124: 1107
50. Heinekey DM, Crocker LS, Gould GJ (1988) J Organomet Chem 342: 243
51. Michels GD, Svec HJ (1981) Inorg Chem 20: 3445

52. Knight Castro HH, Meetsma A, Teuben JH, Vaalburg W, Panek K, Ensing G (1991) J Organomet Chem 410: 63
53. Hieber W, Lux F, Herget C (1965) Z Naturforschg 20b: 1159
54. Alberto R, Herrmann WA, Bryan JC, Schubiger PA, Baumgärtner F, Mihalios D (1993) Radiochim Acta 63: 153
55. Grimm CC, Clark RJ (1990) Organometallics 9: 1118
56. Green MLH, Wilkinson G (1958) J Chem Soc: 4324
57. Raptis K, Dornberger E, Kanellakopulos B, Nuber B, Ziegler M (1991) J Organomet Chem 408: 61
58. Bischof P (1977) J Am Chem Soc 99: 8145
59. Bernal I, Korp JD, Herrmann WA, Serrano R (1984) Chem Ber 117: 434
60. Herrmann WA, Serrano R, Weichmann J (1983) J Organomet Chem 246: C57
61. Hoyano JK, Graham WAR (1982) J Chem Soc, Chem Commun: 27
62. Raptis K, Kanellakopulos B, Nuber B, Ziegler ML (1991) J Organomet Chem 405: 323
63. Kanellakopulos B, Nuber B, Raptis K, Ziegler ML (1991) Z Naturforsch B Chem Sci 46b: 55
64. Nesmeyanov AN, Kolobova NE, Makarov YV, Anisimov KN (1969) Izv Akad Nauk SSSR Ser Khim: 1826
65. Joachim JE, Apostolidis C, Kanellakopulos B, Maier R, Marques N, Meyer D, Müller J, Pires de Matos A, Nuber B, Rebizant J, Ziegler ML (1993) J Organomet Chem 4: 119
66. Joachim JE, Apostolidis C, Kanellakopulos B, Maier R, Meyer D, Rebizant J, Ziegler ML (1993) J Organomet Chem 455: 137
67. Sellmann D (1971) Angew Chem Int Ed Engl 10: 919
68. Sellmann D, Kleinschmidt E (1977) Z Naturforsch Teil B 32: 795
69. Ziegler ML, Weidenhammer K, Zeiner H, Skell PS, Herrmann WA (1976) Angew Chem 88: 761
70. Lorenz B, Findeisen M, Olk B, Schmidt K (1988) Z Anorg Allg Chem 566: 160
71. Kaden L, Findeisen M, Lorenz B, Schmidt K, Wahren M (1991) Isotopenpraxis 27: 265
72. Lorenz B, Findeisen M, Schmidt K (1991) Isotopenpraxis 27: 266
73. Connelly NG, Dahl LF (1979) J Chem Soc Chem Commun: 880
74. Adamov VM, Belyaev BN, Borisova IV, Miroslavov AE, Sidorenko GV, Suglobov DN (1991) Radiokhimiya 33(4): 38
75. Borisova IV, Miroslavov AE, Sidorenko GV, Suglobov DN, Shcherbakova LL (1991) Radiokhimiya 33(4): 27
76. Miroslavov AE, Sidorenko GV, Borisova IV, Legin EK, Lychev AA, Suglobov DN (1990) Radiokhimiya 32(6): 14
77. Miroslavov AE, Sidorenko GV, Borisova IV, Legin EK, Lychev AA, Suglobov DN, Adamov VM (1990) Radiokhimiya 32(4): 6
78. Miroslavovm AE, Sidorenko GV, Borisova IV, Legin EK, Lychev AA, Suglobov DN (1989) Radiokhimiya 31(6): 33
79. Herrmann WA, Alberto R, Bryan JC; Sattelberger AP (1991) Chem Ber 124: 1107
80. Herrmann WA, Mihalios D, Oefele K, Kiprof P, Belmejahed F (1992) Chem Ber 125: 1795
81. Kane-Maguire LAP, Sweigart DA (1979) Inorg Chem 18(3): 700
82. Apostolidis C, Kanellakopulos B, Maier R, Rebizant J, Ziegler ML (1990) J Organomet Chem 396: 315
83. Apostolidis C, Kanellakopulos B, Maier R, Rebizant J, Ziegler ML (1991) J Organomet Chem 411: 171
84. Kaden L, Lorenz B, Schmidt K, Sprinz H, Wahren M (1979) Z Chem 19: 305
85. Kaden L, Findeisen M, Lorenz B, Schmidt K, Wahren M (1992) Inorg Chim Acta 193: 213
86. Abram U, Abram S, Beyer R, Muenze R, Kaden L, Lorenz B, Findeisen M (1988) Inorg Chim Acta 148: 141
87. de Vries N, Cook J, Jones AG, Davison A (1991) Inorg Chem 30: 2662
88. de Vries N, Dewan JC, Jones AG, Davison A (1988) Inorg Chem 27: 1574
89. Wester DW, Coveney JR, Nosco DL, Robbins MS, Dean RT (1991) J Med Chem 34: 3284
90. Baumgärtner F, Fischer EO, Zahn U (1961) Naturwissenschaften 48: 478
91. Krebs B (1969) Angew Chem 81: 328
92. Nugent WA (1983) Inorg Chem 22: 965
93. Herrmann WA, Kuchler JG, Felixberger JK, Herdtweck E, Wagner W (1988) Angew Chem Int Ed Engl 27: 394
94. Herrmann WA, Kühn FE, Fischer RW, Thiel WR, Ramão CC (1992) Inorg Chem 31: 4431
95. Herrmann WA, Alberto R, Kiprof P, Baumgärtner F (1990) Angew Chem Int Ed Engl 29: 189

96. Herrmann WA, Marz DW, Weichselbaumer G, Wagner W, Fischer RW (Jan 27, 1989) GP 3902357.5 Hoechst AG
97. Pearlstein RM, Davison A (1988) Polyhedron 7: 1981
98. Wai JSM, Marko I, Svendsen JS, Finn MG, Jacobsen EN, Sharpless KB (1989) J Am Chem Soc 111: 1123
99. Szyperski T, Schwerdtfeger P (1991) Angew Chem 101: 1271
100. Kanellakopulos B, Nuber B, Raptis K, Ziegler ML (1989) Angew Chem 101: 1055
101. Bryan JC, Burrell AK, Miller MM, Smith WH, Burns CJ, Sattelberger AP (1993) Polyhedron 12(14): 1777
102. Burrell AK, Bryan JC (1992) Organometallics 11: 3501
103. Herrmann W, Weichselbaumer G, Paciello RA, Fischer R, Herdtweck E, Okuda J, Marz DW (1990) Organometallics 9: 489
104. Burrell AK, Bryan JC (1993) Organometallics 12: 2426
105. Burrell AK, Bryan JC (1993) Angew Chem Int Ed Engl 32(1): 94
106. Burrell AK, Clark DL, Smith WH, Sattelberger AP, Bryan JC (1994) manuscript in preparation

Chemistry of Technetium Cluster Compounds

Sergey Vl. Kryutchkov

Institute of Physical Chemistry, Russian Academy of Sciences, Leninsky prosp. 31, Moscow, 117915 Russia

Table of Contents

Topics in Current Chemistry, Vol. 176
© Springer-Verlag Berlin Heidelberg 1996

This review is concerned with the syntheses, electronic and molecular structures, and properties of all types of presently known technetium cluster compounds with acido-ligands. Examination of the literature shows that technetium is not only a typical cluster-forming agent, but also has a number of specific "anomalous" cluster-forming properties. These properties may be interpreted in terms of a greater ability of the outer diffuse 5s(5p)-AO's to participate in additional M–M bonding in technetium acido-clusters, compared to the situation in analogous clusters of other d-transition elements. Theoretical interpretations of the electronic and molecular structures and properties of technetium clusters are supported by experimental data (X-ray diffraction analysis, magnetochemistry, and optical, ESR, X-ray emission, and X-ray photoelectron spectroscopies). The observed increased stability of technetium clusters with an odd number of "metallic" electrons and a decrease in the effective charge on technetium atoms upon formation of M–M bonds are discussed.

1 Introduction

Metal–metal (M–M) bonds, first noted in the early sixties, occur in several thousand transition-metal compounds [1]. Complex technetium compounds and compounds with M–M bonds (clusters) have been studied more extensively than many other classes of inorganic compounds. Increasing interest in technetium compounds is due to the practical uses of the ^{99m}Tc isotope, which ranks first among radioactive isotopes used in nuclear medicine diagnostics [2–4]. On the other hand, technetium clusters are an interesting object for theoretical studies, because until recently, they were the only compounds in which the presence of these "anomalous" chemical bonds was thought possible.

Although there are a lot of publications on the chemistry of technetium [2–4] and transition-metal clusters [1, 5–8], the chemistry of technetium clusters was insufficiently studied until the early eighties [1, 2]. Nevertheless, the available scanty data on the compounds with Tc–Tc bonds inspired hope that interesting results would be obtained in the chemistry of technetium in general, in radiochemistry, and in the chemistry of transition-metal cluster compounds. The anticipated results were actually obtained [9–15] and the conclusion was drawn that technetium had a number of "anomalous" cluster-forming properties [9]. This review looks at the detailed studies of these properties and their interpretation in terms of electronic structure theory.

2 Syntheses, Properties, and Molecular Structures

2.1 The Classification of Technetium Cluster Compounds

Since presently more than 60 compounds with Tc–Tc bonds are known, and their number is likely to increase, it is necessary to classify them. This classification may be done according to the specific molecular and electronic structures of these compounds. In our opinion, the nuclearity[1] of technetium clusters should be the underlying classification principle. Thus all known compounds with technetium–technetium bonds fall into two groups: binuclear complexes (clusters) and polynuclear clusters.

In compliance with the nuclearity principle, polynuclear clusters are subdivided into a number of other subgroups, e.g. hexanuclear, octanuclear, etc. The binuclear clusters of technetium may be classified according to the electronic structure of their Tc–Tc[2] bonds. Then, the d^4–d^4 complexes with quadruple M–M bonds are the "father" of all binuclear complexes with Tc–Tc bonds. The "addition" or "removal" of electrons from Tc–Tc bonds [1, 11] should result in a decrease in the formal multiplicity of M–M bonds. Thus, for instance, the formal multiplicity of Tc–Tc bonds of d^3–d^3 and d^5–d^5 binuclear complexes equals[3] 3, that of d^4–d^5 and d^4–d^3 complexes equals 3.5, etc.

Binuclear complexes may be further classified according to the chemical nature and type of the bonding between ligands. Thus, for example, the subgroup of d^4–d^5 binuclear complexes of technetium may be divided into compounds without bridge ligands and compounds of the "lantern" type with four bridge ligands, which additionally connect the mononuclear fragments together. Further division of the subgroups of polynuclear clusters is according to the specific form of the metal skeleton of the cluster. For example, based on this principle hexanuclear clusters are subdivided into trigonal-prismatic and octahedral clusters.

It should be noted that the above classification system of technetium cluster compounds is not the only possible one. In section 4 another classification is described, which is based on thermal stability and the mechanism of thermal decomposition. Section 2.2 is concerned with the classification based on methods of synthesizing cluster compounds. The classifications based on specific properties of clusters do not at all belittle the advantages of the basic structural classification; they broaden the field of application of the latter, because for a better understanding and explanation of any chemical, physico-chemical and physical properties it is necessary to deal directly or indirectly with the molecular and/or electronic structures of the clusters.

2.2 Methods of Synthesizing Technetium Cluster Compounds

All the current methods of synthesizing of technetium compounds (as well as most compounds of other metals [1, 11]) with M–M bonds involve reductions of complexes of the central atom in the higher oxidation states with further (or

---

[1] The number of metal atoms connected by the direct M–M bonds in the metal skeleton of a cluster.
[2] At present this is the most widespread system of classification of bi- and polynuclear clusters.
[3] In section 5.2.3 it is shown that in contrast to binuclear complexes of other elements [1, 10, 11] for technetium compound this statement is not true.

simultaneous) substitutions by the respective ligands. These methods may be divided into three groups: (1) the reduction of mononuclear complexes using molecular hydrogen as a reducer under a pressure of 3–5 MPa in a hydrohalogenic acid medium at temperatures of 100–220°C; (2) the reduction of mononuclear complexes using other chemical reagents as reducers; (3) syntheses based on the chemical reactions of compounds with M–M bonds.

The first two groups are distinguished by the fact that autoclave syntheses are used mainly by the Soviet (or Russian) cluster school [1, 2, 11, 16–22], while the American and other schools do not use autoclave techniques [1, 23–35]. This distinction is historical and applies not only to technetium compounds, but also to other cluster compounds of d-transition elements [36–41]. Cotton et al. [28] did use an autoclave method once, but it was not further developed [1].

The American scientists believe [1, 28] that the autoclave technique is more labour consuming, and autoclaves are subjected to corrosion and cannot be used for a long time. However, the authors, who developed this method, believe that in the case of rhenium and molybdenum compounds [16–21, 36–41], as well as technetium compounds [2, 9, 10, 15, 22, 42, 43] excessive corrosion may be avoided by using stainless steel and shielding glass test tubes and the high labour consumption is greatly compensated by the advantages of the method. For example, a 0.25 l autoclave made from H18N10T-type stainless steel was used for the synthesis of cluster compounds for about 15 years without repair or parts-replacement [22, 42–44]. During that time about 70% of all the currently known technetium compounds with M–M bonds, including all polynuclear technetium clusters were synthesized. These numbers speak in favour of the autoclave methods of synthesis as do several other advantages: (1) the use of molecular hydrogen as a reducer elimates added steps for formation and precipitation of admixtures; (2) the possibility for high temperatures inside the autoclave accelerates chemical reactions, leading to formation of clusters, and increases considerably the dissolution of reagents in hydrogenhalogenic acids; (3) the high external pressure does not allow the gaseous and volatile mixture to volatilize from the reaction mixture; (4) slow cooling down of the autoclave provides almost ideal conditions for getting well-formed single crystals suitable for X-ray structural analysis[4] and almost all technetium-containing compounds, which at high temperatures are in the dissolved state, crystallize.

With autoclave syntheses a high yield of clusters is achieved, and it is possible for researchers to follow the reaction path in solution by gradually changing (from experiment to experiment) the working parameters of the synthesis (temperature, pressure, exposure at working temperatures, etc). All these advantages of the autoclave technique have resulted in an abundance of new forms of technetium clusters (particularly, polynuclear ones) because it has been possible to develop and improve the method of obtaining these compounds.

Among other widely used reducers for the synthesis of binuclear technetium complexes with M–M bonds, zinc in HCl and H_3PO_2 [1, 11] are most important. These reducers, as a rule, are used for the synthesis of Re clusters of similar structure.

If the first two groups of the methods of synthesis are based on reactions leading to the formation of M–M bonds from mononuclear Tc complexes, the third one is based on the reactions of those compounds, which already have Tc–Tc bonds in their structures. These methods are based on the substitution

---

[4] The specific features of the cluster compounds of technetium are such, that practically each new compound must be studied using single crystal X-ray structural analysis, because their complex structures do not allow the interpretation of the results from other physico-chemical methods of investigation. Therefore, the synthesis of single crystals suitable for X-ray structural analysis is the main and most laborious chemical task.

reactions of ligands [45–49]. Among them there are also autoclave methods with all the above advantages; for example, acetate binuclear technetium clusters, such as [Tc$_2$(CH$_3$COO)$_4$Cl], K[Tc$_2$(CH$_3$COO)$_4$Cl$_2$], [Tc$_2$(CH$_3$COO)$_4$Br] [48, 49], as well as the polynuclear cluster [(CH$_3$)$_4$]$_3$ [Tc$_6$Br$_{14}$] [50] have been synthesized. Other methods of the third group are those based on thermochemical reactions [51], hydrolysis reactions [11, 52], intramolecular rearrangements [44, 53], and some others. Again this group includes autoclave methods [48, 49, 50, 53].

2.3 Binuclear Technetium Cluster Compounds with the d^4–d^5 Electronic Configuration of the Core Atoms

These compounds constitute the most numerous groups of technetium clusters which have been investigated (Table 1, compounds 8–16). In our view this is due to the greater stability of the d^4–d^5 clusters compared with the d^4–d^4 clusters and also the fact that compounds with the [Tc$_2$Cl$_8$]$^{3-}$ anions were obtained initially for technetium [22, 24, 42, 43] and were subsequently employed to synthesize new technetium clusters [11, 60–62].

2.3.1 The Octachloroditechnetates M$_3$[Tc$_2$Cl$_8$] · nH$_2$O

Compounds of this type were first obtained by the reduction of pertechnetate ions by metallic zinc in concentrated hydrochloric acid at approximately 100 °C [24]. Later this method was used with some modifications in other studies also for the preparation of salts of octachloroditechnetate ions [30, 45, 46, 79]. The reduction of pertechnetate ions under the given conditions is as a rule accompanied by the formation of brown solutions, which are rapidly oxidized in air to greenish-blue solutions with $\lambda_{max} \cong 615$ nm [24], which corresponds to the absorption of light by the [Tc$_2$Cl$_8$]$^{3-}$ ions. When salts with the cations M$^+$ = K$^+$, NH$_4^+$, 1/3Y$^+$, or (C$_4$H$_9$)$_4$N$^+$ are added to these solutions, crystals of complexes with the overall general formula M$_3$[Tc$_2$Cl$_8$] · nH$_2$O are formed [11, 30, 45, 46, 79] (Table 1).

Another method of synthesizing the octachloroditechnetates consists in reducing complexes of technetium(IV)–technetium(VII) in concentrated hydrochloric acid with molecular hydrogen under pressure in an autoclave at 120–180 °C [22, 42, 43]. As in the reduction with metallic zinc, the autoclave reduction is accompanied by the formation of brown solutions ($\lambda_{max} = 465$ nm), which are rapidly oxidized in air to greenish-blue solutions ($\lambda_{max} = 645$ nm). The reduction in the autoclave proceeds at an initial hydrogen pressure of ~3 MPa. When technetium(IV) complexes are used as the initial compounds in the synthesis, the working temperature of the autoclave must be higher by 20–40 °C than in the case of pertechnetates, since the hexahalogenotechnetates(IV) are as a rule less soluble than the corresponding pertechnetates. Compounds of the composition $M_3[Tc_2Cl_8] \cdot nH_2O$, where $M^+ = K^+$, Cs^+, NH_4^+, PyH^+, or $QuinH^+$ were obtained by the autoclave method (Table 1) [11, 30, 45, 46, 79].

All the octachloroditechnetates are substances which crystallize in the form of dark hexahedral needles, which give rise to a pale-grey powder on being ground. These complexes dissolve in hydrochloric acid with the formation of turquoise solutions, which rapidly oxidize in air or disproportionate in its absence when the concentration of the $[Tc_2Cl_8]^{3-}$ ions ($\sim 10^{-4}$ M) is low. The complexes are comparatively stable when the concentrations of such ions is high ($\sim 10^{-1}$ M). These compounds are sparingly soluble in polar organic solvents (clusters with organic cations are dissolved more effectively). In non-polar solvents, the octachloroditechnetates do not dissolve, whereas in water and alkaline solutions they rapidly hydrolyze, forming brown colloidal solutions of the hydrated oxide $Tc_4O_5 \cdot nH_2O$ [9, 11, 52, 80].

The autoclave method of synthesizing clusters with the $[Tc_2Cl_8]^{3-}$ anions [22, 42, 43] made it possible to observe an interesting phenomenon [11], which had not been noted by other workers [24, 30, 60–63, 79] and which involves a non-stoichiometric variable content of certain singly-charged cations ($M^+ = K^+$, NH_4^+, or Cs^+) in octachloroditechnetate salts having the general formula $M_3[Tc_2Cl_8] \cdot nH_2O$. It was found that the composition of these clusters depends greatly on the hydrochloric acid concentration in the mother liquor from which they are crystallized. For example, at an HCl concentration of 4–5 M, the K:Tc:Cl ratios in the potassium octachloroditechnetate formed are $\sim 3:2:8$, whereas in concentrated HCl solutions the ratios approach $2:2:8$. The decreased K^+ content in the complexes is accompanied by a decrease in the proportion of crystallization water. It may also be noted [11] that the composition of these compounds depends on other factors as well, in particular on the rate of crystallization. However, X-ray powder analysis of the potassium octachloroditechnetate, obtained under different conditions, showed [11] that the complex undergoes no appreciable crystal-chemical changes: the positions of all the lines on the diffractograms remain unchanged and correspond to the dimensions of the trigonal unit cell ($a = 12.80 \pm 0.05$; $c = 8.30 \pm 0.04$ Å [81, 82]), while the slight variations in intensity of the individual lines is mainly associated with the presence of a texture in the specimens.

It can be assumed a priori [11] that the reason for the variable composition of the octachloroditechnetates may be either a variable oxidation state of technetium or substitution of some of the cations by H_3O^+ ions. The first explanation was proposed [22], when the accuracy (± 0.017 Å) of the X-ray diffraction study by a photomethod [81] precluded the determination of the fine difference between the Tc–Tc distances in the $[Tc_2Cl_8]^{3-}$ and $[Tc_2Cl_8]^{2-}$ anions, which is ~ 0.03 Å according to the latest X-ray diffraction data [11, 35, 60, 63, 82] (Table 1). An attempt was made [11] to determine the H_3O^+ ions in potassium octachloroditechnetate, which led to their detection from the weak characteristic absorption lines in the infrared spectra ($v(OH) = 2965$, 2930, and 2865 cm^{-1}; $\delta(HOH) = 1610$ cm^{-1} [83, 84]). The results [11] did not conflict with the revised X-ray diffraction data [82] according to which the K^+ ions in the unit cell of potassium octachloroditechnetate ($+2.5$) are present in two non-equivalent positions: sixfold (K') and threefold (K''); under these conditions, the sixfold positions were fully occupied by K^+ ions, while the threefold positions were half-occupied.

Summarizing all that has been said above concerning the structures of the octachloroditechnetates ($+2.5$), it may be concluded that their true composition is described by a formula with variable coefficients, namely $M'_6 M''_{3-x}(H_3O)_x[Tc_3Cl_8] \cdot nH_2O$, where x and n vary from 0 to 3. The substitution of some of the M' ions by H_3O^+ ions is possible by virtue of the similarity of the properties of the hydroxonium cation and the alkali metal cations both in solution and in the crystalline state [85, 86].

Knowledge of such fine differences between the compositions of the octachloroditechnetates ($+2.5$) is useful mainly for studying certain solid-phase reactions, since the dissolution of these compounds leads to their dissociation into the M^+, H^+, O^+, and $[Tc_2Cl_8]^{3-}$ ions and the properties of the solutions are then determined by the concentration of the octachloroditechnetate ions. The simplified formula of these complexes $M_3[Tc_2Cl_8] \cdot nH_2O$ was therefore used in subsequent studies [9, 11, 15, 42, 43, 48, 49].

Table 1. Crystal data of the technetium cluster compounds

No	Compound	Space group	a, Å ($\alpha°$)	b, Å ($\beta°$)	c, Å ($\gamma°$)	Interatomic distances, Å				References
						$Tc \equiv Tc$	$Tc-Tc$	$Tc-X_{term}$	$\rho^c-(\mu-X)$	
I	II	III	IV	V	VI	VII	VIII	IX	X	XI
1	$[Tc_2(H_2EDTA)_2(\mu-O)_2] \cdot 5H_2O$	Pna2$_1$	18.41 (1)	10.96 (1)	16.25 (1)	2.331 (1)	–	2.011 (7)-O	1.913 (8)	[54, 55]
2	$[Tc_2(SC_6H_4S)_4] \cdot CHCl_3$	P$\bar{1}$	8.534 (1) (107.02 (2))	8.842 (2) (98.13 (1))	11.192 (3) (100.60 (2))	–	2.591 (3)	2.207 (16)-N 2.2925 (7)-S	2.408 (6)-S	[56]
3	$Ba_2[Tc_2(TCTA)_2(\mu-O)_2]ClO_4 \cdot 9H_2O$	P2$_1$/n	15.128 (1)	18.822 (2) (105.43)	15.582 (1)	2.402 (1)	–	2.07 (2)-O 2.21 (2)-N	1.936 (7)-O	[57]
4	$\{[Tc_2(CH_3COO)_4](TcO_4)_2\}$	P2$_1$/n	8.324 (1)	7.826 (1) (101.81)	14.644 (1)	2.149 (1)	–	2.014 (15) 2.153 (5)-ax	–	[58]
5	$[(C_4H_9)_4N]_2[Tc_2Cl_8]$	P2$_1$/c	10.915 (2)	15.382 (3) (122.37)	16.409 (2)	2.147 (4)	–	2.33 (1)	–	[35]
6	$\{Tc_2[(CH_3)_3CCOO]_4Cl_2\}$	I4/m	11.515 (2)	–	10.625 (3)	2.192 (1)	–	2.032 (4) 2.408 (4)	–	[45]
7	$K_2[Tc_2(SO_4)_4] \cdot 2H_2O$	P$\bar{1}$	7.522 (2) (101.63)	7.666 (2) (100.25)	7.684 (2) (112.32)	2.155 (1)	–	2.02 (2) 2.249-ax	–	[59]
8	$K_6K''_{3-x}(H_3O)_x[Tc_2Cl_8]_3 \cdot nH_2O$	P3$_1$21	12.817 (3)	–	8.185 (2)	2.119 (1)	–	2.363 (14)	–	[11]
9	$K_3[Tc_2Cl_8] \cdot 2H_2O$	P3$_1$21	12.838 (3)	–	8.197 (2)	2.117 (2)	–	2.364 (16)	–	[60]
10	$(NH_4)_3[Tc_2Cl_8] \cdot 2H_2O$	P3$_1$21	13.04 (2)	–	8.40 (1)	2.13 (1)	–	2.36 (3)	–	[61, 62]
11	$Y[Tc_2Cl_8] \cdot 9H_2O$	P42$_1$2	11.712 (2)	7.661 (2)	7.661 (2)	2.105 (2)	–	2.363 (4)	–	[63]
12	$(C_5H_5NH)_3[Tc_2Cl_8]$	P$\bar{1}$	8.323 (2) (78.60 (1))	8.780 (1) (70.02 (2))	8.830 (2) (85.78 (2))	2.1185 (5)	–	2.364 (15)	–	[64]
13	$[Tc_2(o-C_5H_4ON)_4Cl]$	I4/m	11.793 (3)	–	7.454 (1)	2.095 (1)	–	2.087 (3) 2.679 (1)-ax	–	[46]
14	$[Tc_2(CH_3COO)_4Cl]$	C2/c	15.035 (3)	7.801 (1)	12.894 (4)	2.117 (1)	–	2.065 (5) 2.656 (1)-ax	–	[48, 65]

15	K[Tc₂(CH₃COO)₄Cl₂]	P4₂/n	11.9885 (3)	–	11.4443 (2)	2.1260 (5)	–	2.07 (1), 2.589 (1)-ax	[48, 66]
16	[Tc₂(CH₃COO)₄Br]	C2/m	7.102 (3)	14.772 (4)	7.095 (2)	2.112 (1)	–	2.060 (4), 2.843 (1)-ax	[49, 67]
17	K₂ₙ[Tc₂Cl₆]ₙ	Cc	8.287 (2)	13.956 (3) (93.99 (5))	8.664 (2)	2.044 (1)	–	2.39 (7)	2.42 (8)	[49, 68, 69]
18	[(C²H₅)₄N]₂{[Tc₆(μ-Br)₆Br₆]Br₂}	P$\bar{1}$	12.877 (3) (81.25 (2))	13.684 (3) (67.45 (2))	14.613 (4) (63.91 (2))	2.188 (5)	2.66 (2)	2.5J (1), 3.03 (9)-ax	2.49 (1)	[50]
19	[(CH₃)₄N]₃{[Tc₆(μ-Br)₆Br₆]Br₂}	P6₃/mmc	13.781 (3)	–	1.801 (4)	2.154 (5)	2.702 (2)	2.514 (4), 3.142 (5)-ax	2.485 (4)	[50]
20	[(CH₃)₄N]₃{[Tc₆(μ-Cl)₆Cl₆]Cl₂}	Pcmn	11.583 (2)	13.527 (3)	24.387 (3)	2.16 (1)	2.69 (1)	2.38 (1), ~ 3.0-ax	2.38 (1)	[11, 70-72]
21	[(CH₃)₄N]₂[Tc₆(μ-Cl)₆Cl₆]	P$\bar{1}$	11.614 (4) (69.66)	11.633 (4) (65.89)	14.017 (6) (60.13)	2.22 (1)	2.57 (1)	2.40 (1), ~ 3.2-ax	2.35 (1)	[11, 71, 72]
22	{[Tc₈(μ-Br)₈Br₄]Br}·2H₂O	P2₁/n	7.573 (2)	13.428 (3) (103.05)	12.661 (2)	2.146 (2)	2.521 (2), 2.687 (23)	2.509 (8), 3.0-ax	2.504 (22)	[44, 53, 73]
23	[H(H₂O)₂]{[Tc₈(μ-Br)₈Br₄]Br}	P2₁/n	7.561 (1)	13.553 (5) (102.62)	12.620 (3)	2.155 (3)	2.531 (2), 2.70 (2)	2.53 (2), 3.0-ax	2.51 (2)	[44, 53, 73]
24	[H(H₂O)₂]₂{[Tc₈(μ-Br)₈Br₄]Br₂}	P2₁/a	22.151 (9)	9.026 (3) (105.94)	9.396 (4)	2.152 (9)	2.520 (9), 2.69 (1)	2.52 (1), 2.90-ax	2.52 (1)	[44, 53, 74]
25	[(C₄H₉)₄N]₂{[Tc₈(μ-Br)₄(μ-I)₄-Br₂I₂]I₂}	P2₁/c	16.875 (6)	23.842 (5) (111.11)	17.539 (3)	2.162 (9)	2.507 (2), 2.704 (10)	2.70 (2), 3.10-ax	2.62 (5)	[75]
26	[(C₄H₉)₄N]₂{[Tc₆Br₆(μ₃-Br)₅]	P2₁/c	9.999 (3)	14.072 (2)	21.127 (4)	–	2.59 (2)	2.53 (1)	2.45 (8)	[44, 53, 76]
27	[H₃O(H₂O)₃]₂[Tc₆Br₆(μ₃-Br)₅]	P2₁/c	9.258 (4)	9.211 (3) (101.09)	17.437 (7)	–	2.59 (1)	2.55 (1)	2.46 (5)	[44, 53]
28	[Fe(C₅H₅)₂]₃{[Tc₆I₆(μ-I)₆]I₂}	P6/m(P6)	15.343 (2)	–	12.704 (2)	2.17 (1)	2.67 (1)	2.71 (1)	2.61 (2)	[77]
29	[Tc₂(CO)₁₀]	P2₁/c	14.65 (1)	7.18 (1)	14.93 (1)	3.04 (1)	–	2.00 (1), 1.90 (1)-ax	–	[78]

The structures of the ammonium [61, 62], potassium [60, 81, 82], pyridinium [64] and yttrium [63] salts of the anion $[Tc_2Cl_8]^{3-}$ (Fig. 1a; Table 1) have been investigated by single crystal X-ray diffraction analyses. The characteristic features of the molecular structure of this anion are its eclipsed conformation (D_{4h} symmetry) and short Tc–Tc distance, much shorter than the analogous distance in metallic technetium (~ 2.70 Å [2]). It is striking that the change of the cation leads to an alteration of the Tc–Tc distance by ~ 0.04 Å (Table 1). The influence of the cation nature on the metal–metal distance in the rhenium and molybdenum complexes of the type $[M_2X_8]^{n-}$ (where M = Re or Mo and X = Cl or Br) with a quadruple M–M bond is weaker (the distance changes by ~ 0.02 Å [1]). In the case of technetium the true change in the M–M bond length on replacing the cations in the octachloroditechnetates is apparently somewhat smaller than the unduly high experimental value, owing to the low accuracy of the X-ray diffraction studies [81, 82].

2.3.2 The Octabromoditechnetates M_3 $[Tc_2Br_8] \cdot nH_2O$

When the potassium and ammonium halide complexes of technetium(VII) to technetium(IV) are reduced with molecular hydrogen under pressure (~ 3 MPa) in an autoclave in concentrated hydrobromic acid at 120–140 °C, the corresponding octabromoditechnetates(+2.5) of the type $M_3[Tc_3Br_8] \cdot 2H_2O$ (where M = NH_4, or K), which are isostructural with the analogous cluster technetium chlorides [9, 15] are formed. Since the $[Tc_2Br_8]^{3-}$ ions are much less stable than $[Tc_2Cl_8]^{3-}$ [9, 49, 80], the formation of the octabromoditechnetates is observed only in a mixture with technetium(II) and technetium(IV) bromide complexes. This mixture can be conveniently resolved under the microscope using the distinctive characteristics of the octabromoditechnetates (+2.5) (large hexahedral needles or conglomerates of the latter, whose comminution yields a pale-brown powder). The octabromoditechnetates (+2.5) are sparingly soluble in polar and non-polar organic solvents, hydrolyze rapidly in water and alkalis, and dissolve in concentrated hydrobromic acid solutions with formation of the $[Tc_2Br_8]^{3-}$ ions, which rapidly oxidize to $[TcBr_6]^{2-}$ or, in the absence of oxidants, disproportionate to $[Tc_2Br_6]^{2-}$ and $[TcBr_6]^{2-}$ [9, 80, 87].

2.3.3 Binuclear Cluster Compounds with the "Lantern" Type Structure

$\{[Tc_2(o\text{-}C_5H_4ON_4)_4]Cl\}$. Green crystals of $\{[Tc_2(o\text{-}C_5H_4ON_4)_4]Cl\}$ sublimed in vacuo ($\sim 10^{-9}$ Pa) at 350 °C without decomposition have been obtained by the reaction of ammonium octachloroditechnetate with o-hydroxypyridine in a nitrogen atmosphere at 150 °C [46]. The compound is sparingly soluble in organic and inorganic solvents. Fig. 1b illustrates the molecular structure of this cluster, which has D_{2h} symmetry. In contrast to $[Tc_2Cl_8]^{3-}$, the Tc–Tc bond in the oxopyridine cluster is additionally strengthened by four $o\text{-}C_5H_4ON$ bridging ligands, which have a contracting effect and lead to structures of the "lantern" type. The shortening of the Tc–Tc bond in the oxopyridine binuclear

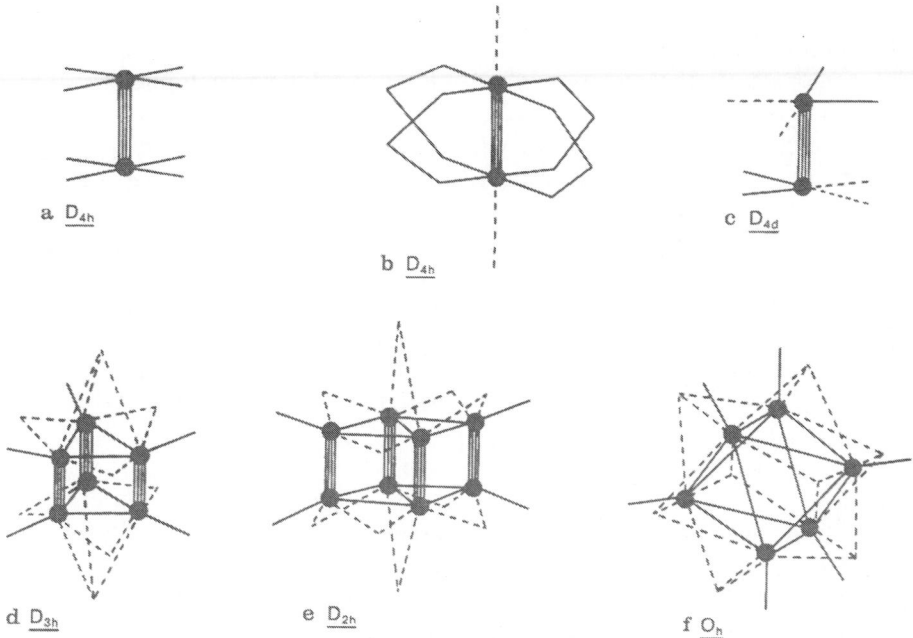

Fig. 1a–f. The main schematic structures of technetium acido-clusters [13]: **a** $[Tc_2X_8]^{n-}$ (X = Cl, Br; n = 2, 3); **b** $[Tc_2L_4]X_2]^{n-}$ (L = SO_4^{2-}, CH_3COO^-, $(CH_3)_3CCOO^-$, X = Cl, Br); **c** $[Tc_2X_6]^{2-}$ (X = Cl, Br); **d** $[Tc_6X_{12}]^{n-}$ (X = Cl, Br; n = 0, 1, 2); **e** $[Tc_8X_{12}]^{n+}$ (X = Br, I; n = 0, 1); **f** $[Tc_6(\mu_3-Br)_5Br_6]^{2-}$

cluster relative to the octachloroditechnetates (+2.5) amounts to ~ 0.015 Å and is close to the analogous value for binuclear rhenium(III) clusters [88].

Acetate complexes. When $K_3[Tc_2X_8] \cdot 2H_2O$ or $K_2[Tc_2X_6] \cdot 2H_2O$ (where X = Cl or Br) interacts with glacial acetic acid in an autoclave at 120–200 °C in an atmosphere of argon or molecular hydrogen (3–5 MPa), green crystals of the acetate binuclear complexes having a structure of the "lantern" type are formed: $K_2\{[Tc_2(CH_3COO)_4)]Cl_2\}$ and $\{[Tc_2(CH_3COO)_4)]X\}$ (where X = Cl or Br) [48, 49] (Table 1, Fig. 1b). The compounds are readily soluble in polar organic solvents and water and give rise to pale-green solutions, which gradually darken in air. On dissolution in concentrated hydrogen halide acid solutions, the corresponding octahalogenoditechnetates(+2.5) are formed. Technetium acetate complexes without inorganic cations are volatile at ~ 400 °C like the oxopyridine complex which has a similar structure.

2.3.4 Compounds Which Have Not Been Investigated by Single Crystal X-ray Diffraction Analyses

$Tc_4O_5 \cdot nH_2O$. A black powder, whose composition can be expressed by the overall formula $Tc_4O_5 \cdot 14H_2O$, has been obtained by the hydrolysis of

$K_3[Tc_2Cl_8] \cdot 2H_2O$ in argon [9, 52]. The freshly prepared hydrated oxide dissolves in concentrated hydrochloric acid with the formation of $[Tc_2Cl_8]^{3-}$ ions. This demonstrates the retention of M–M bonds of high multiplicity in the hydrated oxide. The aging of $Tc_4O_5 \cdot 14H_2O$ or its thermal dehydration is accompanied, according to ESCA, infrared spectroscopic, and X-ray powder diffraction data [9] by irreversible structural changes. The final dehydration product is the black oxide Tc_4O_5. The X-ray photoelectron and infrared spectra of Tc_4O_5 indicate structural non-equivalence of both technetium and oxygen atoms, but this is not associated with the formation of a metallic technetium phase, TcO_2, or other known technetium oxides on dehydration of $Tc_4O_5 \cdot 14H_2O$. According to magnetic measurements of the hydrated and anhydrous oxides [9, 52] there is approximately one unpaired electron per every two technetium atoms, as in all the binuclear d^4–d^5 clusters [1, 11]. The exact structure of both compounds is unknown, but their chemical and physico-chemical properties suggest the presence of multiple Tc–Tc bonds in their structures.

$[Tc_2(C_5H_4N)_2Cl_5]$. The thermal decomposition of pyridinium octa-chloroditechnetate (+2.5) in an argon atmosphere at 250–280 °C yielded the compound $[Tc_2(C_5H_5N)_2Cl_5]$, the chemical analysis of which as well as optical and X-ray photoelectron spectra, made it possible to ascribe to it a binuclear structure with a multiple Tc–Tc bond and with pyridine and chloride ligands [51]. The molecular structure of the compound is so far unknown. It has been isolated in the form of a dark-brown powder soluble in pyridine and unstable when dissolved in inorganic acids and alkalis. Treatment with nitric acid or a peroxide-alkali mixture as well as prolonged storage in air lead to its oxidation to technetium(IV) in the form of the stable compound $[Tc(C_5H_5N)_2Cl_4]$. In water the complex $[Tc_2(C_5H_5N)_2Cl_5]$ rapidly hydrolyzes, whereas in alkalis it dissolves with decomposition.

2.4 Binuclear Technetium Cluster Compounds with the d^4–d^4 Electronic Configuration of the Core Atoms

Compounds 4–7 (Table 1) belong to this group. The physicochemical properties of the given binuclear clusters are close to those of the analogous rhenium and molybdenum compounds with the quadruple M–M bonds and are as a rule isostructural with them [1, 89, 90].

2.4.1 Binuclear Cluster Compounds with the "Lantern" Type Structure

$\{Tc_2[(CH_3)_3CCOO]_4\}Cl_2$. The first binuclear d^4–d^4 technetium complex was obtained by Cotton and Gage [45] by oxidative substitution of the chloride ions in the coordination sphere of $[Tc_2Cl_8]^{3-}$ by trimethylacetate ions. The substitution reaction was carried out by heating (t = 150 °C) a suspension of $(NH_4)_3[Tc_2Cl_8] \cdot 2H_2O$ in pivalic acid under an argon atmosphere for 36 h. After the reaction mixture had been cooled, fine red crystals of $\{Tc_2[(CH_3)_3CCOO]_4\}Cl_2$ were precipitated; one of them was subjected to single crystal X-ray diffraction (Fig. 1b, Table 1), and the results demonstrated the similarity of the crystalline and molecular structures to binuclear technetium(III) and rhenium(III) pivalate complexes.

$\{[Tc_2(CH_3COO)_4]Cl_2\}$. The reduction of $[TcO_4]^-$ in a mixture of acetic and hydrochloric acids by molecular hydrogen under pressure yielded red crystals of $\{[Tc_2(CH_3COO)_4]Cl_2\}$ [47]. The structure of the compound was not investigated by single crystal X-ray diffraction, but, in view of the fact that the technetium and rhenium [47,91] acetate complexes are isostructural, presumably it too has a structure of the "lantern" type with four bridging acetate ions in the equatorial planes and two axial chlorine atoms on either side of the quadruple Tc–Tc bond, i.e. their structure is similar to that of $\{Tc_2[(CH_3)_3-CCOO]_4\}Cl_2$ (Fig. 1b). This structure of the acetate complex has also been confirmed from its diamagnetism and infrared spectroscopic data.

$\{[Tc_2(CH_3COO)_4](TcO_4)_2\}$. This compound was obtained by the slow oxidation of alcoholic solutions of the intermediate products of the synthesis of $[Tc_2(CH_3COO)_4Cl]$ and $K_2[Tc_2(Ac)_4Cl]_2$ in air [58]. The oxidation yields several red transparent crystals of the regular octahedral form.

The compound is built up of double-charged cations of tetra(μ-acetato)-ditechnetium(III) and pertechnetate anions, axially coordinated to the binuclear cation (Figs. 1b, 2; Table 1). The pertechnetate anions are bonded with the technetium atoms of the binuclear cluster by means of the bridge oxygen atoms of the $[TcO_4]^-$ group. The shortening of the Tc–Tc distance compared with the pivalate complex is due to the lower trans-effect of axial pertechnetate ions compared with chloride ions (see section 5.2.2). Pertechnetate ions like most pertechnetates with organic cations [12] are subjected to the pseudo-Jahn–Teller C_{3v}-distortion; the bond lengths of technetium atoms with the bridging oxygen atoms have the maximum values, whereas the lengths of the three remaining Tc–O bonds are the minimum. This type of distortion of pertechnetate ions corresponds to the case of $[(C_2H_5)_4N][TcO_4]$ [92,93], however the values of the mean Tc–O distances in tetra(μ-acetato)ditechnetium pertechnetate better corresponds to analogous distances in the structures of caesium and tetramethylammonium salts [92,94,95]. Thus the structure of $\{[Tc_2(CH_3COO)_4](TcO_4)_2\}$ again confirms the fact that pertechnetate ions are subject to the pseudo-Jahn–Teller C_{3v}-distortion.

Sulphate complexes. Small green transparent octahedral crystals of $K_2[Tc_2(SO_4)_4] \cdot 2H_2O$ were obtained [59] through the interaction of $K_3[Tc_2Cl_8] \cdot 2H_2O$ with a 10M solution of sulphuric acid at 100 °C. The compound dissolves readily in non-complexing mineral acids and hydrolyses slowly in water. The solutions are greenish with λ_{max} in the region of 700 nm and they slowly oxidize in air.

The $K_2[Tc_2(SO_4)_4] \cdot 2H_2O$ is made up of $[Tc_2(SO_4)_4]^{2-}$ anions, water molecules axially coordinated to them and K^+ cations (Fig. 1b, Table 1) [12][5]. The Tc–Tc distance in the sulphate complex has an intermediate value between the analogous distance in pivalate and acetate dimers, which is due to the intermediate force of the trans-effect of coordinated water molecules on the Tc–Tc bonds compared with chloride and pertechnetate ions in the respective compounds. In fact, in the analogous rhenium sulphate dimer the $Re-O_{ax}$ has a somewhat higher value (2.28 Å), practically equal with $[Tc_2(So_4)_4]^{2-}$ $M-O_{eq}$ bond lengths (2.01–2.02 Å), while the Re–Re distance is much lower in $\{Re_2(SO_4)_4(H_2O)_2\}^{2-}$ (2.214(2) Å) [96] than in $\{[Re_2(\mu-C_3H_7COO)_4](ReO_4)_2\}$ (2.251(2) Å) [97] though the structure of the latter is very close to that of tetra-(μ-acetato)ditechnetium pertechnetate, and the $M-O_{ax}$ distance in it (2.18 Å) is a little longer than in the respective technetium dimer (Table 1). Thus, from the above data, it follows that water molecules which are axially coordinated to dimers at distances of ~ 2.25 Å have the greater trans-effect on the M–M bond lengths than the pertechnetate groups with the distance of $Tc-O_{ax} \sim 2.15$ Å.

In paper [59] a sulphate complex with the molecular formula $K_4(H_3O)_2$-$[Tc_2(SO_4)_6]$ was synthesized. This compound is formed in concentrated sulphuric acid followed by further salting out from the solution by diethyl ester. The compound is a greenish-brown powder sparingly soluble in sulphuric acid and water, it is even less soluble in alcohol and other polar organic solvents. The freshly prepared compound dissolves in concentrated hydrochloric acid and gives a greenish-blue solution with $\lambda_{max} = 620$ nm, corresponding to the $[Tc_2Cl_8]^{3-}$ ions. When the compound is no longer fresh, its reaction with HCl does not lead to the formation of $[Tc_2Cl_8]^{3-}$ ions, but yields only greenish-brown solutions without characteristic absorption maxima. The compound possesses temperature-independent paramagnetism. The IR-spectra of $K_3(H_3O)_2[Tc_2(SO_4)_6]$ indicate the presence in its structure of H_3O^+ cations and, in contrast to the $K_2[Tc_2(SO_4)_4] \cdot 2H_2O$, sulphate anions monodentately bonded with technetium. The single crystal X-ray structural analysis of the given compound was not made; however the physico-chemical characteristics point to the presence of M–M bonds, though a polynuclear structure of the complex is not at all excluded.

2.4.2 The Octahalogenoditechnetates $M_2[Tc_2X_8]$

$[(C_4H_9)_4N]_2[Tc_2Cl_8]$. In contrast to $[Mo_2Cl_8]^{4-}$ and $[Re_2Cl_8]^{2-}$, the octa-chloroditechnetate ions have been obtained and characterized by means of chemical [32], spectroscopic [34], and single crystal X-ray diffraction [35] analytical methods comparatively recently. This is due to the lower stability of the $[Tc_2Cl_8]^{2-}$ ion, compared with $[Tc_2Cl_8]^{3-}$, in hydrochloric acid solutions

---

[5] The X-ray structural analysis was made by P.A. Kozmin et al. on the single crystal synthesized by the authors of paper [59].

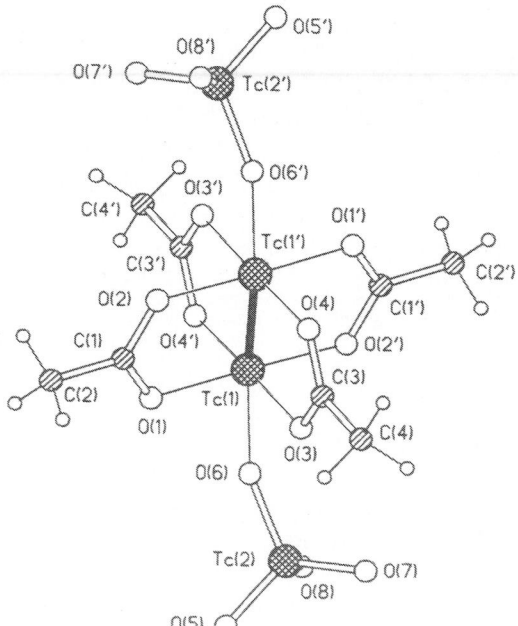

Fig. 2. Molecular structure of $[Tc_2(\mu\text{-}CH_3COO)_4]$ $[TcO_4]_2$ [58]

in which binuclear cluster chlorides are as a rule formed on reduction of technetium(VII)-technetium(IV) by various reductants. However, subsequent investigations showed that the $[Tc_2Cl_8]^{2-}$ ions are much more stable in organic solvents than $[Tc_2Cl_8]^{3-}$ ions.

This property of the octachloroditechnetate ions has been used to develop an optimum method for the synthesis of compounds with the $[Tc_2Cl_8]^{2-}$ anion [9]. For this purpose, the oxo-, oxochloro-, or chloro-complexes of technetium(VII)–technetium(IV) were reduced in concentrated hydrochloric acid (H_3PO_2, metallic zinc, or molecular hydrogen under a pressure of 3–5 MPa in an autoclave at $\sim 140°C$ can be used as reductants). An excess of a hydrochloric acid solution of tetrabutylammonium chloride is added to the resulting solution of the mixture of $[Tc_2Cl_8]^{2-}$, $[Tc_2Cl_8]^{3-}$, and $[Tc_2Cl_6]^{2-}$ ions and the solution is then extracted with a mixture of chloroform, carbon tetrachloride, and methylene chloride. This procedure gives a satisfactory separation of the organic and aqueous phases. In order to increase the solubility of technetium ions in the organic phase, acetone may also be added to the extraction mixture. When the solution is coloured an intense brown or turquoise, which corresponds to the presence in the mixture of a high content of $[Tc_2Cl_6]^{2-}$ or $[Tc_2Cl_8]^{3-}$ ions respectively, it is necessary to pass air through the extraction mixture until it becomes fully bright green in colour, but an yellow colour, indicating the formation of the $[TcCl_6]^{2-}$ ions, is not obtained. After this, the organic phase is separated and allowed to stand in air in order to evaporate the solvent slowly. Crystallization yields bright green $[(C_4H_9)_4N]_2[Tc_2Cl_8]$ dendrites, which are less soluble than $[(C_4H_9)_4N]_2[TcCl_6]$. After two recrystallizations from acetone or methyl chloride, a pure tetrabutylammonium octachloroditechnetate (III) is produced. By repeated employment of the evaporated mother liquors, formed in all stages of the experiment, as the starting materials for the reduction of technetium chloride complexes to technetium(III), the overall yield of tetrabutylammonium octachloroditechnetate can be increased almost to $\sim 100\%$.

The preparation of single crystals of tetrabutylammonium octachloroditechnetate(III) has been described by Cotton et al. [35]. The crystals were studied by

X-ray diffraction, which showed that the structure of the $[Tc_2Cl_8]^{2-}$ ions is analogous to that of the isoelectronic cluster rhenium and molybdenum chlorides and bromides with a quadruple M–M bond [1, 89, 90] (Fig. 1a, Table 1).

Table 1 shows that the replacement of the chloride ions in $[Tc_2Cl_8]^{2-}$ by trimethylacetate ions $[(CH_3)_3CCOO]^-$ leads to an appreciable increase in the Tc–Tc distance, which is caused by the trans-influence of the axial chlorine atoms in $\{Tc_2 [(CH_3)_3CCOO]_4Cl_2\}$ on the length of the quadruple M–M bond. The mechanism of the trans-influence is based on in the fact that the formation of the Tc–Cl_{ax} bonds in the carboxylate complexes involve mainly the $4d_82$, 5s, and $5p_z$ AO's of the technetium atoms and the $3p_z$ AO's of the chlorine atoms. Since the $4d_z2$, 5s, and $5p_z$ AO's of technetium also participate in the formation of the σ-components of the Tc–Tc bonds, there is a mutual correlation between the Tc–Cl_{ax} and Tc–Tc bond lengths. Analogous effects have been observed for the binuclear complexes of rhenium, molybdenum, chromium, platinum, etc. [1, 11, 89, 90].

$[(C_4H_9)_4N]_2[Tc_2Br_8]$. When several drops of hydrobromic acid are added to an acetone solution of $[(C_4H_9)_4N]_2 [Tc_2Cl_8]$ and the reaction mixture is then heated briefly to boiling, a carmine-red solution is formed; this colour corresponds to that of the $[Tc_2Br_8]^{2-}$ ions [34]. When the solution is evaporated, carmine-red $[(C_4H_9)_4N]_2[Tc_2Br_8]$ dendrites are formed, but their exact molecular structure has not been established yet.

2.5 Binuclear Technetium Clusters Compounds with the d^5–d^5 Electronic Configuration of the Core Atoms

Six complexes with this electronic structure, having the general formula $M_2[Tc_2X_6] \cdot nH_2O$, where M = K, NH_4, or $(C_4H_9)_4$ N; X = Cl or Br, and $n = 0$ or 2, have been described [9, 49, 68] (Table 1, compound 17). These compounds are formed under the conditions of far-reaching reduction of the technetium halide complexes in concentrated solutions of the corresponding hydrogen halides. The $[Tc_2X_8]^{3-}$ ions are the direct precursors of $[Tc_2X_6]^{2-}$. Atomic hydrogen (metallic zinc or aluminum in concentrated HCl) or molecular hydrogen (under a 3–5 MPa pressure in an autoclave at 140–180 °C) can be used as the reductant. The compounds are dark-brown cryptocrystalline powders or conglomerates of acicular crystals formed into flat and three-dimensional stars. The complexes are sparingly soluble in organic polar and non-polar solvents (the $[(C_4H_9)_4N]_2[Tc_2X_6]$ salts constituting an exception, since they are readily soluble in polar organic solvents), easily dissolve in hot solutions of the corresponding hydrogen halides forming brown solutions ($\lambda_{max} = 460$ nm), which are rapidly oxidized in air first to $[Tc_2X_8]^{3-}$ and then to $[TcX_6]^{2-}$. All compounds of this type are diamagnetic.

Crystals of $K_2[Tc_2Cl_6]$ suitable for X-ray diffraction analysis were obtained by the reduction of $K[TcO_4]$ in hydrochloric acid by molecular hydrogen at about 3 MPa in an autoclave at 140 °C by our reported procedures [50, 58, 69].

We noted that most of the $K_2[Tc_2Cl_6] \cdot 2H_2O$ crystals obtained on the bottom and walls of the test tube, covered by the mother liquor, are unsuitable for X-ray single crystal structural analysis. They are bulky fused star-shaped masses consisting of fine needles which could not be mechanically separated without damage. For the structural analysis, we selected the phase found on the walls of the test tube above the upper edge of the meniscus of the mother liquor. This phase is composed of compact, planar star-shaped concretions of nonhydrated $K_3[Tc_2Cl_8]$ and $K_2[Tc_2Cl_6]$. One of six crystals obtained by their mechanical separation under a microscope met the requirements for X-ray structural analysis. This crystal was a wedge-shaped needle, ~ 0.08 mm thick at the base, ~ 0.02 mm thick at the apex, and ~ 0.5 mm in length.

These crystals consist of potassium cations and polymeric $[Tc_2Cl_6]_n^{2n-}$ anions. The polymeric $[Tc_2Cl_6]_n^{2n-}$ anions consist of binuclear (Tc_2Cl_8) fragments linked by common bridging chlorine atoms Cl(1) and Cl(3) into infinite zigzag chains. Figs. 1c and 3 schematically show the connection of the (Tc_2Cl_8) binuclear fragments with each other in the structure of $K_2[Tc_2Cl_6]$.

The binuclear (Tc_2Cl_8) fragment (Figs. 1c, 3) is a four-sided prism formed by eight chlorine atoms which is strongly distorted relative to ideal D_{4d} symmetry and twisted by 39–45°. This prism contains two technetium atoms connected by a short Tc–Tc bond. Each technetium atom is surrounded by four chlorine atoms. We noted that the lengths of the bonds of the technetium atoms with the bridging chlorine atoms Cl(1) and Cl(3) hardly differ from the Tc–Cl(n) bonds with the terminal chlorine atoms. The difference between the Tc(1)–Cl(n) and Tc(2)–Cl(n) bonds is more significant. In addition, the mean Tc–Cl distances in the Cl(2)′–Tc(1)′–Cl(1)–Tc(2)–Cl(6) chains are about 0.05 Å greater than the analogous distances in the Cl(5)′–Tc(1)′–Cl(3)–Tc(2)–Cl(4) chains (Fig. 3, Table 1), i.e., the D_{4d} symmetry of the (Tc_2Cl_8) fragment is highly distorted.

The infinite $[Tc_2Cl_6]_n^{2n-}$ zigzag chains in the structure of $K_2[Tc_2Cl_6]$ are elongated along the Z-axis of the crystal and the technetium atoms forming these chains are located in the unit cell in two parallel planes with $y \sim (0 \pm 0.05)b$ and $Y \sim (0.5 \pm 0.05)b$. The K^+ ions are found between the $[Tc_2Cl_6]_n^{2n-}$ layers in two nonequivalent K(1) and K(2) positions. Each potassium cation has seven or eight contacts at distances from 3.0 to 3.5. All the contacts are equal or greater than the sum of the van der Waals radii, with the exception of K(2) $\cdots$ Cl(6) = 3.008(6) Å.

These studies [68, 69] showed that $K_2[Tc_2Cl_6]$ has the following major structural features: (1) the shortest Tc–Tc bond length[6] of all the analogous values reported (Table 1) [11, 12]; and (2) staggered conformation of the major structural fragment (Tc_2Cl_8). The formation of an M–M $(\sigma^2 2\pi^2 \delta^2 \delta^{*2})$ triple bond is observed in binuclear rhenium and molybdenum d^5–d^5 clusters [1]. The

---

[6] According to [98, 99], the compound $\{Tc_2O_3[C_5(CH_3)_5]_5\}_n$ has a polymeric structure with R(Tc–Tc) = 1.867(4) Å. However, these authors do not report the details of the X-ray diffraction experiment and the atomic coordinates. Therefore, these data seem doubtful.

Tc–Tc bond in the $[Tc_2Cl_6]^{2-}$ complex (Table 1) has a higher multiplicity since the M–M distance is about 0.1 Å shorter than the analogous distance in $[Tc_2Cl_8]^{2-}$ with a quadruple Tc–Tc bond [35] (see also part 5.2.3).

2.6 Binuclear Technetium Cluster Compounds with the d^4–d^3, d^3–d^3, and d^3–d^2 Electronic Configurations of the Core Atoms

$[Tc_2(\mu\text{-O})_2(H_2EDTA)_2] \cdot 5H_2O$. This compound has been obtained by reducing $[TcO_4]^-$ with an excess of $NaHSO_3$ in the presence of an excess of Na_2H_2EDTA with heating (70–80 °C) of the reaction mixture for several days [54, 55]. After recrystallization from 0.01 M $HClO_4$, the complex consisted of a diamagnetic cryptocrystalline red-brown powder. In contrast to all the binuclear technetium clusters examined above, the ethylenediaminetetra-acetate complex, with two bridging oxygen atoms additionally binding the technetium atoms, has the idealized C_{2v} symmetry. The ethylenediaminetetra-acetate ions behave as tetradentate ligands, coordinating around each technetium atom with the two nitrogen atoms and the two oxygen atoms of the acetate group (Fig. 4); the coordination polyhedra of the technetium atoms formed in this process consist of two distorted octahedra with a common edge made up of bridging oxygen atoms.

The compounds $K_2[Tc_2(\mu\text{-O})_2(NTA)_2]$ [54, 55], $Ba_2[Tc_2(\mu\text{-O})_2(TcTA)_2]$-$(ClO)_4 \cdot 9H_2O$ [54], $Ba_3[Tc_2(\mu\text{-O})_2(TCTA)_2]_2 \cdot 8H_2O$ [57], $Na_3[Tc_2(\mu\text{-O})_2$ $(TCTA)_2 \cdot 2CH_3OH \cdot 2H_2O$ [57], and $Ba[Tc_2(\mu\text{-O})_2(TCTA)_2] \cdot 3/4Ba(ClO_4)_2 \cdot$ $8H_2O$ [57] (where NTA is nitrylotriacetate, TCA is 1,4,7-triazacyclononano-N,N',N''-triacetate) have analogous structures. The single crystal X-ray structural analysis was done only for the first two of the above compounds. Table 1 shows the results, from which it follows, that in the case of the dimer Tc(III/IV) ($[Tc_2(\mu\text{-O})_2(TCTA)_2]^{3-}$) the Tc–Tc distance lengthens a little compared with the dimers Tc(IV/IV) ($Tc_2(\mu\text{-O}_2)_2(H_2EDTA)_2]$ and $[Tc_2(\mu\text{-O})_2(NTA)_2]^{2-}$). This is probably related to the "additional" unpaired electron in the Tc(III/IV) being located in one of the antibonding MO's (see part 5.2.4). However the authors [57] believed that this lengthening of the bond may also be due to substitution of the ligands.

All the above dimers are readily formed from the respective monomer complexes, e.g., TcO(EG) (TCTA) (where EG is ethylglicolate). When $NaBH_4$ in aqueous solution is added the dimeric TCTA-complex of Tc(IV) is formed. The Tc(III/IV) compounds are blue ($\lambda_{max} \sim 600$ nm), while the Tc(IV/IV) compounds are pink ($\lambda_{max} \sim 500$ nm) [54–57]. All the complexes are rather stable in aqueous and alcohol solutions. The Tc(III/IV) may be oxidized electrochemically, to the respective dimers Tc(IV/IV) or by the action of $K_2S_2O_8$. The reverse reduction easily occurs in the presence of hydrazine. The Tc(III/IV) compounds are paramagnetic and in the polycrystalline state at -100 °C give a broad ESR signal with a weak hyperfine structure, composed of 18 bands, which are due to two equivalent technetium atoms with a nuclear spin of 9/2.

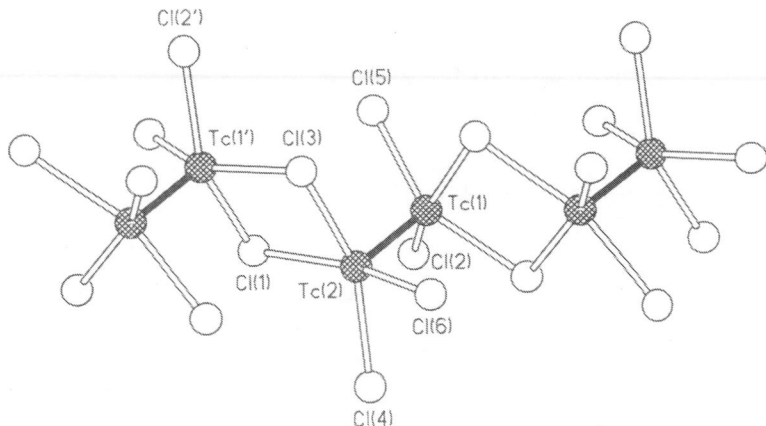

Fig. 3. Fragment of the structure of $K_2[Tc_2Cl_6]$ [68, 69]

TcO_2.[7] Technetium dioxide has a structure close to that of the ethylene-diaminetetra-acetate complex described above. According to the results of X-ray diffraction analysis of the TcO_2 powder [100] the technetium dioxide has a distorted rutile structure: six oxygen atoms are located at the vertices of a distorted octahedron in the centre of which is the technetium atom. The TcO_6 octahedra are linked to one another along the edges made up of oxygen atoms forming an infinite three-dimensional network. The technetium atoms in the octahedra have been brought closer together in pairs so that formation of Tc–Tc bonds is expected, but their strength is much lower (and their length much greater – see Table 1) than in the ethylenediaminetetra-acetate complex. This is shown indirectly by the paramagnetism of technetium dioxide, indicating the presence of approximately one unpaired electron per each technetium atom [9].

$K_3[Tc_2Cl_8O_2]$. This compound is hitherto the only representative of binuclear technetium clusters with the d^3–d^2 electronic configuration of the core atoms. The complex has been obtained by refluxing a solution of $K_3[Tc_2Cl_8] \cdot 2H_2O$ in methylethylketone in air [9, 15]. It is a yellow-green powder, sparingly soluble in methylethylketone and other organic solvents, but readily soluble in inorganic acids. The complex is resistant to the action of oxidants and reductants both in the solid phase and in solutions. It can be oxidized to $[TcO_4]^-$ only after prolonged refluxing with strong oxidants such as Ce^{4+} in 0.4 M H_2SO_4. In presence of complex-forming agents (for example, concentrated hydrochloric acid), the $[Tc_2Cl_8O_2]^{3-}$ ions decompose comparatively rapidly with the formation of initially equivalent amounts of $[TcCl_6]^{2-}$ and $[TcOCl_5]^{2-}$.

---

[7] The methods of synthesis and properties of TcO_2 have been described in detail in the monograph [2].

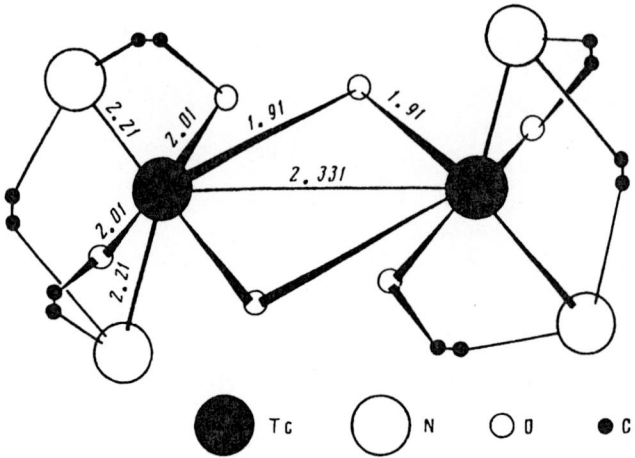

Fig. 4. Schematic presentation of nearest ligand surrounding the binuclear d^3–d^3 and d^3–d^2 technetium complexes (according to [11])

Infrared spectroscopy of $K_3[Tc_2Cl_8O_2]$ indicates the presence of two types of bonds between the oxygen and technetium atoms in the structure of this complex: $\nu(Tc{=}O) = 1020$ cm^{-1} (strong); $\nu(Tc{-}O) = 680$ cm^{-1} (medium); $\delta(Tc{-}O{-}Tc) = 475$ cm^{-1} (medium). The X-ray photoelectron spectra of this compound [101, 102] indicate the absence of bridging chlorine atoms in its structure. The complex is paramagnetic with approximately three unpaired electrons per two technetium atoms. Thus, although there are no direct data concerning the structure of $K_3[Tc_2Cl_8O_2]$, the physicochemical properties of this complex suggest that it has a binuclear structure with a weak metal–metal bond.

2.7 Polynuclear Technetium Cluster Compounds

Until recently, there had been no data on polynuclear ($n > 4$) clusters of either technetium or the other elements of group VIIb [1, 11], although such compounds had been synthesized [1, 8] for transition metals of groups V, VI, and VII. The elements of the vanadium and chromium subgroups formed clusters with mainly weak crystal field ligands, whereas platinum metals formed clusters with weak crystal field ligands. Thus, the possibility of formation of polynuclear clusters of the group VIIb elements and, particularly, the possibility of formation of technetium clusters with weak crystal field ligands are of fundamental interest to general chemistry in the light of the development of Mendeleev's concept about the periodic character of elements.

The first communications about the synthesis [9, 49] and structure [9, 70, 101] of polynuclear technetium halides appeared in 1982–83. These studies demonstrated the superior capacity of technetium to form polynuclear clusters with metal–metal bonds. Furthermore, the compounds isolated constituted a type of cluster compounds with conjugated quadruple metal–metal bonds, previously unknown for other elements. It is therefore useful to consider in greater detail the results of the studies on the chemistry of polynuclear technetium clusters that we have carried out.

In the deliberate search for ways to form polynuclear technetium clusters, we were guided by the following postulates [9]: (1) the stability and probability of the formation of polynuclear clusters should be greater in the case of a smaller effective charge on the technetium atoms, i.e. for complexes with ligands of lower electronegativity; (2) it is most likely that the polynuclear technetium clusters constitute a kind of intermediate link in the reduction of binuclear clusters with multiple Tc–Tc bonds to the metallic state; (3) other conditions being equal, the probability of the formation of polynuclear clusters should be greater at high concentrations of technetium ions in solutions; (4) in the presence of a large assortment of reductants capable of reducing binuclear technetium clusters to polynuclear clusters, it is preferable to employ reductants which do not contaminate the final products.

It was found [9, 49] that all the postulates enumerated above are satisfied by the autoclave method for the reduction of technetious acid in concentrated hydrogen halide solutions by molecular hydrogen under a pressure of 3–5 MPa at 140–220 °C. A series of experiments showed that the final product of the reduction of H [TcO_4] under these conditions is a mixture of outwardly similar crystalline substances with similar physico-chemical properties. The composition of the mixture can be described by the general overall formula [$TcX_{1.8 \pm 0.3} \cdot m(H_2O, OH^-, H_3O^+)]_n$, where X = I or Br and $n > 2$.[8]

According to X-ray photoelectron data [9, 102, 103], the technetium atoms in the compounds constituting this mixture are in an average oxidation state of 1.5–2.0 (a single narrow peak with $\varepsilon(Tc_{3d5/2}) = 254.7 \pm 0.3$ eV for X = Br and $\varepsilon(Tc_{3d5/2}) = 254.5 \pm 0.3$ eV for X = I is observed in the spectra), while the halogen atoms in the structures of these compounds belong to at least two types: terminal $(\varepsilon(Br_{3p3/2}) = 181.1 \pm 0.3$ eV; $(\varepsilon(I_{4d5/2}) = 48.4 \pm 0.4$ eV) and bridging $(\varepsilon(Br_{3p3/2}) = 182.8 \pm 0.3$ eV; $\varepsilon(I_{4d5/2}) = 49.8 \pm 0.4$ eV). Magnetic data indicate a weak paramagnetism of the $(TcX_2 \cdot 0.5H_2O)_n$ complexes, which include ~1 unpaired electron for 4–6 technetium atoms.[9]

The results of the studies on the physico-chemical properties of $(TcX_2 \cdot 0.5H_2O)_n$ described above suggested a polynuclear structure for the compounds constituting this mixture. Subsequently, this was confirmed by the single-crystal X-ray diffraction analysis of a series of compounds forming part of a mixture of polynuclear technetium bromides [9, 70, 101]. Using the differences between the outer appearances and the physico-chemical properties of the crystals, initially investigated by methods of microcrystal chemical analysis, it was later possible to achieve an initial separation of the mixture into its constituent parts under the microscope. Such separation made it possible to investigate thoroughly the physico-chemical properties of each phase in [$TcBr_2 \cdot 0.5H_2O]_n$ and to develop optimum methods for the synthesis of these compounds. Methods for synthesizing hexanuclear tetramethylammonium chloride cluster compounds of technetium were developed analogously.

2.7.1 Polynuclear Cluster Bromides

Figure 5 shows a scheme for synthesizing all the bromides of polynuclear technetium clusters presently known. Table 2 summarizes details of their ap-

---

[8] For brevity and also in view of the difficulty of determining the exact composition of this mixture, the above formula was expressed initially [9, 49, 102] in the form [$TcX_2 \cdot 0.5H_2O]_n$ and is used as such in this review.

[9] The values of μ_{eff} for [$TcX_2.0.5H_2O]_n$ have been determined with a low accuracy because the diamagnetic and paramagnetic contributions to the total magnetic susceptibility are comparable and the exact diamagnetic corrections for technetium compounds of this kind are unknown. For this reason, it was initially assumed [49] that the constituent compounds of this mixture were diamagnetic; only subsequent more accurate magnetic susceptibility data confirmed their weak paramagnetism [9], the nature of which has not yet been established.

pearance and main properties, while Table 1 lists their main crystallo-chemical and structural data.

Compounds (I), (II), (III), and (IV) (Table 2) have similar crystal and molecular structures (Fig. 1e), characterized by the following: the crystals of these compounds are made of octanuclear cluster fragments $[Tc_8Br_4(\mu\text{-}Br_8]^{n+}$ ($n = 1, 0$), Br^- anions and $[H(H_2O)_2]^+$ cations (in compound (I), H_2O molecules). The $[Tc_8Br_4(\mu\text{-}Br)_8]^{n+}$ fragments have an identical structure in all the compounds (Fig. 1e): eight technetium atoms form a prism, the base of which is a rhomb; the short diagonal of the rhomb corresponds to formally single M–M bonds; the sides of the rhomb correspond to the M–M bonds with formal multiplicity of about 0.5; and side edges of the prism correspond to formally quadruple ones.

In compounds (II) and (III) (Table 2) a very short O–O' distance is observed, which is due to the formation of strong hydrogen bonds in $[H(H_2O)_2]^+$ cations, whereas in compound (I) the O–O' distance is about 0.3 Å longer (Table 1 and papers [44, 73, 74, 101]). In compounds (I) and (II) in contrast to (III), the Br_{ax} atoms are simultaneously bonded by way of an electrostatic interaction with two octanuclear fragments, thus forming in the crystals of these compounds endless chains.

In compounds (III) and (IV), the fragments $\{[Tc_8X_4(\mu\text{-}X)_8]X_2\}^{2-}$ are structurally isolated; in the crystals, they are bonded to $[H(H_2O)_2]^+$ cations by means of hydrogen bonds and by weak electrostatic interactions such as Br...Br'.

The main difference between compounds (II), (III) and (IV) on the one hand and (I) on the other is related to a difference in the formal oxidation degree of technetium: in compounds (II), (III) and (IV) the value is $+1.5$, and in (I) it is $+1.625$. This leads to a slight difference in the length of the Tc–Tc and Tc–Br bonds (except Tc–Br_{ax}). For example, when passing from (I) to (II), the average lengthening of the Tc–Tc and Tc–Br bonds is ~ 0.01 Å, and the average shortening of the Tc–Br_{ax} bonds is ~ 0.005 Å (Table 1 and paper [49]). It should be noted that in the case of chloride hexanuclear clusters, $[Tc_6Cl_6(\mu\text{-}Cl)_6]^-$ and $[Tc_6Cl_6(\mu\text{-}Cl)_6]^{2-}$, the similar addition of only one electron leads to more considerable changes in the length of the Tc–Tc and Tc–X bonds [44, 50].

The main difference between the structures of compound (IV) and compounds (I), (II), and (III) is that in compound (IV) half of the bridging and terminal bromine atoms, and all the axial bromine atoms, are substituted by iodine atoms, with partial substitution being statistical [75]. Thus, according to the X-ray diffraction data, the effective Tc–Tc distances are intermediate between analogous distances in bromide and iodide complexes of technetium(IV) (Table 1) [104, 105]. We observed that the substitution of I for Br has almost no effect on the Tc–Tc distances.

Based on its method of synthesis and some of its spectroscopic properties, compound (X) belongs to the group of octanuclear clusters (I)–(IV). However, its exact structure is still undefined. According to most of its physico-chemical properties, compound (X) is similar to $[Fe(C_5H_5)_2]_3\{[Tc_6(\mu\text{-}Cl)_6Cl_6]Cl_2\}$ and

Table 2. Appearances and main properties of polynuclear bromide technetium clusters [11, 44, 53, 72, 76]

No	Compound	Apperance and properties
I	II	III
I	$\{[Tc_8(\mu\text{-}Br)_8Br_4]Br\}\,2H_2O$	Black oblong needle-like crystals with a square cross-section and intensive metallic lustree. Insoluble in hot HBr (concentr.). Grinding yields black powder.
II	$[H(H_2O)_2]\{[Tc_8(\mu\text{-}Br)_8Br_4]Br\}$	Black oblong needle-like crystals with a square cross-section and intensive metallic lustre, sligtly soluble in hot HBr (concentr.) forming a green solution. Grinding yields black powder.
III	$[H(H_2O)_2]_2\{[Tc_8(\mu\text{-}Br)_8Br_4]Br_2\}$	Black oblong (up to 5 mm) needle-like crystals with a square cross-section. No metallic lustre is observed. Very soluble in HBr (concentr.), alcohol and other polar solvents, forming reddish-brown solutions. Grinding yields reddish-brown powder.
IV	$[(C_4H_9)_4N]_2\{[Tc_8(\mu\text{-}Br)_4(\mu\text{-}I)_4 Br_2I_2]I_2\}$	Fine black rhombic crystals. Very soluble in polar organic solvents, forming black-brown solutions. Grinding yields black powder with a brown tinge.
V	$[(C_2H_5)_4N]_2\{[Tc_6(\mu\text{-}Br)_6]Br_6]Br_2\}$	Black-brown dendrite-like crystals consisting of fine needles. Soluble in hot HBr (concentr.), slightly soluble in polar organic solvents forming brown solutions. Grinding yields black-brown powder.
VI	$[(CH_3)_4N]_3\{[Tc_6(\mu\text{-}Br)_6Br_6]Br_2\}$	Black-brown plate- and needle-like crystals. Soluble in hot HBr (concentr.), forming black-brown solutions. Grinding yields brown powder with a violet tinge
VII	$[H_3O(H_2O)_3]_2[Tc_6(\mu_3\text{-}Br)_5Br_6]$	Black with red tinged crystals of a distorted octahedral form. Very soluble in alcohol, HBr (concentr.), acetone and other polar solvents with formation of red-brown solutions. Grinding yields red-brown powder.
VIII	$[(C_4H_9)_4N]_2[Tc_6(\mu_3\text{-}Br)_5Br_6]$	Red transparent octahedral crystals very soluble in polar organic solvents (less soluble in inorganic and nonpolar organic solvents), forming reddish-brown solutions. Grinding yields red powder.
IX	$[H_3O(H_2O)_3]_2[Tc_6(\mu_3\text{-}Br)_5Br_6]\cdot 4H_2O$	Black with brown tinged needle-like crystals very soluble in HBr (concentr.), alcohol and other polar solvents, forming brown solutions. Grinding yields brown powder.
X	$[Fe(C_5H_5)_2]_2\{[Tc_8(\mu\text{-}Br)_8Br_4]Br_2\}$	Black fine needle-like crystals with metallic lustre. Insoluble in orgaic and inorganic solvents but does not decompose. Grinding yields amorphous black powder.

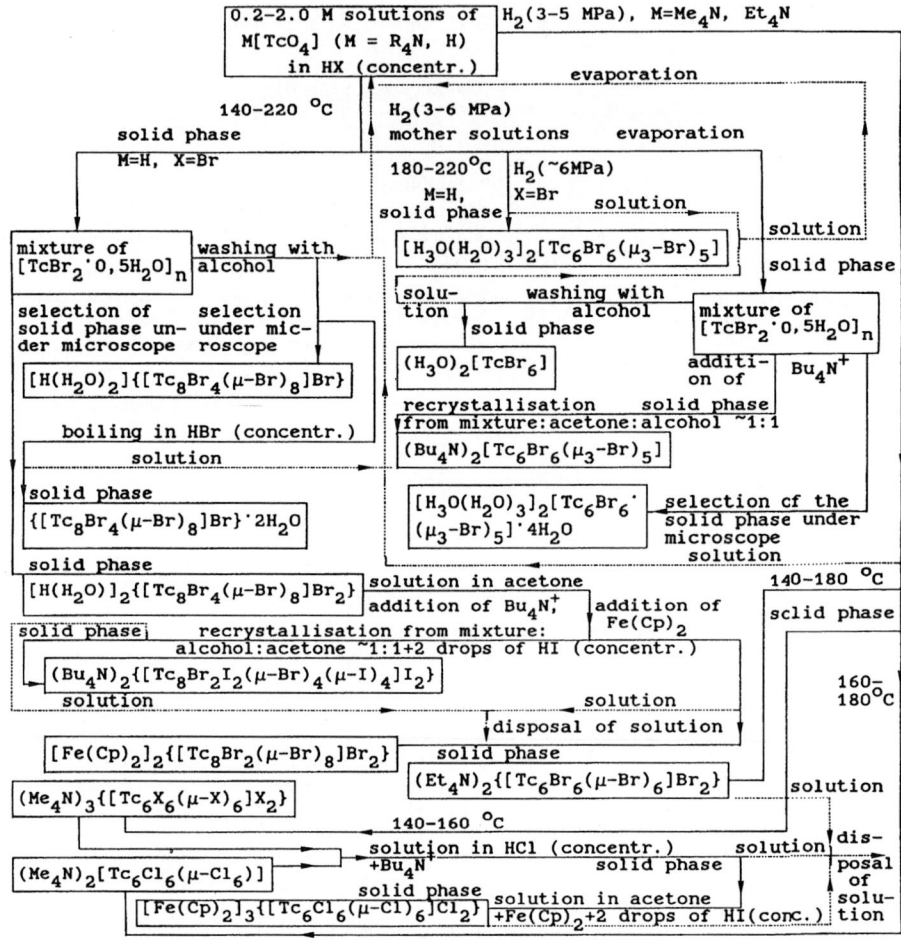

Fig. 5. Scheme of the synthesis of polynuclear technetium clusters [13, 44]

$[Fe(C_5H_5)_2]_3\{[Tc_6(\mu\text{-}I)_6I_6]I_2\}$. For example, the Mössbauer ^{57}Fe spectra of all three ferricinium-containing compounds are characterized by the presence of one absorption peak with the averaged parameters of Fe^{2+} and Fe^{3+} peaks at room temperature and the appearance of peaks characteristic of ferrocene and ferricinium at liquid nitrogen temperature. The properties of these compounds are due to a fast electron exchange between the ferricinium cations and the cluster anions. This electron exchange results from the overlapping of 4s-AO's of the iron atoms and 5s(5p)-AO's of the technetium atoms.

The crystals of (V) and (VI) (Table 1) are built from tetraalkylammonium cations and cluster anions $\{[Tc_6(\mu\text{-}Br)_6Br_6]Br_2\}^{n-}$, where $n = 2$ for (V) and $n = 3$ for (VI). The metal core of the cluster anions $\{[Tc_6(\mu\text{-}Br)_6Br_6]Br_2\}^{3-}$ (Fig. 1d) is a regular trigonal prism formed from technetium atoms. The vertical edges of this prism are consistent with the M–M bonds with a formal multiplicity of ~ 4.0 (according to [106, 107], the multiplicity of these Tc–Tc bonds is 3.0) and the horizontal edges are consistent with M–M bonds with a formal multiplicity of ~ 0.5 (according to [106, 107], 1.0). In horizontal planes, the technetium atoms are additionally bridged in pairs through bromine atoms and each technetium atom is bonded to the terminal bromine atoms. Over each trigonal base of the prism, there is an axial bromine atom simultaneously bonded to three technetium atoms through a weak electrostatic interaction. Thus, the structures of compounds (V) and (VI) are similar to those of $[(CH_3)_4N]_3\{[Tc_6(\mu\text{-Cl})_6Cl_6]Cl_2]\}$ and $[(CH_3)_4N]_2[Tc_6(\mu\text{-Cl})_6Cl_6]$ [70, 72, 74].

The molecular structures of compounds (VII), (VIII), and (IX) also have much in common (Fig. 1f, Table 1): all three compounds consist of $[TC_6Br_6(\mu_3\text{-}Br)_5]^{2-}$ anions and M^+ cations where $M^+ = (C_4H_9)_4N^+$ or $[H_3O(H_2O)_3]^+)$. For compound (IX) such a structure is established indirectly, because compound (VIII) is easy to obtain by adding $[(C_4H_9)_4N]^+$ cations both to the solutions of compound (IX) and to those of compound (VII). Besides, compound (IX) possesses chemical and physico-chemical properties similar to those of compounds (VII) and (VIII).

Figure 1f shows the molecular structure of the $[Tc_6Br_6(\mu_3\text{-}Br)_5]^{2-}$ anion. In general, the structure of this anion is similar to the structure of the well-known octahedral halogenide clusters of molybdenum and tungsten $[M_6X_8]^{4+}$ ($X = Cl, Br, I$) [5, 8]. The principal difference is that the eight equivalent positions of bridging bromine atoms in the technetium clusters are not fully occupied.

The proposed composition and structure of the cluster anion $[TC_6Br_6(\mu_3\text{-}Br)_5]^{2-}$ is confirmed by experimental facts [44, 53, 76]: chemical analysis data [44, 76]; the intensity of the electron density peaks corresponding to the bridging bromine atoms in the Fourier synthesis for compound (V), which constitute about 50–60% of the intensity of the peaks corresponding to the terminal bromine atoms; the anomalously high B_{iso}^{equ}-values observed for the μ_3-Br atoms (10–15 Å^2), which for all other atoms are quite normal and do not change during further least-squares refinement of the structure; anomalously high values of R_1-factors (13.3% in the isotropic approximation and 12.0% in the anisotropic one) and the absence in the Fourier synthesis of the electron density peaks, which could be attributed to unrevealed atoms; further full-matrix least-squares refinement of the structure of compound (VIII) together with the refinement of the multiplicity of the positions of the bridging bromine atoms which result in a marked decline of the R-factor ($R_1 = 4.52\%$, $R_w = 5.94\%$) accompanied by the removal of all the above-mentioned drawbacks of the solution of the structure of compound (VIII). It should be noted that the multiplicity of positions of the bridging bromine atoms markedly decreased and the most probable variant of the structure of compound (VIII) is $[TC_6Br_6(\mu_3\text{-}Br)_5]^{2-}$, because it has the lowest values of the R-factors of all the stoichiometric compositions of this compound. This variant is in accordance with the chemical analysis data and does not contradict the experimentally found paramagnetism of compound (VIII) [44, 76].

It should be noted that for compound (VII) analogous phenomena are observed both in the determination of its structure and in its final variant after refinement of the multiplicity of the positions of bridging bromine atoms. However, the authors of the structural study [70], performed on the crystals synthesized by us using the method described in paper [44], believe that in compound (VII) some of the μ_3-Br atoms are replaced by μ_3-OH groups. According to the same authors of [70] a considerable shortening of some of the Tc-(μ_3-Br) distances to the value of ~ 2.4 Å agrees with this fact. Thus they represent the composition of the octahedral cluster anion in compound (VII) as $[Tc_6Br_6\mu_3\text{-}(Br,OH)_6]^{2-}$.

However, from our point of view [44, 76], the shortening of part of the Tc-(μ_3-Br) distances in the $[Tc_6Br_6(\mu_3\text{-}Br)_5]^{2-}$ anion can be accounted for without introducing additional atoms into the assumed compositions of the given anion. The presence of these atoms is confirmed neither by chemical nor by physico-chemical investigation methods. In fact, the shortening of these bonds is observed only for μ_3-Br atoms whose multiplicity of positions is 1.5 times lower than that for other μ_3-Br atoms [44, 76]. In other words, the positions corresponding to the first of the bridging Br atoms are characterized either by stronger bonds between the bromine and technetium atoms, or by the absence of bromine atoms in these positions. Thus, in the $[Tc_6Br_6(\mu_3\text{-}Br)_5]^{2-}$ ion a new type of distortion of octahedral clusters is observed. The reasons for this distortion which are related to the electron structure of the given cluster ion are considered in part 5.3.3.

2.7.2 Polynuclear Cluster Chlorides

The reduction of $[(CH_3)_4N]_2 [TcCl_6]$ or $[(CH_3)_4N] [TcO_4]$ in concentrated hydrochloric acid by molecular hydrogen under pressure (3–8 MPa) in an autoclave at 140–220 °C leads to the formation of a mixture of dark-brown almost black crystals of different geometrical shapes (tetrahedral rods, dendrites, and planar layer polyhedra). Using methods analogous to those described for $(TcBr_2 \cdot 0.5H_2O)_n$, it was possible to separate this mixture and to find the optimum conditions for synthesizing crystals of the two compounds: $[(CH_3)_4N]_3\{[Tc_6Cl_6(\mu\text{-}Cl)_6]Cl_2\}$ and $[(CH_3)_4N]_2[Tc_6Cl_6(\mu\text{-}Cl)_6]$ [11, 72].

$[(CH_3)_4N]_3\{[Tc_6Cl_6(\mu\text{-}Cl)_6]Cl_2\}$. This is formed with a maximum yield at 140–150 °C under an initial hydrogen pressure in the autoclave of 3–5 MPa. Crystals of the compound have the form of elongated tetragonal rods or are otherwise frequently conglomerates of such rods in the form of dendrites. After grinding, the crystals give rise to a dark-brown powder that is soluble in hydrochloric acid, water, alcohol, and other polar solvents, giving rise to brown solutions [72].

The molecular structure of the compound is presented in Fig. 1d and Table 1 [70, 71]. In broad outline, the $\{[Tc_6Cl_6(\mu\text{-}Cl)_6]Cl_2\}^{3-}$ structure resembles that of the octanuclear technetium bromide cluster, but in the structure of the hexanuclear complex the Tc atoms are located at the vertices of a prism whose base consists of an equilateral triangle and not a rhomb as, for example, in the $[H(H_2O)_2] \{[Tc_8Br_4(\mu\text{-}Br)_8]Br\}$ structure. The vertical edges of this prism are much shorter than the horizontal edges and they correspond to quadruple Tc–Tc bonds, while the horizontal edges correspond to M–M bonds with a lower formal multiplicity. In addition, the technetium atoms are linked in the horizontal planes by bridging chlorine atoms. Furthermore, in the $[(CH_3)_4N]_3\{[Tc_6Cl_6(\mu\text{-}Cl)_6]Cl_2\}$ molecule there are six terminal chlorine atoms and two chlorine atoms of the ionic type, which constitute a kind of axial ligand. The compound is paramagnetic with μ_{eff} corresponding to the presence of approximately one unpaired electron per molecule.

$[(CH_3)_4N]_2[Tc_6Cl_6(\mu\text{-}Cl)_6]$. This is formed under analogous but more severe conditions (150–180 °C) in an autoclave by the reduction of $[(CH_3)_4N]_2$-$[TcCl_6]$ or $[(CH_3)_4N][TcO_4]$ in concentrated hydrochloric acid [72]. Under the microscope the crystals of this compound resemble anthracite and are orthorhombic layered plates. After grinding, a dark-brown powder with a violet tint is produced. It is soluble in hydrochloric acid, water, alcohol, and other polar solvents, giving rise to brown solutions. The complex is slightly paramagnetic and has an ESR spectrum [15, 72].

According to single crystal X-ray diffraction data, obtained by Koz'min and co-workers [71], the structure of the cluster is close to that of $[(CH_3)_4N]_3\{[Tc_6Cl_6(\mu\text{-}Cl)_6]\,Cl_2\}$ (Fig. 1d, Table 1). The differences occur in the length of the Tc–Tc bonds. For example, in the case of $[(CH_3)_4N]_2\,[Tc_6Cl_6(\mu\text{-}Cl)_6]$ the three Tc–Tc bonds located in the vertical planes are lengthened at the expense of the shortened M–M bonds in two horizontal planes. Furthermore, in the structure of this hexanuclear complex the terminal chlorine atoms in the neighboring polynuclear fragments take the role of axial ligands.

$[(C_4H_9)_4N]_3\{[Tc_6Cl_6(\mu\text{-}Cl)_6]Cl_2\}$. Similar to $[(CH_3)_4N]_3\{[Tc_6Cl_6(\mu\text{-}Cl)_6]Cl_2\}$, the compound was synthesized in an autoclave by reduction of tetrabutylammonium pertechnetate in a concentrated hydrochloric acid solution by molecular hydrogen under pressure (~ 5 MPa) at a temperature of 150 °C. The initial product was a black hemispheric particle. After being dissolved in acetone and further salted out by diethyl ester, a brown powder was obtained, whose chemical analysis, magnetic properties ($\mu_{eff} = \sim 1.7\ \mu_B$), and IR and X-ray photoelectron spectra indicate that its composition may be described by the formula $[(C_4H_9)_4N]_3\{[Tc_6Cl_6(\mu\text{-}Cl)_6]Cl_2\}$ [15, 102].

$[Fe(C_2H_5)_2]_3\{[Tc_6Cl_6(\mu\text{-}Cl)_6]Cl_2\}$. This compound was obtained from the previous one through reaction with ferrocene in an acetone solution [15, 77, 108]. This compound is a black fine-crystalline powder, which contains needle-like crystals with a metallic lustre. The compound is insoluble in most organic and inorganic solvents, possesses a metallic type of temperature dependence with respect to conductivity, anomalously high paramagnetism and unusual Mössbauer ^{57}Fe spectra.

Thus, although only the first steps have been made in elucidating the chemistry of polynuclear clusters, these steps are rather important: (1) the new type [11, 12, 13] of polynuclear clusters with acido-ligands in which formally triple and quadruple M–M bonds alternate with formally single M–M bonds were synthesized and structurally characterized; (2) hexanuclear trigonal-prismatic rhenium clusters of an analogous structure were synthesized and comprehensively characterized [109–112]; (3) the synthesized compounds confirmed the "anomalous" cluster-forming properties of technetium namely, the increased ability of technetium to form mixed-valency, paramagnetic, and acido-clusters with excess electrons and strong M–M bonds [10, 15].

3 Chemical Properties in Solutions

3.1 Mechanism of the Metal–Metal Bond Formation

In complexing acid media (e.g. hydrohalogenic acids HX) the reduction of pertechnetate ions proceeds in a different way. In the works [11, 13, 15] it was shown that such a reduction occurred according to the scheme (1).

$$[TcO_4]^- \Rightarrow [TcOX_4]^- \Rightarrow [TcX_6]^{2-} \Leftrightarrow [Tc_2X_8]^{2-} \Leftrightarrow [Tc_2X_8]^{3-}$$

$$\Leftrightarrow [Tc_6X_{14}]^{2-} \Leftrightarrow [Tc_6X_{14}]^{n-} \Leftrightarrow [Tc_6X_{12}]^{2-} \Rightarrow [Tc_6X_{11}]^{2-}$$

$$\Leftarrow [Tc_8X_{14}]^{n-} \tag{1}$$

In the process of their reduction, technetium ions unite to form increasingly large clusters. Fig. 1 shows the molecular structures of these clusters, Fig. 6 shows the results of the calculation of their electronic structure in terms of the EHT method [113].

From the results presented it follows that the "driving force" behind the growth of technetium clusters in the process of their reduction is a decrease in the total electron energy of the ions due to the formation of M–M bonds. In fact, as is shown in Fig. 6, if the M–M bonds were absent the total electron energy of technetium complexes would be considerably higher and the complex would be unstable. However, besides purely thermodynamic reasons leading to the cluster formation, there should also be kinetic possibilities for these processes to take place. This aspect of technetium cluster formation is partially considered below.

A number of studies [42, 43, 52, 114] have shown that $[TcX_6]^{2-}$ entered into reduction reactions more readily in hydrohalogenic acids at a concentration of 4–7 M than in concentrated hydrohalogenic acids. It was assumed [9, 11, 52] that the reduction of $[TcCl_6]^{2-}$ was promoted by their acidic hydrolysis (anation reaction) (2).

$$[TcCl_6]^{2-} + H_2O \Rightarrow [TcCl_5(H_2O)]^- + Cl^- \tag{2}$$

The resulting Tc(IV) aqua-ion is reduced at a high rate, e.g. under the action of pressurized molecular hydrogen [42, 43], hydrazine [114] or other reductants (3) [115].

$$[TcCl_5(H_2O)]^- + e^- \Rightarrow [TcCl_5(H_2O)]^{2-} \tag{3}$$

The Tc(III) aqua-ion is rather unstable and enters one of the following reactions: (1) oxidation to Tc(IV) [11, 80, 87]; (2) dimerization with the formation of $[Tc_2X_8]$ [80, 87]; (3) disproportionation with the formation of Tc(IV) and Tc(II) [80, 87]. It should be noted that $[Tc_2Cl_8]^{2-}$ is rather unstable in aqueous solutions (the lifetime is several minutes [30, 115]); however, in organic solvents it is stable and can be obtained from them in a solid state, e.g. in the form of $[Bu_4N]_2[Tc_2Cl_8]$ [11, 32, 34, 35].

Further reduction of Tc(III) may occur by the disproportionation mechanism as well as according to the reaction (4) [87].

$$[Tc_2Cl_8]^{2-} + e^- \Rightarrow [Tc_2Cl_8]^{3-} \tag{4}$$

The stability of the $[Tc_2X_8]^{3-}$ ions forming in hydrohalogenic acids is determined by the sum of the competing processes (5–7) [80, 87].

$$[Tc_2X_8]^{3-} + 4H_2O \Rightarrow [Tc_2X_8\text{-}(nH_2O)_n]^{(3-n)-} + nX^- \tag{5}$$

$$[Tc_2X_8]^{3-} + X^- \Rightarrow [Tc_2X_9]^{4-} \tag{6}$$

$$[Tc_2X_9]^{4-} + HX \Rightarrow Tc(II) + Tc(III) \tag{7}$$

Reaction (5) describes the acidic hydrolysis process (the anation reaction) and does not lead to rupture of the M–M bond, its final product being $Tc_4O_5 \cdot nH_2O$ [80]. Reaction (6) is the complex formation reaction. The initially formed $[Tc_2X_9]^{4-}$ ion is extremely unstable; it disproportionates with the rupture of M–M bonds according to reaction (7). The rate of the equilibrium complex formation reaction (6) limits the rate of the total disproportionation reaction of $[Tc_2X_8]^{3-}$ with the rupture of the Tc–Tc bond and increases down the sequence HCl, HBr, HI, and in proportion to the concentration of these acids [80, 87]. In the absence of oxidants a total mobile equilibrium is attained and in their presence (e.g. oxygen of the air) the reduction to $[TcCl_6]^{2-}$ takes place [80, 87].

Under stronger reducing conditions (e.g. Zn + HCl during heating or under pressurized molecular hydrogen at 3–5 MPa and a temperature of 140–180 °C)

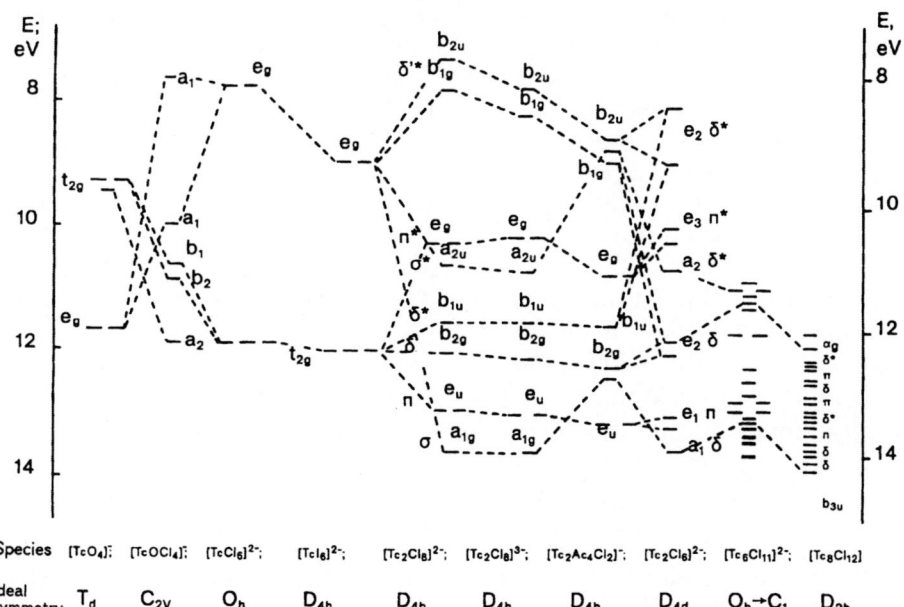

Fig. 6. Results of the calculation of the electronic structure of technetium oxo- and halogenocomplexes using EHT method [13, 113]

further reduction of $[Tc_2Cl_8]^{3-}$ is also possible (8–11) [80]:

$$[Tc_2Cl_8]^{3-} + e^- \Rightarrow [Tc_2Cl_6]^{2-} + 2Cl^- \tag{8}$$

$$Tc(III) + e^- \Rightarrow Tc(II) \tag{9}$$

$$Tc(II) + Tc(II) \Rightarrow [Tc_2Cl_6]^{2-} \tag{10}$$

$$[Tc_2Cl_6]^{2-} + e^- \Rightarrow [Tc_2Cl_6]^{3-} \tag{11}$$

It is noteworthy that all these reactions are reversible, but under reducing conditions the equilibrium shifts to the right-hand side, and under oxidizing conditions it shifts to the left-hand side. For all resulting binuclear ions at rather low concentration in solutions, reactions take place, accompanied by the rupture of the M–M bond and the formation of unstable mononuclear complexes, which are readily oxidized even by the solvent. At high concentrations of binuclear ions cycloaddition reactions of multiple M–M bonds (these reactions are considered in section 3.5) occur to form stable, under the given conditions, polynuclear clusters.

3.2 The Behaviour of the $[Tc_2X_8]^{3-}$ Ions in Solutions of Hydrogen Halides

In order to elucidate the causes of the increased stability of the hydrolyzed cluster ions compared with the unhydrolyzed ions, further studies were made of the behaviour of $[Tc_2X_8]^{3-}$ (where X = Cl, Br, or I) in solutions of hydrogen halides [43, 52, 80, 87]. The studies were performed mainly in relation to the most stable and most readily synthesized $[Tc_2Cl_8]^{3-}$ ion (Fig. 1a); kinetic methods with optical recording were employed. The identity of the reaction products was in most cases confirmed by their isolation in the solid phase. The studies showed that the stability of the $[Tc_2X_8]^{3-}$ ions (where X = Cl, Br, or I) in aqueous solutions is determined by the sum of competing processes: acid hydrolysis complex formation with subsequent disproportionation and dissociation of the M–M bonds, and oxidative addition of atmospheric oxygen to the Tc–Tc multiple bond.

Analysis of reactions (5–7) shows that there is a narrow range of HX and technetium ion concentrations in which there is a possibility of prolonging the existence of $[Tc_2X_8]^{3-}$; the limits of this range become markedly narrower in the order $Cl > Br > I$[10]. For example, at an initial concentration $[Tc_2Cl_8]^{3-} > 10^{-2}$ M and for ~3 M < [HCl] < ~6 M, the solutions of the octachloroditechnetate can exist indefinitely without visible changes in the

---

[10] The paper electrophoresis experiments carried out to study the mobility of polynuclear technetium clusters in aqueous solutions of HX of varying acidity, as a mobile phase, showed that these clusters were also characterized by reversible reactions such as (5) without leading to destruction of M–M bonds. On the other hand, an autoclave recrystallization of the polynuclear clusters at 200–220 °C in an atmosphere of argon from concentrated solutions of HX led to a partial destruction of M–M bonds and the formation of binuclear complexes $[Tc_2X_8]^{3-}$ and $[Tc_2X_6]^{2-}$. This indirectly shows that reactions (6) and (7), leading to the destruction of M–M bonds, are likely in solutions of polynuclear clusters [15].

absence of atmospheric oxygen, while the octaiododitechnetates decompose almost at the instant of their formation in hydroiodic acid. The octa-bromoditechnetates, which can be obtained at a high (~ 1 M) concentration of technetium ions and a high temperature in the autoclave (when the autoclave is cooled, crystals of $M_3[Tc_2Br_8] \cdot nH_2O$ are precipitated and virtually only the $[TcBr_6]^{2-}$ and $[Tc_2Br_6]^{2-}$ ions remain in the mother liquor), occupy an intermediate position.

Thus partial evaporation of HCl during the synthesis of $[Tc_2Cl_8]^{3-}$ by the method of Eakins et al. [24] and by the autoclave method [22, 42, 43] creates favorable conditions (from the standpoint of the acidity of the solution and the concentration of technetium ions in it) for the formation of the octachloro-ditechnetates (+2.5), since, on the one hand, the rate of reduction of the technetium(IV) ions increases owing to their hydrolysis [42] and, on the other hand, the stability of the hydrolyzed cluster ions formed increases in relation to the reactions involving disproportionation and oxidation by atmospheric oxy-gen [9, 52, 80, 87]. We may note that under the conditions of more pronounced hydrolysis, the rate of reduction of technetium(IV) increases so much that the formation of metallic technetium becomes possible.

3.3 Oxidation–Reduction Reactions

In solutions with a high $[Tc_2Cl_8]^{3-}$ concentration the octachloroditechnetate ions can enter into reactions in which oxygen undergoes oxidative addition to the Tc–Tc multiple bond (12) [9, 80].

$$[Tc_2Cl_8]^{3-} + O_2 \Rightarrow [Tc_2Cl_8O_2]^{3-} \tag{12}$$

The Tc–Tc bond in dinuclear Tc(IV/V) complexes is greatly weakened (Table 1) and the complexes readily decompose in HCl solutions to the mononuclear Tc(V) and Tc(IV) complexes (13) [9].

$$[Tc_2Cl_8O_2]^{3-} \Rightarrow [(TcO(OH)Cl_4]^{2-} + [TcCl_6]^{2-} \tag{13}$$

In the absence of complex-forming agents and in aprotic solvents the $[Tc_2Cl_8O_2]^{3-}$ ions are comparatively stable and can be isolated in the form of the salt $K_3[Tc_2Cl_8O_2]$ (Table 1). We noted that the analogous reactions involving the addition of oxygen to the quadruple M–M bond in the $[Tc_2Cl_8]^{2-}$ complexes lead to the formation of mononuclear oxohalogenotech-netium(V) complexes even in aprotic solvents (14) [9].

$$[Tc_2Cl_8]^{2-} + O_2 \Rightarrow [Tc_2Cl_8O_2]^{2-} \Rightarrow [TcOCl_4]^- \tag{14}$$

Oxidative addition reactions have been observed also for the binuclear clusters of other d-transition elements with multiple M–M bonds [1, 116, 117]. The multiplicity of the M–M bonds must decrease in these reactions. It is known from organic chemistry that similar reactions are extremely characteristic of unsaturated organic compounds. We believe that the capacity for oxidative addition reactions is a common property of homeopolar multiple bonds [1, 9, 11, 15].

Apart from reactions involving the oxidative addition of oxygen, the usual oxidation–reduction reactions have also been observed for binuclear technetium clusters with multiple M–M bonds (in this case they must be accompanied by a change in the multiplicity of the Tc–Tc bonds up to their complete dissociation)[11] (15, 16) [11, 80].

$$[Tc_2Cl_6]^{2-} + 2Cl^- \Rightarrow [Tc_2Cl_8]^{3-} + e^- \tag{15}$$

$$[Tc_2Cl_8]^{3-} \Rightarrow [Tc_2Cl_8]^{2-} + e^- \tag{16}$$

Reaction (15) is very rapid and takes place in hydrochloric acid solution. Reaction (16) is slower and is characteristic for aprotic solvents since the $[Tc_2Cl_8]^{2-}$ ions rapidly decompose in hydrochloric acid solutions via reaction types (6), (7). In the presence of an excess of the reductants, reactions (15) and (16) can also proceed in the reverse direction.

3.4 Ligand Substitution Reactions

Three types of ligand substitution reactions are now known for technetium clusters.

3.4.1 Reversible Substitution of Monodentate Ligands by Other Monodentate Ligands

$$[Tc_2Cl_8]^{3-} + 8X^- \Rightarrow [Tc_2X_8]^{3-} + 8Cl^- \tag{17}$$

$$[Tc_2Cl_8]^{2-} + 8X^- \Rightarrow [Tc_2X_8]^{2-} + 8Cl^- \tag{18}$$

$$[Tc_6Cl_{14}]^{3-} + 14X^- \Rightarrow [Tc_6X_{14}]^{3-} + 14Cl^- \tag{19}$$

The equilibria of these reactions are significantly shifted to the right-hand side for X = Br or I [11, 34, 49, 50, 75]. In organic solvents, these reactions occur at a slower rate. Therefore, for polynuclear clusters, it is possible to synthesize mixed-ligand $\{[Tc_8Br_2I_2(\mu\text{-}Br)_4(\mu\text{-}I)_4]I_2\}^{2-}$-type complexes [75]. The axial atoms are substituted first, followed by the bridging and terminal atoms. Unlike classical octahedral halide clusters, there is not much difference in the rates of substitution of the bridging and terminal ligands in the case of trigonal- and tetragonal-prismatic clusters. No formation of mixed-ligand binuclear complexes was observed.

3.4.2 Reversible Substitution of Monodentate Ligands by Bidentate Ligands

$$[Tc_2Cl_8]^{3-} + 4CH_3COO^- \Leftrightarrow [Tc_2(CH_3COO)_4Cl_2]^- + 6Cl^- \tag{20}$$

$$[Tc_2(CH_3COO)_4Cl_2]^- \Leftrightarrow [Tc_2(CH_3COO)_4Cl] + Cl^- \tag{21}$$

---

[11] Trigonal- and tetragonal-prismatic clusters are also characterized by redox reactions (15) and (16) changing the formal multiplicity of the M–M bonds in them (Sects 3.5 and 5.3).

$$[Tc_2Cl_8]^{3-} + 4o\text{-}C_5H_4(OH)N \Leftrightarrow [Tc_2(o\text{-}C_5H_4ON)_4Cl]$$
$$+ 7Cl^- + 4H^+ \qquad (22)$$

For polynuclear clusters similar reactions were not obtained. These reactions proceed in the forward direction only at high temperatures ($\sim 120\,^\circ$C) and in the presence of an excess of the bidentate complex-forming agent, whereas the reverse reactions (20–22) also take place at room temperature in the presence of excess HCl [11, 46, 49].

3.4.3 Oxidative Substitution of Monodentate Ligands by Bidentate Ligands

$$[Tc_2Cl_8]^{3-} + 4(CH_3)_3CCOOH \Rightarrow \{Tc_2[(CH_3)_3CCOO]_4Cl_2\}$$
$$+ e^- + 6Cl^- + 4H^+ \qquad (23)$$

$$[Tc_2X_6]^{2-} + 4CH_3COOH \Rightarrow [Tc_2(CH_3COO)_4X] + 5X^-$$
$$+ e^- + 4H^+ \qquad (24)$$

$$4[Tc_2(CH_3COO)_4Cl_2]^- + 7O_2 + 2H_2O \Rightarrow$$
$$2\{[Tc_2(CH_3COO)_4](TcO_4)_2\} + 8Cl^- + 4H^+ \qquad (25)$$

These reactions are irreversible; organic acid ions apparently behave in the reactions as oxidants, being converted into the corresponding alcohols and their esters [11, 48]. Reactions (23) and (24) take place only at high temperatures ($> 150\,^\circ$C) and in the presence of a large excess of organic acid [11, 48, 58]. These reactions are also possible for polynuclear clusters, because redox reactions occur readily and reversibly for them [50].

Disproportionation reactions of type (11) with the formation of $[Tc_2X_6]^{2-}$ and $[TcX_6]^{2-}$ take place simultaneously with the ligand substitution reactions (17), (18), and (20), involving the $[Tc_2X_8]^{3-}$ complexes. However, the analogous complexes with bidentate bridging ligands (the d^4–d^5 complexes with the "lantern" structural type, see Fig. 1b) do not undergo disproportionation reactions. This fact provides additional confirmation that the disproportionation of $[Tc_2X_8]^{3-}$ is accompanied by the dissociation of the Tc–Tc bond, while the bidentate bridging acetate ions stabilize these bonds. The binuclear complexes with Tc–Tc bonds also behave analogously in thermal decomposition reactions (see Sect. 4.2).

3.5 Polynuclear Cluster Compound Reactions

The cluster formation reactions considered in this review occur in technetium-concentrated solutions of hydrohalogenic acids. From comparison of the data reported in [9, 11, 44, 49, 50, 53, 68, 72, 73, 75, 76, 118] it follows that the ions $[Tc_2X_6]_n^{2n-}$ (X = Cl, Br) are immediate precursors to polynuclear technetium clusters. The given ions enter into the composition of the diamagnetic compounds of the type $M_2[Tc_2X_6] \cdot nH_2O$.

In HBr solutions the $[Tc_2Br_6]_n^{2n-}$ ions exist in the form of $[Tc_2Br_8]^{4-}$. However, experiments on the composition of the mother solutions over the polynuclear technetium clusters being formed, show that paramagnetic ions giving the ESR signals are always present in them along with the diamagnetic $[Tc_2Br_8]^{4-}$ ions. Since none of the polynuclear clusters being formed under these conditions can give a similar ESR signal (neither in form nor intensity) and the possible admixture of binuclear $[Tc_2Br_8]^{3-}$ would give another ESR spectrum [113], the given signal can be expected to belong to $[Tc_2Br_8]^{5-}$ ions, which are the products of the one electron reduction of the $[Tc_2Br_8]^{4-}$ ions. Auxiliary experiments measuring ESR spectra of the frozen glass solutions in hydrobromic acid of the compounds $[Tc_2(CH_3COO)_4Br]$, $K_2[Tc_2Br_6] \cdot 2H_2O$ and compounds (I)–(XII) (Table 2) support this supposition.

Thus, it can be ascertained that polynuclear technetium clusters are formed from the binuclear ions $[Tc_2X_8]^{4-}$ and $[Tc_2X_8]^{5-}$ with the M–M bonds of formal order 5.0 and, probably, 4.5, respectively, i.e. under the conditions of reduction reactions at high concentrations of technetium ions the cycloaddition of multiple Tc–Tc bonds takes place according to (26).

$$(26)$$

The yield of the hexa- and octanuclear prismatic clusters formed by reaction (26) is about proportional to the square of the concentration of $[Tc_2X_6]^{n-}$ ions in solution [15]. This fact indicates that the first stage in the addition reaction of $[Tc_2X_6]^{n-}$ ions is their dimerization. Although individual tetranuclear clusters were not synthesized in the studied system, the basic possibility of their formation was confirmed by synthesis of tetranuclear right-angle molybdenum and tungsten clusters, $M_4Cl_8[P(C_2H_5)_3]_4$ and $[M_4Cl_{14}]^{3-}$ [119, 120].

Since the clusters formed, $\{[Tc_8Br_4(\mu\text{-}Br)_8]Br\} \cdot 2H_2O$ and $[H(H_2O)_2]$-$[[Tc_8Br_4(\mu\text{-}Br)_8]Br\}$, have a very low solubility, they are precipitated from the reaction solution. On the other hand $\{[Tc_8Br_4(\mu\text{-}Br)_8]Br_2\}^{2-}$ possesses high solubility, so it can be accumulated in solution and enter into further reactions. Thus, for example, if $[H(H_2O)_2]_2\{[Tc_8Br_4(\mu\text{-}Br)_8]Br_2\}$ is taken as the initial compound for the synthesis in the autoclave, then at a temperature $\sim 180\,^\circ C$, compounds (VII) and (IX) (Table 2) and ions $[Tc_2Br_8]^{5-}$ will be three of the products (27) [44].

$$\{[Tc_8Br_4(\mu\text{-}Br)_8]Br_2\}^{2-} \xrightarrow{+\,5Br} [Tc_6Br_6(\mu_3\text{-}Br)_5]^{2-} + [Tc_2Br_8]^{5-} \quad (27)$$

The formation of the $[Tc_6Br_6(\mu_3\text{-}Br)_5]^{2-}$ ions is also possible in the cycloaddition reaction (28) [44].

(28)

It should be noted that trigonal-prismatic clusters in the system $H_2[TcBr_6]$–HBr–H_2 were not obtained, whereas under similar conditions for the system $[TcX_6]^{2-}$–$[R_4N]^+$–HX–H_2 (where X = Cl, Br) hexanuclear prismatic halogenide technetium clusters of the composition $\{[Tc_6X_6(\mu\text{-}X)_6]X_2\}^{n-}$ (where $n = 2, 3$) (A) and $[Tc_6Cl_6(\mu\text{-}Cl)_6]^{2-}$ (B) were obtained [11, 72]. Thus, it can be supposed that the $H_2[TcBr_6]$–HBr–H_2 system, hexanuclear clusters of (A) and (B) compositions are also initially formed, but they appear to be less stable and are transformed into the octahedral $[Tc_6Br_6(\mu_3\text{-}Br)_5]^{2-}$ cluster. For such a transformation the triangular bases of the trigonal prisms of the metallic skeletons should be turned at 60° with respect to each other round the thirdfold axis and the verticle edges should be lengthened (see Section 5.3.3).

A specific feature of reactions occurring in the autoclave is that the least soluble compounds are always precipitated from the homogeneous phase of the reaction. As a result, the equilibrium of the reaction is always shifted to the formation of these very insoluble compounds. Thus, it becomes clear that by varying the composition of the reaction mixture (mainly due to the introduction of new cations and anions) practically all types of the cluster forms being generated in the given system can be obtained in the solution. This is a clear advantage of the hydrothermal technique for cluster synthesis in the autoclave.

The chemical properties of the technetium clusters are similar to those of the usual mononuclear complexes (hydrolytic reactions, reactions involving the substitution of monodentate ligands by other monodentate ligands, complex formation reactions, and oxidation–reduction processes) as well as typical polynuclear clusters of other d-transition elements (reactions accompanied by a change in the multiplicity of the M–M bonds up to their complete dissociation and by the substitution of monodentate ligands by bidentate ligands with formation of complexes having a structure of the "lantern" type). However, some of the reactions of the dinuclear and polynuclear technetium clusters are characteristic also of unsaturated organic compounds with multiple homeopolar bonds: i.e. reactions involving oxidative addition to multiple bonds and reactions in which multiple bonds undergo cycloaddition with the formation of closed cyclic systems.

In a number of cases, "anomalies" were observed in the properties of technetium cluster compounds compared with the clusters of other d-transition elements, for example an increased capacity for the formation of stable paramagnetic clusters or clusters with complex systems of M–M bonds, including several quadruple Tc–Tc bonds. The possibility cannot be excluded that these "anomalies" are caused solely by the lack of adequate experimental data on the clusters of other d-transition elements and that in the future similar properties will also be observed for clusters of other metals. However, it is also possible that, by virtue of its "crossing" position in the Periodic Table, that technetium should simultaneoulsy possess the capacity for forming both (a) binuclear clusters with weak crystal field ligands and M–M bonds having a high multiplicity and (b) polynuclear clusters with strong and weak crystal field ligands. It may be that the "anomalous" properties of technetium clusters are due precisely to this factor, together with the fact that the theory of clusters is as yet inadequate to explain unambiguously and to predict the wide range of properties of these compounds.

4 Thermal Properties and Mechanism of Thermal Decomposition in the Solid Phase

In addition to research on the structure and other physicochemical properties in the solid state and solution, research into the thermal stability of technetium clusters is also of interest. However, there are few publications [11, 13, 51, 53, 59, 121–124] devoted to this aspect of technetium cluster chemistry. This section summarizes results taken from the literature and our own results on the thermal stability of technetium clusters.

Table 3 lists the main thermal stability characteristics of all the technetium cluster compounds studied to date. The compounds are divided into two large groups according to the structural principle: bi- and polynuclear clusters. The binuclear clusters are again subdivided, according to the structural principle, into clusters having only terminal ligands and clusters of the "lantern" structure with four μ-bridging ligands additionally bonding mononuclear fragments into a cluster. Moreover, in the first group of binuclear clusters there is a subgroup of clusters with organic cations and ligands. Polynuclear clusters are divided into two subgroups: with organic and inorganic cations.

Such a classification of technetium cluster compounds, in our opinion, reflects the relationship between the thermal stability and structure of the clusters quite well. Moreover, on the basis of this classification it is easier to follow the mechanism of the main thermochemical transitions of technetium clusters, such as: (1) dehydration; (2) disproportionation and related processes occurring without changes or with only small changes in mass; (3) one-stage processes of thermolysis. We shall now consider these main mechanisms of the thermochemical reactions of technetium clusters in greater detail.

4.1 Dehydration

In all clusters the dehydration process is an endothermic one; and it does not lead to the destruction of the cluster, the only exception being $K_4(H_3O)_2$-$[Tc_2(SO_4)_6]$ [59], whose dehydration process occurs simultaneously along with the thermal decomposition (29).

$$2K_4(H_3O)_2[Tc_2(SO_4)_6] \rightarrow 3K_2SO_4 + 2K[TcO_4]$$
$$+ 2Tc + 9SO_2 + O_2 + 6H_2O \qquad (29)$$

The temperature range of the hydrate stability is very wide (60–400 °C) and it depends on the character of water bonding in the crystals and the hydration energies of the ions. The most weakly bonded water molecules are removed from the compounds at 60–130 °C. These compounds include: $K_2[Tc_2(SO_4)_4 \cdot 2H_2O]$, where the water molecules act as axial ligands with a weak Tc–O bonds (Tc–Tc ~ 2.16 Å, Tc–O ~ 2.4 Å [59]); $\{[Tc_8Br_4(\mu$-$Br)_8]Br\} \cdot 2H_2O$, where water molecules are bonded to each other by means of hydrogen bonding (O–O ~ 2.7 Å) and are located in crystalline holes formed by large cluster particles [44]. It is noteworthy that at temperatures of 80–130 °C the first stage of dehydration of $[H_3O(H_2O)_3]_2[Tc_6Br_6(\mu_3$-$Br)_5] \cdot 4H_2O$ occurs. At this stage the four molecules of cystallization water are removed from the compound. The character of their bonding in the crystal lattice is unknown, but probably close to the compounds described above [44, 53].

Table 3. Thermal stability of the technetium cluster compounds

No.	Compound	Temperature ranges of destruction stages (°C)	Identified products of destruction (stoichiometric coefficient)		Weight loss; obs./calc. (%)	[References] (reaction number in text)
			Gaseous	Solid		
I	II	III	IV	V	VI	VII
1	$K_3[Tc_2Cl_8] \cdot 2H_2O$	130–160 370–450	(2)H_2O –	$K_3[Tc_2Cl_8]$ (5/4)$K_2[TcX_6]$; (3/4)Tc; (1/2) KCl	5.7/5.67 0/0	[121] (32)
2	$(NH_4)_3[Tc_2Cl_8] \cdot 2H_2O$	140–150* 160–170* 280–340* 360–460*	(1)H_2O (1)H_2O NH_4Cl; Cl_2 NH_4Cl; HCl	$(NH_4)_3[Tc_2Cl_8] \cdot H_2O$ $(NH_4)_3[Tc_2Cl_8]$ (1)Tc; (1)$(NH_4)_2[TcCl_6]$ TcNCl }	3.1/3.15 3.1/3.15 16.0/15.56 46.5/41.08	[123] (33, 34)
		400–460* 600–800	Cl_2 (1/2)N_2	TcN Tc	2.0/2.44	
3	$(PyH)_3[Tc_2Cl_8] \cdot 2H_2O$ (Py-C_5H_5N)	130–150 250–280 400–430	(2)H_2O PyHCl; HCl Py; (5/2)Cl_2	$(PyH)_3[Tc_2Cl_8]$ $Tc_2(Py)_2Cl_5$ Tc	4.0/4.75 24.2/24.87 43.9/44.53	[11] (37, 38)
4	$(NH_4)_3[Tc_2Br_8] \cdot 2H_2O$	60–90* 260–430*	(2)H_2O NH_4Br; Br_2	$(NH_4)_3[Tc_2Br_8]$ $(NH_4)_2[Tc_2Br_8]$.	4.0/3.88	[13, 124] (33–35)
		430–470* 600–800	NH_4Br; Br_2 N_2	$(NH_4)_2[TcBr_6]$; Tc } Tc(TcN) Tc	74.2/74.76 2.0/1.51	
5	$K_2[Tc_2Cl_6] \cdot 2H_2O$	130–150 640–680	(2)H_2O –	$K_2[Tc_2Cl_6]$ $K_2[TcCl_6]$; Tc	6.2/6.85 0/0	[13, 124] (32)
6	$K_2[Tc_2Br_6] \cdot 2H_2O$	130–150 550–650	(2)H_2O –	$K_2[Tc_2Br_6]$ $K_2[TcBr_6]$; Tc	4.2/4.55 0/0	[13, 124] (32)
7	$[(C_4H_9)_4N]_2[Tc_2Cl_8]$	245–275*	$[(C_4H_9)_3N]$; Cl_2; NH_4Cl	Tc(TcC)	76.1/79.50	[11]
8	$[(C_2H_5)_4N]_2[Tc_2Cl_6]$	295–305*	$[(C_4H_9)_3N]$; Cl_2; NH_4Cl	Tc(TcC)	69.5/70.49	[51]
9	$Tc_2(CO)_{10}$	220–260	CO	Tc(TcC)	59.3/58.58	[13](40)
10	$K_2[Tc_2(SO_4)_4] \cdot 2H_2O$	90–110 250–450	(2)H_2O SO_2; Tc_2O_7	$K_2[Tc_2(SO_4)_4]$ (1) $KTcO_4$; (1) $Tc(TcO_2)$; (1/2)K_2SO_4	5.0/5.17 49.0/44.2	[59, 124] (36)

No.	Compound	Temperature (°C)	Reagents	Products	% (found/calc.)	Ref.
11	K$_4$(H$_3$O)[Tc$_2$(SO$_4$)$_6$]	350–500	SO$_2$; H$_2$O	(1) KTcO$_4$; (1) Tc(TcO$_2$); (3/2) K$_2$SO$_4$	38.1/41.91	[59, 124] (29)
12	K[Tc$_2$(CH$_3$COO)$_4$Cl$_2$]	400–460	(CH$_3$CO)$_2$O; Cl$_2$	(1) KCl; (2) Tc	56.5/57.09	[11] (39)
13	[Tc$_2$(CH$_3$COO)$_4$Cl]	360–420	(CH$_3$CO)$_2$O; Cl$_2$	(2) Tc	58.3/57.83	[11] (39)
14	[Tc$_2$(CH$_3$COO)$_4$Br]	400–460	(CH$_3$CO)$_2$O; Br$_2$	(2) Tc	62.3/61.47	[11] (39)
15	{[Tc$_8$(μ-Br)$_8$Br$_4$]Br} · 2H$_2$O	80–130	(2)H$_2$O	{[Tc$_8$(μ-Br)$_8$Br$_4$]Br}	2.0/1.93	[53]
		300–600	(13/2)Br$_2$	(8)Tc	55.5/55.64	
16	[H(H$_2$O)$_2$]{[Tc$_8$(μ-Br)$_8$Br$_4$]Br}	260–380	(2)H$_2$O	{[Tc$_8$(μ-Br)$_8$Br$_4$]Br}	2.0/1.93	[53] (31)
		400–600	(13/2)Br$_2$	(8)Tc	56.0/55.64	
17	[H(H$_2$O)$_2$]$_2${[Tc$_8$(μ-Br)$_8$Br$_4$]Br$_2$}	250–390	(4)H$_2$O	{[Tc$_8$(μ-Br)$_8$Br$_4$]Br$_2$}	3.6/3.63	[53] (31)
		610–620	melting	{[Tc$_8$(μ-Br)$_8$Br$_4$]Br$_2$}	0/0	
		620–820	(7)Br$_2$	(8)Tc	57.2/56.36	
18	[H$_3$O(H$_2$O)$_3$]$_2$[Tc$_6$(μ_3-Br)$_5$Br$_6$]	140–200	(6)H$_2$O	[H$_3$O]$_2$[Tc$_6$(μ_3-Br)$_5$Br$_6$]	6.9/6.67	[53] (30)
		260–400	(2)H$_2$O	[Tc$_6$(μ_3-Br)$_5$Br$_6$]	2.2/2.22	
		540–800	(11/2)Br$_2$	(6)Tc	55.0/54.29	
19	[H$_3$O(H$_2$O)$_3$]$_2$[Tc$_6$(μ_3-Br)$_5$Br$_6$] 4H$_2$O	80–130	(4)H$_2$O	[H$_3$O(H$_2$O)$_3$]$_2$[Tc$_6$Br$_{11}$]	4.8/4.26	[53] (30)
		150–190	(6)H$_2$O	[H$_3$O]$_2$[Tc$_6$Br$_{11}$]	6.2/6.38	
		260–390	(2)H$_2$O	[Tc$_6$(μ_3-Br)$_5$Br$_6$]	2.0/2.13	
		550–800	(11/2)Br$_2$	(6)Tc	52.3/51.98	
20	[(C$_2$H$_9$)$_4$N]$_2$[Tc$_6$(μ_3-Br)$_5$Br$_6$]	250–400	(11/2)Br$_2$; [(C$_4$H$_9$)$_3$N]	Tc(TcC)	68.0/69.65	[53]
21	[(CH$_3$)$_4$N]$_2$[Tc$_6$(μ-Cl$_6$)Cl$_6$]	265–400	{ (6)Cl$_2$; [(CH$_3$)$_3$N]	Tc(TcC)	~ 48/49.14	[11]
		370–375[a]				
22	[(CH$_3$)$_4$N]$_3${[Tc$_6$(μ-Cl$_6$)Cl$_6$]Cl$_2$	260–400	{ (7)Cl$_2$; [(CH$_3$)$_3$N]	Tc(TcC)	53.1/54.76	[11]
		365–370[a]				
23	[(C$_2$H$_5$)$_4$N]$_2${[Tc$_6$(μ-Br$_6$)Br$_6$]Br$_2$}	260–390	(7)Br$_2$; (C$_2$H$_5$)$_3$]N	Tc(TcC)	68.2/69.89	[11]
24	[(CH$_3$)$_4$N]$_3${[Tc$_6$(μ-Br$_6$)Br$_6$]Br$_2$}	270–390	(7)Br$_2$; (C$_2$H$_5$)$_3$]N	Tc(TcC)	68.9/69.29	[11]

[a] Thermolysis was carried out under quasi-isothermal conditions (in other cases under dynamic conditions with a heating rate of 10°C/min).

In the compounds $M_3[Tc_2X_8] \cdot 2H_2O$ and $M_2[Tc_2X_6] \cdot 2H_2O$ ($M = NH_4$, K; $X = Cl$, Br) dehydration occurs at 130–170 °C (except for $(NH_4)_3[Tc_2Br_8] \cdot 2H_2O$, which is dehydrated at 60–90 °C). In a number of cases (Table 3) under quasi-isothermal conditions, dehydration occurs in two stages, which testifies to small structural differences in the coordination of water molecules in the crystals of these compounds [11, 121, 123]. It should be noted that the X-ray structural analysis of potassium hexachloroditechnetate [68] showed the absence of crystallization water molecules, while the data on thermal dehydration indicated the presence of crystallization water molecules in this compound. Therefore, actually both hydrated and dehydrated forms of this compound exist [49, 68].

Of special interest is the dehydration of polynuclear technetium bromide clusters, which contain hydroxonium cations with different numbers of hydration water molecules. Analysis of the results obtained leads us to conclude that at 140–200 °C dehydration occurs with a partial decomposition of the $[H_3O(H_2O)_3]^+$ cations (30).

$$[H_3O(H_2O)_3]^+ \rightarrow 3H_2O + [H_3O]^+ \qquad (30)$$

The final dehydration of the given compounds takes place only at 250–400 °C, when the more stable cations $[H_3O]^+$ and $[H(H_2O)_2]_2^+$ are decomposed. It is noteworthy that the last dehydration stage is accompanied by the oxidation of the clusters (31).

$$[H(H_2O)_2]_2\{[Tc_8Br_4)(\mu\text{-}Br)_8]Br_2\} \rightarrow 4H_2O$$
$$+ H_2\{[Tc_8Br_4(\mu\text{-}Br)_8]Br_2\} \qquad (31)$$

The appearance of two unpaired electrons per cluster resulting in its paramagnetism with $\mu_{eff} \sim 2.7\mu_B$ (the initial compound $[H(H_2O)_2]_2\{[Tc_8Br_4(\mu\text{-}Br)_8]Br_2\}$ does not have any unpaired electrons) further confirms the latter statement [53].

4.2 Disproportionation and Related Processes Occurring Without Changes or with Only Small Changes in Mass

These processes include: (1) phase transitions; (2) disproportionation reactions with rupture of metal–metal bonds; (3) the Anderson reaction.

The majority of cluster technetium compounds are subject to thermal decomposition topochemically (i.e. their decomposition reaction occurs in the solid phase), $[H(H_2O)_2]_2\{[Tc_8Br_4(\mu\text{-}Br)_8]Br_2\}$ being an exception. This compound melts before decomposition (at 610–620 °C), which is good evidence in favour of the molecular crystalline structure of its dehydrated form $\{[Tc_8Br_4(\mu\text{-}Br)_8]Br_2\}$.

Clusters with K^+ cations decompose according to the mechanism of the disproportionation reaction (10) [11, 121].

$$4K_3[Tc_2Cl_8] \rightarrow 5K_2[TcCl_6] + 3Tc + 2KCl$$

$$K_2[Tc_2X_6](X = Cl, Br) \rightarrow K_2[TcX_6] + Tc \tag{32}$$

Binuclear halogenide clusters with ammonium cations under quasi-isothermal conditions also decompose by the disproportionation mechanism (33) [51, 123].

$$2(NH_4)_3[Tc_2Cl_8] \rightarrow 2Tc + 2(NH_4)_2[TcCl_6] + 2NH_4Cl + Cl_2 \tag{33}$$

However, this process is accompanied by the formation of volatile products; the resultant hexachlorotechnetate(IV) of ammonium is unstable under these conditions [121, 122] and decomposes further according to (34) and (35).

$$(NH_4)_2[TcCl_6] \rightarrow TcNCl + NH_4Cl + 4HCl \tag{34}$$

$$2TcNCl \rightarrow 2TcN + Cl_2 \tag{35}$$

Under dynamic conditions the two decomposition stages merge into one.

The thermal decomposition of binuclear technetium sulfate clusters also occurs according to the disproportionation mechanism, but in this case, (a) other technetium-containing products are formed, and (b) a weight loss due to the evolution of gaseous products is also observed (36) [59].

$$2K_2[Tc_2(SO_4)_4] \rightarrow K_2[TcO_4] + 2Tc + K_2SO_4 + 7SO_2 + 3O_2 \tag{36}$$

According to the mechanism of its thermal decomposition, the pyrolysis of $(PyH)_3[Tc_2Cl_8] \cdot 2H_2O$ [11, 51] belongs to the type of processes under consideration. The peculiar features of this process are the Anderson reaction (the migration of pyridine into the inner coordination sphere of the complex) and the M–M bond preservation. This mechanism is supported by the IR $(\nu(Tc-N) = 243$ and $258 \, cm^{-1})$ and ESCA spectra $(\varepsilon_{Tc3d5/2} = 255.3 \, eV)$ and by measurements of static magnetic susceptibility $(\mu_{eff} = 1.8 \, \mu_B)$. The compound formed at 250–280 °C (37) further decomposes (38) yielding metallic technetium.

$$(PyH)_3[Tc_2Cl_8] \rightarrow (PyH)Cl + [Tc_2(Py)_2Cl_5] + 2HCl \tag{37}$$

$$2[Tc_2(Py)_2Cl_5] \rightarrow 2Tc + 4Py + 5Cl_2 \tag{38}$$

It should be noted [122] that the thermal decomposition of $2(NH_4)_3[Tc_2Cl_8]$ [reaction (33)] obviously occurs according to a similar mechanism, but in this case the initially formed Tc(IV) ammoniacate is unstable and decomposes to nitride [125, 126].

4.3 One Stage Processes of Thermolysis

These processes include the thermal decomposition of all polynuclear clusters and also binuclear clusters with organic cations and ligands (Table 3). As a rule, these processes are accompanied by ample gas evolution and, for a given compound, they take place in narrow temperature ranges. The final decomposition products are metallic technetium and volatile technetium-free gaseous products. In the case of compounds with stable inorganic cations (e.g. $K[Tc_2(CH_3COO)_4Cl_2]$), KCl, nonvolatile under these conditions, is formed. For the compounds containing a large amount of carbon, as a rule, formation of metallic technetium along with its carbide TcC is observed [11, 51]. The occurrence of the thermolysis reactions under quasi-isothermal conditions in a semi-closed vessel increases the decomposition temperature of the clusters and the yield of technetium carbide. This favours the complex reaction mechanism, leading to the formation of TcC. In a number of cases (e.g. $Tc_2(CH_3COO)_4X$ and $Tc_2(CO)_{10}$) the thermal decomposition of the clusters is preceded by their sublimation, and the forming gaseous products decompose (39, 40) on the walls of the reaction vessel [11, 51].

$$2Tc_2(CH_3COO)_4X \ (X=Cl, Br) \rightarrow 4Tc + 4(CH_3CO)_2O$$

$$+ X_2 + 2O_2 \tag{39}$$

$$Tc_2(CO)_{10} \rightarrow 2Tc + 10CO \tag{40}$$

All the compounds considered in this section can be divided into three groups according to their thermal stability: (a) compounds whose thermal stability is determined by the stability of the cluster anions (all bromide polynuclear clusters with hydroxonium cations); (b) compounds, whose thermal stability is determined by the stability of organic cations (bi- and polynuclear clusters with tetra-alkylammonium cations); (c) compounds with organic ligands, whose thermal stability is determined by the strength of the coordination bonds of the technetium atoms with the ligands (acetate and carbonyl binuclear complexes). The most stable are the compounds of group (a) (Table 3). Their thermal decomposition occurs in the range of 600–800 °C, as a rule, with a small heat release. The compounds of group (b) decompose in the range of 260–430 °C (the heat effect is negligible). It follows from [9, 13, 15] that in this temperature range mononuclear technetium complexes with tetra-alkylammonium cations also decompose. For group (c) compounds, thermal stability strongly depends on the nature of the organic ligands, e.g. $Tc_2(CO)_{10}$ decomposes exothermally in the range of 220–260 °C (like the mononuclear carbonyl halogenides $Tc(CO)_5Cl$ and $Tc(CO)_5Br$), while acetate binuclear complexes decompose endothermally in the range of 360–460 °C (reactions (39, 40)).

Thus, the investigations performed have shown that cluster fragments in technetium compounds are thermally rather stable. However, in a number of compounds thermal decomposition occurs at comparatively low temperatures, which is mainly related to the instability of organic cations and the weak bonding of technetium with organic ligands.

As far as the mechanism of thermal decomposition is concerned, halogenide binuclear clusters with multiple M–M bonds differ considerably from other clusters in their stronger tendency to undergo disporportionation reactions to

Tc(IV) [or Tc(VII)] and Tc. An analogous tendency was observed for these clusters in aqueous solutions [11, 80, 87]. Polynuclear technetium clusters decompose with the formation of metallic technetium, i.e. the system of M–M bonds in this case appears to be more stable than the bonding of technetium with ligands.

5 Electronic Structure of Technetium–Technetium Bonds

Because magnetic, spectroscopic and other physicochemical properties of technetium clusters have been discussed in detail in our earlier reviews [9–15, 77, 103, 108, 113, 127–129], primary attention is given in this paper to a discussion of the electronic structure of technetium acido-clusters.

5.1 Electronic Structure of Tc_2^{n+} Dimers

The simplest case of M–M bonds was observed in M_2^{n+} dimers without surrounding coordinate ligands. These compounds (with $n = 0$) were synthesized for a number of metals by fast condensation of metal vapors in inert-gas matrices at low temperatures [130–135].

The M–M bond distances and dissociation energies D_e of the metal dimers were determined experimentally. The theoretical calculations were carried out by the SCF-X_α–SW, SCF-X_α–DVM, Hartree–Fock–Slater methods, with allowance made for the configuration interactions (CI's), and by other methods [1, 136–142]. It should be noted that the theoretical values do not always completely agree with the observed M–M bond distances and dissociation energies D_e. The results obtained by the SCF-X_α methods are in better agreement with experimental results than those obtained by the Hartree–Fock–Slater method with CI's, which is, obviously, due to the fact that the Hartree–Fock–Slater method fails to allow for all configurations. Among the SCF-X_α methods, the relativistic methods give more adequate results, even for comparatively light metal atoms.

Figure 7 presents the results of the calculation by the SCF-X_α–SW method for the Tc_2^{n+} dimer. The energy gap between the $2\sigma_{g\alpha}$-molecular orbital formed by mainly 5s(5p) atomic orbitals and the group of molecular orbitals formed by mainly the 4d orbitals significantly decreases when n decreases from 6 to 2. At $n < 2$, there is always a critical M–M distance (Fig. 8), for which, the $2\sigma_{g\alpha}$ and $1\delta_{u\beta}$ molecular orbitals have equal energies. Thus, the problem of determining the possibility that the technetium outer 5s(5p) atomic orbitals participate in the formation of M–M bonds is confined to determining the effective charges in real technetium clusters: if Z_{eff} is less than $1.0 +$, the technetium 5s(5p) atomic orbitals will participate in the formation of M–M bonds, provided that the Tc–Tc bond distances are shorter than the critical value (r_{cr}).

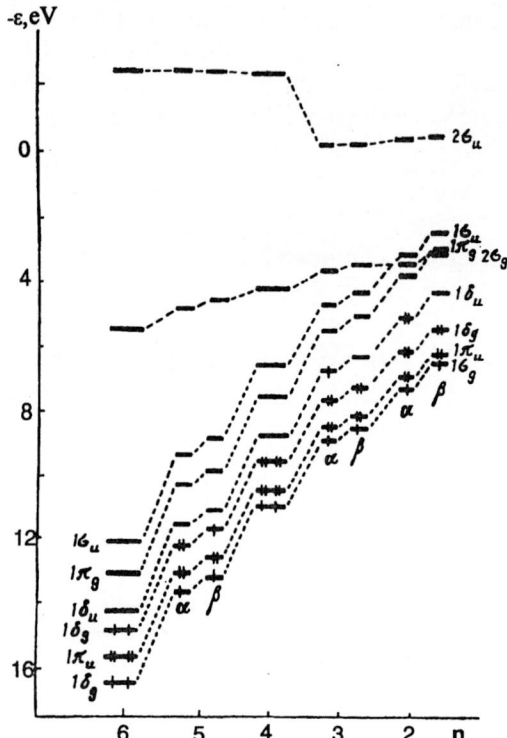

Fig. 7. Results of the SCF-X_α-SW calculation for the M_2^{n+} dimers ($n = 2$–6) with Tc–Tc distances of 4.00 a.u. [142]

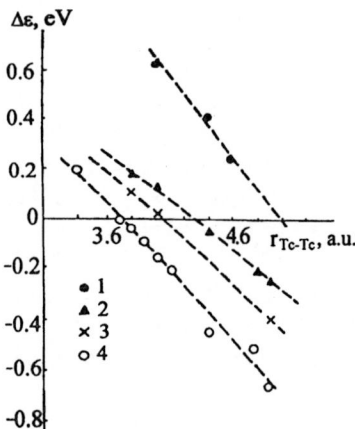

Fig. 8. Dependence of the energy difference $\Delta\varepsilon$ between the $2\sigma_{g\alpha}$ and $1\delta_{u\beta}$ molecular orbitals of the Tc_2^{n+} dimers on the Tc-Tc distance according to the results of the SCF-X_α-SW calculation for different values of n: (1) — 0.5; (2) — 1.0; (3) — 1.2; (4) — 1.5 [142]

Table 4 presents the calculated results of the effective charges on technetium atoms in technetium compounds, arrived at by using various theoretical approximations. In technetium compounds with M–M bonds and formal technetium oxidation states 2.0 + and lower, Z_{eff} is less than 1.0 + , whatever the

Table 4. The effective charges on the atoms of technetium in its compounds [103, 127, 142]

No.	Compound	Z_{ENP}	Z_{EHT}	Z_{eff}
1	$MTcO_4\{M = NH_4; K; (C_6H_5)_4 As; Na; (C_4H_9)_4N\}$	1.60–1.40	3.90–3.25	1.73–1.51
2	$(C_4H_9)_4 NTcOCl_4$	1.05 (5)	2.40 (5)	0.82 (13)
3	$(C_4H_9)_4 NTcOBr_4$	0.85 (5)	2.0 (1)	0.54 (5)
4	$M_2TcCl_6\{M = K^*;(CH_3)_4N;(C_2H_5)_4 N;(C_4H_9)_4N\}$	0.90 (5)*	1.80–1.45	0.57–0.47
5	$M_2TcBr_6\{M = K^*; NH_4;(CH_3)_4N\}$	0.70 (5)*	1.6 (1)*	0.33–0.11
6	$M_2TcI_6\{M = K^*;(C_2H_5)_4 N\}$	0.5 (1)*	1.2 (3)*	0.09–0.08
7	$[(C_4H_9)_4N]_2Tc_2X_8 (X = Cl^*, Br)$	0.55 (5)*	1.45 (1)*	0.56 (8)
8	$M_3Tc_2Cl_8 \cdot nH_2O (M = K^*, C_5H_5NH)$	0.40 (5)*	1.07–0.80	0.12–0.02
9	$K_2Tc_2Cl_6$	0.30 (5)	0.70 (5)	− 0.14 (2)
10	$[(C_2H_5)_4N]_2[Tc_6Br_{14}]$	0.20 (5)	0.55 (5)	− 0.02 (2)
11	$[(CH_3)_4N]_3[Tc_6Br_{14}]$	0.25 (5)	0.65 (5)	− 0.03 (2)
12	$[(CH_3)_4N]_3[Tc_6Cl_{14}]$	0.15 (5)	0.35 (5)	− 0.03 (2)
13	$[(CH_3)_4N]_2[Tc_6Cl_{12}]$	0.20 (5)	0.60 (1)	0.24 (5)
14	$[Tc_8Br_{13}] \cdot 2H_2O$	0.20 (5)	0.55 (5)	− 0.07 (1)
15	$[H(H_2O)_2]_2[Tc_8Br_{14}]$	0.20 (5)	0.55 (5)	− 0.04 (2)
16	$[H_3O(H_2O)_3]_2[Tc_6Br_{11}]$	0.40 (5)	1.0 (1)	0.24 (1)

Note: Data are given for the particular salt that is marked with an asterisk. Standard errors are given in round brackets.

approximation used. Thus, it may be claimed that, in all technetium acido-clusters with multiple M–M bonds and the formal oxidation states of technetium at $2.0 +$ and lower, the 5s(5p) atomic orbitals participate in the formation of Tc–Tc bonds.

It was mentioned earlier [9, 10] that technetium was significantly different from other typical cluster-forming metals. One of the reasons for this difference is a low effective charge Z_{eff} on the technetium atoms within technetium clusters. According to [9, 10, 142], the low value of Z_{eff} for technetium atoms in clusters is due to the fact that it decreases when Tc–Tc bonds are formed. For other typical cluster-forming metals, Z_{eff} either increases, as in the case of molybdenum [139, 143], or remains unchanged, as in the case of rhenium [9, 10, 53]. We believe that the behavior of technetium is different because of the greater participation of the ns(np) atomic orbitals in the formation of M–M bonds in technetium clusters. As a result, shielding of the positive charge on the nucleus increases, leading to a decrease in Z_{eff}.

We next consider the participation of metallic outer ns(np) atomic orbitals in the formation of M–M bonds for close neighbors of technetium in the Periodic Table: Mo, Re, and Ru. Fig. 9 presents calculated results of the electronic structures of M_2^{n+} dimers. For Ru, the $2\sigma_{g\alpha}$ molecular orbital obviously has a significantly higher energy than the $1\delta_{u\beta}$ molecular orbital, and it is therefore always unoccupied whatever the bond distance r_{M-M} and the charge n. In the case of Mo, only the $2\sigma_{g\alpha}$ molecular orbital has a lower energy value than that of the $1\delta_{u\beta}$ molecular orbital for Mo_2^0, and at $n > 1 +$, which is consistent with the values of Z_{eff} in real clusters, the $2\sigma_{g\alpha}$ molecular orbital becomes unoccupied, owing to the "deficiency" of the valence electrons in molybdenum atoms. In the case of Re, a somewhat greater similarity in the behavior of the $2\sigma_{g\alpha}$ molecular orbital to its behavior in the technetium clusters is observed. There are, at least, three reasons why the probability of the participation of the ns(np) atomic orbitals in the formation of Re–Re bonds becomes smaller. (1) In the rhenium clusters, the M–M distance is 0.05–0.1 Å longer than that in the corresponding technetium clusters [1, 11, 12, 77]. (2) In rhenium clusters, Z_{eff} is higher than that in technetium clusters [9, 10, 31]. (3) Calculations show that the necessary conditions for the participation of ns(np) atomic orbitals in M–M bonding only exist for Re_2^0 and Re_2^+ (for $r_{cr} < 2.22$ Å).

Thus, the SCF–X_α–SW calculations for the metal dimers confirm the "anomaly" of the technetium clusters that was experimentally observed in previous studies [9–11]. It should be noted that because the actual structure of the

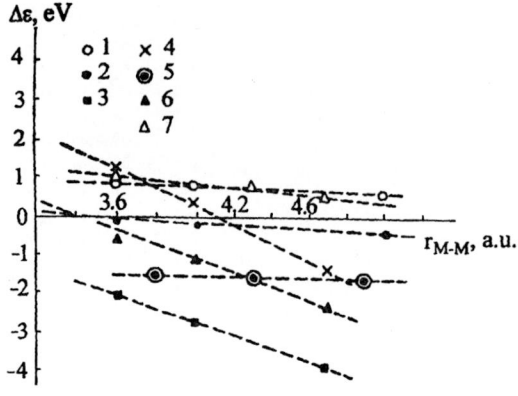

Fig. 9. Dependence of the energy difference $\Delta\varepsilon$ between the $2\sigma_{g\alpha}$ and $1\delta_{u\beta}$ orbitals of the M_2^{n+} dimers on the M–M distance according to the results of the SCF-X_α-SW calculation: $1 - Mo_2^0$; $2 - Mo_2^+$; $3 - Re_2^{3+}$; $4 - Re_2^+$; $5 - Ru_2^0$ and Ru_2^+; $6 - Re_2^{2+}$; $7 - Re_2^0$ [142]

clusters may significantly change the molecular orbital scheme in question, these calculations only demonstrate an increasing probability for the participation of the outer ns(np) atomic orbitals in the M–M bonding according to the series Ru < Mo < Re ≪ Tc and they only hold for clusters with a similar structure.

According to [1, 141, 142, 144, 145], the $2\sigma_{g\alpha}$-MO is of diffuse character. Its main electron density is beyond the atomic spheres of Tc atoms. Therefore, the energy and composition of this MO is not greatly dependent on either the M–M bond distance or Z_{eff}. However, from the data reported in [142], in contrast to [141], it follows that the 2σ-bonding is only somewhat weaker than the 1σ-bonding and comparable to the total δ-bonding. Thus, the $2\sigma_g$-MO may make a considerable contribution to the total M–M bonding (up to 18%) at Z_{eff} on Tc less than $1.0 +$. The other results in [136, 138, 140, 142, 144, 145] are not mutually contradictory.

Thus, the calculations show that the outer ns(np) atomic orbitals can play a significant role in the formation of M–M bonds in transition metal acido-clusters. The probability that these atomic orbitals will participate in the formation of M–M bonds is maximal for elements of Group 7, particularly, for technetium, in whose clusters Z_{eff} for technetium atoms is the lowest of those observed in all known acido-clusters.

5.2 Electronic Structure of Binuclear Technetium Complexes

5.2.1 Octachloroditechnetates

Calculations of the electronic structure of the complexes of this type were first carried out for $[Tc_2Cl_8]^{3-}$, using the semi-empirical EHT method [146]. It was shown that a strong quadruple M–M bond is formed in both the d^4–d^4 binuclear transition-metal complexes and d^4–d^5 technetium complexes. The addition of an "excess" electron does not decrease the Tc–Tc bond strength, because the nonbonding a_{2u}-MO is occupied. These calculations were generally confirmed both experimentally [30, 147, 148] and theoretically [10, 149]. However, the character of the HOMO in these complexes has been discussed in the literature for some time [113, 141, 150].

The calculations of the electronic structure of $[Tc_2Cl_8]^{3-}$ by the SCF–X_α–SW method were carried out in [149]. It was shown that $b_{1u}(\delta^*_{M-M})$-MO acted as a HOMO in this complex. The calculations were confirmed by the optical [151] and ESR [10, 11, 30, 148, 149] spectroscopic data. However, these investigations posed new problems [10, 11, 148]. In terms of the current theoretical concepts [1], the introduction of an "excess" electron, in addition to the electronic configuration for a quadruple M–M bond, into the $b_{1u}(\delta^*_{M-M})$ level should disturb the eclipsed conformation of the complex or, at least, increase the metal–metal bond distance. The following mechanism was suggested for this phenomenon [10]: the presence of an "excess" electron initially results in the loss of half of the M–M δ bond energy. The other half of the δ bond is insufficient to compensate for the energy of the electrostatic repulsion between the negatively charged ligands, when the complex has an eclipsed conformation. As a result, the complex is transformed into a complex with a staggered conformation, in which the δ component of the M–M bond is absent. In this case, the multiplicity of the M–M bond should decrease, and its length should appreciably increase. It should be noted that even in the case where the eclipsed conformation of the complex persists, the Tc–Tc bond distance should still increase, because the transition from $[Tc_2Cl_8]^{2-}$ to $[Tc_2Cl_8]^{3-}$ would have changed the multiplicity of the M–M bond from 4.0 to 3.5 [10].

The experimental results are completely at variance with this theoretical explanation. First, the stability of the $[Tc_2Cl_8]^{3-}$ anion is considerably higher than that of the $[Tc_2Cl_8]^{2-}$ anion. Second, the Tc–Tc bond distance in the $[Tc_2Cl_8]^{2-}$ anion is longer than in the $[Tc_2Cl_8]^{3-}$ [1, 12, 77]. Third, the greater Tc–Tc bond strength in all technetium d^4–d^5 complexes is higher than that in d^4–d^4 [1, 12, 77]. Fourth, it was shown experimentally, using ESR and static magnetic susceptibility measurements [1, 30, 113, 147, 148] that the "excess" electron in the d^4–d^5 complexes actually lies on the $b_{1u}(\delta^*_{M-M})$ and that the formal multiplicity of the M–M bond in them is 3.5. This proof is particularly important, because according to the literature data, the "excess" electron lies on the $a_{2u}(\sigma'_{M-M})$ [146] or $a_{1g}(\sigma''_{M-M})$ MO [150], and the multiplicity of the Tc–Tc bonds in d^4–d^5 complexes is 4.0 or 4.5, respectively.

Modern theoretical methods do not allow very accurate calculation of the energy involved in weak interactions. This is the reason for all the discrepancies in the theoretical and experimental results for compounds with δ bonds. Attempts have been made to improve the accuracy of theoretical predictions, using the current methods of calculation. For example, the valence bond theory in combination with the X_α–SW method allowed [152] a more accurate prediction of the bands of the low-energy δ–δ^* transitions in the optical spectrum of $[Mo_2Cl_8]^{4-}$. The allowance made for the CI's in the nonempirical self-consistent-field methods also allowed the interpretation of these absorption bands for $[M_2X_8]^{n-}$ ions with M–M quadruple bonds [153, 154]. Fairly good agreement between theory and experiment was observed when the CI's were taken into account for complexes with a Cr–Cr quadruple bond, whose energy and length change under the action of various factors [154]. Calculations with the CI's also allow an explanation of the greater stability of binuclear technetium clusters with the 3.5 M–M bond multiplicity compared to that in complexes with a Tc–Tc quadruple bond [10, 113, 148, 153]. The greater stability of these clusters follows from the Brillouin theorem [155], according to which singly excited states do not make any contribution to the ground electronic state, when the CI's are taken into account in self-consistent-field methods. Therefore, "dilution" of the quadruple bond will occur in d^4–d^4 complexes, owing to the doubly excited $\sigma^2 2\pi^2 \delta^0 \delta^{*2}$ state with an M–M double bond, while the low-lying singly excited $\sigma^2 2\pi^2 \delta^0 \delta^{*1}$ state does not make any contribution to the total energy of the M–M bond in d^4–d^5 complexes owing to the symmetry prohibition. All doubly excited states are characterized by higher energies and make only small contributions to the ground state.

Whatever the advantages of the SCF calculations with the CI's, they do not always give good results, as is the case for M_2 molecules with quintuple and sextuple M–M bonds [135, 156]. Nevertheless, some deviations from the simple MO scheme can be explained using simpler qualitative arguments. For example, the increased ability of technetium to form d^4–d^5 complexes and their greater stability in comparison to that of the d^4–d^4 complexes are explained on the basis of a model of the electrostatic repulsion of M atoms with like charges in a binuclear cluster [10, 90, 150] or on the basis of different diffusivities of $\sigma, \sigma', \pi, \delta$ and δ^* "metallic" MO's of the clusters [63, 141, 157].

5.2.2 Carboxylate Complexes with the "Lantern"-Type Structure

The electronic structure of carboxylate complexes with the "lantern"-type structure was studied [58] using EHT calculations (Fig. 10, Table 5).

Figure 10 shows the MO schemes for $[Tc_2(\mu\text{-}CH_3COO)_4][TcO_4]_2$ and $[Tc_2(\mu\text{-}CH_3COO)_4]Cl_2$, and $[TcO_4]^-$ ions with the Tc–O bond distances determined as a result of an X-ray diffraction study of $[(CH_3)_4N]$ $[TcO_4]$ [92, 95] and $[Tc_2(\mu\text{-}CH_3COO)_4]$ $[TcO_4]_2$ [58]. In the HOMO–LUMO region

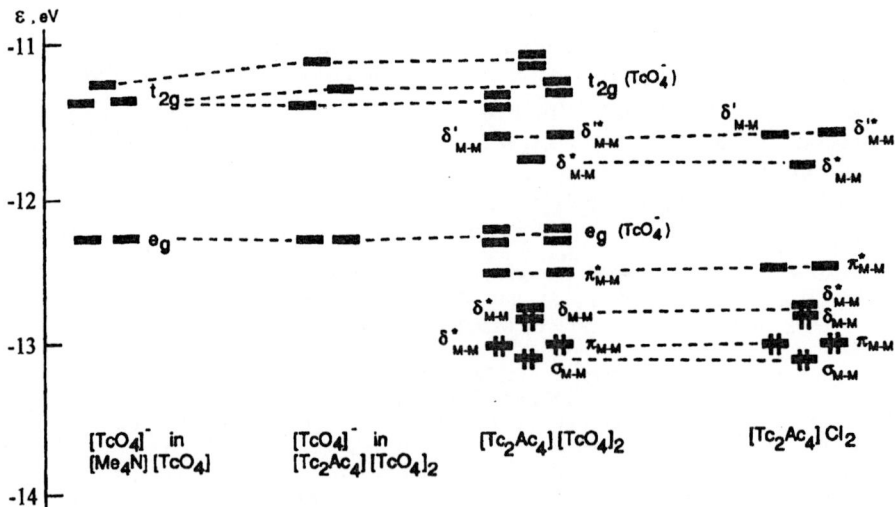

Fig. 10. Schemes of MO's for technetium compounds according to the results of extended Huckel calculations [58]

of $[Tc_2(\mu\text{-}CH_3COO)_4]$ $[TcO_4]_2$, there are no MO's belonging to pertechnetate ions over a wide energy range. This provides evidence for the significantly higher oxidation–reduction stability of pertechnetate ions than that of carboxylate cations, whose oxidation–reduction activity is due mainly to high-lying δ_{M-M}, δ^*_{M-M}, π_{M-M} and π^*_{M-M} MO's.

Calculations [58] showed that Z_{eff} and bond orders in $[Tc_2(\mu\text{-}CH_3COO)_4]$-$[TcO_4]_2$ were similar to those of $[Tc_2(\mu\text{-}CH_3COO)_4]Cl_2$ with a Tc–Cl bond distance of 2.6 Å. In other words, the axial coordination of the pertechnetate ions has the same influence on the tetraacetatoditechnetium cation as that of the chloride ions in normal dichlorotetraacetate complexes with Tc–Cl$_{ax}$ distances equal to ~ 2.6 Å. According to [1, 12, 77], short M–Cl$_{ax}$ distances are observed for analogous d^4–d^4 rhenium complexes and d^4–d^5 chloroacetatoditechnetium complexes. However, in the technetium d^4–d^4 complex $\{Tc_2[(CH_3)_3 CCOO]_4\}Cl_2$, the Tc–Cl$_{ax}$ bond distances are considerably shorter [45, 58]. Therefore, the Tc$\equiv$Tc bond distances are considerably longer than those in $[Tc_2(\mu\text{-}CH_3COO)_4]$ $[TcO_4]_2$.

The Tc–Cl$_{ax}$ distance of ~ 2.6 Å is the optimal distance, when there is no significant *trans*-effect on the Tc$\equiv$Tc bond. In fact, a comparison between the M–M interatomic distances in $[Tc_2(\mu\text{-}CH_3COO)_4]$ $[TcO_4]_2$ and in other d^4–d^5 and d^4–d^4 complexes [1, 12, 77] supports this fact. Thus, a slight increase in the Tc–Tc bond distance in $[Tc_2(\mu\text{-}CH_3COO)_4]$ $[TcO_4]_2$ in comparison with that in d^4–d^5 acetatotechnetium complexes is assigned to the effects discussed in the literature (see section 5.2.1). According to the latest experimental and theoretical results [10, 11, 15, 148, 153], this increase in the Tc–Tc bond distances in d^4–d^4 complexes with M–M quadruple bonds in comparison to those in analogous d^4–d^5 complexes with formal bond orders of 3.5 is assigned to the CI effects.

All these facts show [58] that both the covalent component and the ionic component make contributions to the binding of the pertechnetate ions with the binuclear fragment in the structure of $[Tc_2(\mu\text{-}CH_3COO)_4][TcO_4]_2$. According to the values of the mean Tc–O distances [12, 58, 77], the total influence of the dimeric acetato cations on the pertechnetate ions is intermediate between those in pertechnetates with very small aklali–metal cations and pertechnetates with large tetrabutylammonium cations, i.e., the total influence of the dimeric cations is closer to that in the case of $[(CH_3)_4N]$ $[TcO_4]$. For this reason, it would be interesting to compare the structure of the pertechnetate ions in $[Tc_2(\mu\text{-}CH_3COO)_4][TcO_4]_2$ and $[(CH_3)_4N][TcO_4]$. The small differences in the electronic structure of these anions lead to different values of Z_{eff} for technetium and oxygen atoms and in the order of the Tc–O bonds (Table 5). However, these small differences cause considerable differences in the character of the distortion of the pertechnetate ions. In fact, while pertechnetate ions in $[(CH_3)_4N]$ $[TcO_4]$ have one short $\{1.589(11)$ Å$\}$ and three long $\{1.719(9)$ Å$\}$ Tc–O bonds, pertechnetate ions in $[Tc_2(\mu\text{-}CH_3COO)_4][TcO_4]_2$ have one long $\{1.732(5)$ Å$\}$ (Tc–O) and three short $(1.670$–1.694 Å) Tc–O bonds. $[Tc_2(\mu\text{-}CH_3COO)_4][TcO_4]_2$ is more similar to $[(C_4H_9)_4N]$ $[TcO_4]$ with respect to the character of the distortion of the pertechnetate ions [12, 58, 77].

Table 5. Average values of bonds order and effective charges (Z_{eff}) in pertechnetates and binuclear technetium acetates calculated by the extended Hückel method [58]

Compound fragment	Formal bond order				Z_{eff}				
	Tc≡Tc	Tc–X	Tc–O; TcO$_4$	Tc–O; Ac	Tc; Ac	Tc; TcO$_4^-$	O^a; Ac	O^a; TcO$_4^-$	X
I	II	III	IV	V	VI	VII	VIII	IX	X
$\{[Tc_2Ac_4][TcO_4]_2\}$	0.254	–	0.606 0.537*	0.296 0.197*	1.60	4.02	0.89	1.21 1.20*	–
$\{[Tc_2Ac_4][TcO_4]_2\}^-$	0.251	–	0.606 0.537*	0.294 0.197*	1.12	4.02	0.89	1.21 1.20*	–
$\{[Tc_2Ac_4][TcO_4]_2\}^{2-}$	0.248	–	0.606 0.537*	0.291 0.197*	0.64	4.01	0.90	1.21 1.20*	–
$\{[Tc_2Ac_4][TcO_4]_2\}^{4-}$	0.204	–	0.606 0.538*	0.289 0.196*	−0.33	4.00	0.90	1.21 1.20*	–
$[TcO_4]^-_{compound\ 1}$	–	–	0.604	–	–	4.03	–	1.26 1.36*	–
$[(CH_3)_4N][TcO_4]$	–	–	0.617	–	–	3.9	–	1.24	–
$[Tc_2Ac_4]^{2+}$	0.243	–	–	0.305	1.72	–	0.88	–	–
$[Tc_2Ac_4]^+$	0.240	–	–	0.303	1.24	–	0.88	–	–
$\{[Tc_2Ac_4]Cl_2\}^a$ 2.7	0.253	0.19	–	0.295	1.61	–	0.89	–	0.85
$\{[Tc_2Ac_4]Cl_2\}^-$ 2.7	0.250	0.19	–	0.293	1.13	–	0.89	–	0.85
$\{[Tc_2Ac_4]Cl_2\}^a$ 2.5	0.258	0.27	–	0.290	1.56	–	0.89	–	0.78
$\{[Tc_2Ac_4]Cl_2\}^-$ 2.5	0.255	0.27	–	0.287	1.08	–	0.90	–	0.78
$\{[Tc_2Ac_4]Cl_2\}^a$ 2.3	0.264	0.37	–	0.283	1.49	–	0.90	–	0.70
$\{[Tc_2Ac_4]Cl_2\}^-$ 2.3	0.261	0.37	–	0.281	1.01	–	0.90	–	0.70
$\{[Tc_2Ac_4]Cl_2\}^a$ 2.2	0.266	0.42	–	0.280	1.45	–	0.90	–	0.64
$\{[Tc_2Ac_4]Cl_2\}^-$ 2.2	0.263	0.42	–	0.277	0.97	–	0.91	–	0.64
$\{[Tc_2Ac_4]Cl_2\}^a$ 2.1	0.056	0.51	–	0.277	1.22	–	0.90	–	0.45
$\{[Tc_2Ac_4]Cl_2\}^-$ 2.1	0.155	0.50	–	0.274	0.82	–	0.91	–	0.52
$\{[Tc_2Ac_4]Cl_2\}^a$ 2.0	0.054	0.56	–	0.273	1.14	–	0.90	–	0.36
$\{[Tc_2Ac_4]Cl_2\}^-$ 2.0	0.439	0.56	–	0.272	0.67	–	0.90	–	0.39

^{a} The Tc–Cl distances (Å) are given alongside the large curly brackets. The values for the bridging and oxygen atoms of the pertechnetate groups are indicated with an asterisk. For O and Cl atoms, absolute values of negative Z_{eff} are given.

A hypothesis, in which these distortions in pertechnetate ions were attributed to the pseudo-Jahn–Teller effect, was advanced in [12, 13, 158]. According to this hypothesis, there are two energy-equivalent variants of C_{3v} distortions for pertechnetate ions, and a particular variant occurs as a result of a finer crystal effect. Thus, the coordination of one of the oxygen atoms of each pertechnetate ion to the binuclear cluster fragment in $[Tc_2(\mu\text{-}CH_3COO)_4]$-$[TcO_4]_2$ initiates the distortion of $[TcO_4]^-$ according to the pattern of "one long Tc–O bond and three short Tc–O bonds". An absolutely different phenomenon takes place in the case of $[(CH_3)_4N][TcO_4]$. Nevertheless, these two distortions are virtually identical in energy (Table 5 and Fig. 10).

Thus, the investigations performed [58, 77] showed that $[Tc_2(\mu\text{-}CH_3COO)_4]$ $[TcO_4]_2$ combines the properties of two classes of technetium complexes: those of pertechnetates and those of binuclear carboxylates with multiple M–M bonds. On the other hand, the anionic and cationic fragments of this compound are characterized by a mutual effect which affects their electronic and molecular structures.

5.2.3 Hexahalogenoditechnetates(II)

Among the most interesting and unusual technetium complexes with M–M bonds, as regards electronic and molecular structures, are hexahalogenoditech-netates(II): binuclear clusters with the d^5–d^5 electronic configuration of the central atoms. A study of the molecular structure of K_2 $[Tc_2Cl_6]$, which is a representative compound of this class, was carried out in [68, 69]. According to the classical concepts [1], the addition of two electrons to the stable d^4–d^4 electronic configuration of the central atoms may lead to the formation of triple M–M bonds ($\sigma^2 2\pi^2 2\sigma^2 \delta^{*2}$-configuration). The occupation of the δ^*-MO should lead to the rotation of the mononuclear fragments around the axis passing through the M atoms in the binuclear complex and result in the formation of a staggered conformation of the dimer. This rotation is due to the absence of the bridging ligands supporting the eclipsed conformation. Moreover, the M–M bond distance should increase and become approximately equal to the bond distances in the d^3–d^3 complexes with triple M–M bonds. In fact, these assumptions are confirmed by the results obtained for some d^5–d^5 transition–metal complexes (Table 6).

However, X-ray diffraction data show [68, 69] that in $K_2[Tc_2Cl_6]$, the Tc–Tc bond distance is 0.1 Å shorter than that in an analogous d^4–d^4 chloride technetium complex, although its main structural fragment $[Tc_2Cl_8]^{4-}$ has a staggered structural conformation. That is why a study of the electronic

Table 6. M–M bond length ($d_{M–M}$) and twist angle (χ) relative to "eclipsed" conformation of d^5–d^5 complexes

No.	Compound	d_{M-M} (Å)	χ (°)	References
1	$Re_2Cl_4[P(C_2H_5)_3]_4$	2.232 (6)	~0	[159]
2	$Re_2Cl_4[P(CH_3)_2(C_6H_5)]_4$	2.241 (1)	~0	[160]
3	$Re_2Cl_4(dppe)_2$ [a]	2.244 (1)	~45	[161]
4	$Re_2(C_3H_5)_4$	2.225 (7)	–	[162]
5	$Mo_2[F_2PN(CH_3)PF_2]_4Cl_2$	2.457 (1)	21	[163]
6	$Os_2(o\text{-}C_5H_5NO)_4Cl_2$	2.344 (2)	5.5	[164, 165]
7	$Os_2(o\text{-}C_5H_5No)_4Cl_2 \cdot 2CH_3CN$	2.357 (1)	17.5	[164, 165]
8	$[Os_2X_8]^{2-}$ $\left(\begin{array}{l} X = Cl \\ X = Br \end{array} \right)$	2.182 (1) 2.196 (1)	49.0 46.7	[166]

[a] dppe = $(C_6H_5)_2PC_2H_4P(C_6H_5)_2$.

structure of $[Tc_2Cl_6]^{2-}$ is very important for understanding the M–M bonding in dimeric transition–metal complexes.

Several reasons for the specific structure of $[Tc_2Cl_6]^{2-}$ were suggested: (1) the participation of 5s-AO's of technetium atoms in the additional σ'_{M-M} bonding (according to [142], the probability that 5s(5p)-AO's of technetium atoms will participate in the M–M bonding is high for hexa-chloroditechnetates(II), because they have the smallest Z_{eff} of all technetium complexes, see Table 4); (2) the distortion of the idealized D_{4d} symmetry of the complexes as a result of the pseudo-Jahn–Teller effect [15]; (3) a decrease in the Coulomb repulsion of negatively charged ligands in the staggered conformation of the complex [15]; (4), chain effects [15]; (5) a different effect of the technetium formal oxidation state on the extent of contraction (or diffusivity) of σ-, π- and δ-MO [69]. The EHT calculations [15, 113] fail to directly confirm any of these effects, and we cannot find any other explanation for the structural specificity of this compound. Also, the EHT calculations, with iteration of the ionization potentials and Z_{eff} for all or some of the atoms in this complex, do not give the desired results. Attempts to improve the EHT calculation parameters by taking into account the experimental X-ray photoelectron spectra and the SCF–X_α–SW calculations have been unsuccessful. From the above considerations it follows that a more reliable explanation of the specific structure of the $[Tc_2Cl_6]^{2-}$ requires further experiments and more accurate calculations.

Nevertheless, a probable quantitative explanation of the $[Tc_2Cl_6]^{2-}$ structural specificity was given in [69]. In fact, some years ago Cotton [167] showed that when changes of δ bond order were accompanied by changes in metal atom oxidation numbers, the net change in the M–M bond distance can be as much or more influenced by the latter than the former, because δ bonding is weak. On the other hand [69], the very strong aggregate bonding effect of the $\sigma^2\pi^4$ configuration can be altered by changes in effective charge on the metal atoms, because these changes in charge slightly expand or contract the d orbitals. Thus, even though the δ bond order goes from 1 to 0 from $[Tc_2Cl_8]^{2-}$ to $[Tc_2Cl_6]^{2-}$, the change in the oxidation state from Tc_2^{6+} to Tc_2^{4+} so enhances the σ and π bonding that a substantial contraction in the Tc–Tc distance occurs. The changes from an eclipsed to a staggered rotational orientation about the Tc–Tc axis would be expected to allow an additional small reduction in bond length.

5.2.4 Other Binuclear Technetium Complexes

Literature data are available on the electronic structures of two more binuclear technetium complexes: $[(NH_3)_2(OH)_2Tc(\mu\text{-}O)_2Tc(OH)_2(NH_3)_2]$ (a hypothetical complex with the structure and composition analogous to those of the ethylen-diamminetetra-acetate complex [54, 55]) and $Tc_2(CO)_{10}$ (a binuclear complex with strong crystal field ligands [168, 169]. We shall consider the results of these calculations in greater detail.

The first compound belongs to d^3–d^3. In terms of the classical scheme of the multiple M–M bonding [1] and the idealized D_{2h} symmetry this compound is expected to have triple M–M bonds [11]. However, the calculations by the EHT method show [54] that the $[(NH_3)_2(OH)_2Tc(\mu\text{-}O)_2Tc(OH)_2(NH_3)_2]$ complex has single Tc–Tc bonds with the $\sigma^2\pi^2\delta^{*2}$ electronic configuration. The authors of [54, 55] believe that the reverse energy order of δ_{M-M} and δ^*_{M-M} MO's is due to the destabilizing effect of 2p-AO's of the bridging oxygen atoms, which explains the longer Tc–Tc bond distance in the ethylendiamminetetra-acetat complex (~ 2.33 Å) compared to that (~ 2.22 Å) resulting from extrapolation of the X-ray data [11, 15, 77].

In this regard it would be interesting to consider the electronic structure of some binuclear complexes of other elements. For example, an analogous effect of the bridging ligands on the reverse energy order of δ_{M-M} and δ^*_{M-M} MO's was found in [170]. In [171], this decrease in the δ_{M-M} and δ^* bond energy was used in the explanation of the effect of the chemical nature of the cation on the magnetic and structural properties of nonachlorodimolybdates(III).

Although the experimental and theoretical data showed the reverse energy order of δ_{M-M} and δ^*_{M-M} MO's in binuclear transition–metal complexes, we doubted this fact [11]. If the reverse energy order actually takes place, for example, in ethylendiamminetetra-acetat technetium complexes, it is not so pronounced as that assumed in [54, 55]. Thus, as seen from [11, 12, 77], the expected "normal" triple M–M bond length is ~0.1 Å shorter than that in ethylendiamminetetra-acetat technetium complexes. The "normal" single Tc–Tc bond length should be equal to ~2.25 Å, which is ~0.23 Å longer than that in $[Tc_2(\mu\text{-}O)_2(H_2EDTA)_2]$ and $[Tc_2(\mu\text{-}O)_2(NTA)_2]^{2-}$ [54]. Moreover, the addition of one electron which, according to the calculations in [55], should get into the δ_{M-M}-MO, increase the Tc–Tc distances to 2.402 Å. Therefore, the difference between the energies of δ_{M-M}- and δ^*_{M-M}-MO's should be positive (but not negative, as was proposed in [54]), and its absolute value should be higher than that of the electron repulsion energy, because otherwise the $[Tc_2(\mu\text{-}O)_2(H_2EDTA)_2]$ and $[Tc_2(\mu\text{-}O)_2(NTA)_2]^{2-}$ complexes would be paramagnetic with two unpaired electrons, as is the case with TcO_2 [113].

Thus, the formal multiplicity of the Tc–Tc bonds in $[Tc_2(\mu\text{-}O)_2(M_2EDTA)_2]$ and $[Tc_2(\mu\text{-}O)_2(NTA)_2]^{2-}$ is most likely equal to 3.0 $(\sigma^2\pi^2\delta^2)$. However, the δ-components of these bonds are close to zero. Obviously, the δ-level is more destabilized in TcO_2 and $[Tc_2Cl_8O_2]^{3-}$. This results in their paramagnetism [113]. Moreover, μ_{eff} is consistent with the presence of two or more unpaired electrons per molecule.

5.3 Electronic Structure of Polynuclear Technetium Clusters

5.3.1 Trigonal-Prismatic Clusters

Qualitative descriptions of the electronic structure of the trigonal-prismatic technetium clusters were given in the first papers on their synthesis [11, 72] and structure [11, 70–72]. According to these descriptions, multiple M–M bonds in trigonal-prismatic technetium clusters are mainly due to $4d_z2(\sigma_{M-M})^-$, $4d_{xz}$-, and $4d_{yz}(\pi_{M-M})$-AO's, whereas $4d_{xy}$- and $4d_{x^2-y^2}$-AO's are simultaneously involved in both the multiple M–M bonds (δ_{M-M}) and single M–M bonds in horizontal directions. Moreover, the latter lead to the formation of a system of M–M bonds delocalized along the whole metal skeleton of the cluster.

The calculation of the electronic structure of these clusters by the EHT method was carried out in [106, 107]. These calculations confirmed the primary role of $4d_z2(\sigma_{M-M})$-, $4d_{xz}$-, and $4d_{yz}(\pi_{M-M})$-AO's in the formation of multiple M–M bonds. However, the $4d_{yz}(\pi_{M-M})$-AO's also play a significant role (but less significant than that of $4d_{xy}$- and $4d_{x^2-y^2}$-AO's) in the formation of a system of delocalized M–M bonds. A satisfactory description of the specific structure of trigonal prismatic technetium clusters is given in [106, 107]. However, there are at least two specific features of the electronic and molecular structures of these technetium clusters [113, 142] that cannot be explained using the results of these investigations. In fact, according to [15, 106, 107], the HOMO in $[Tc_6(\mu\text{-}Cl)_6Cl_6]^{2-}$ is represented by the a''_2-MO having the π^*-character with respect to the vertical M–M bonds and the σ-bonding character with respect to horizontal Tc–Tc bonds. It is apparent that the consecutive removal of one or two electrons, which really occurs in the $[Tc_6(\mu\text{-}Cl)_6Cl_6]^{n-}$ clusters $(n = 1, 0)$, should decrease the vertical M–M bond distances and increase the horizontal ones. However, a somewhat different change in the M–M bond distances is observed in the $[Tc_6(\mu\text{-}Cl)_6Cl_6]^{n-}$ clusters (where $n = 2, 1, 0$). Also, the weak paramagnetism of $[Tc_6(\mu\text{-}Cl)_6Cl_6]^{2-}$ and the ESR spectra observed in polycrystalline state are the second experimental fact that does not quite agree with the results of [15, 106, 107]. Actually, according to [106, 107, 113], a comparatively large energy gap between the HOMO and LOMO (~0.7 eV) implies that $[Tc_6(\mu\text{-}Cl)_6Cl_6]^{2-}$ is in a spin-paired ground electronic state.

The electronic structure of trigonal-prismatic technetium clusters was discussed in [142]. The main results obtained in this work are considered below.

Figure 11 shows the results of calculations on the electronic structure of technetium trigonal-prismatic clusters, carried out in [12, 15, 106, 107, 113]. As seen in Fig. 11, a significant relative energy rearrangement of MO's occurs in $[Tc_6Cl_{14}]^{3-}$ and $[Tc_6Cl_{12}]^{2-}$ in the region of HOMO–LUMO. This change cannot be due to the difference in the calculation methods and parametrization, because the calculations of $[Tc_6Cl_{12}]^{2-}$ made in [106, 107] and [12, 15, 113] yield essentially the same results. Most probably, the difference in the calculations of $[Tc_6Cl_{14}]^{3-}$ and $[Tc_6Cl_{12}]^{2-}$ is due to the actual difference in the interatomic distances of these clusters [12, 15, 77]. According to the calculations in [15], the formal multiplicities of short M–M bonds for $[Tc_6Cl_{14}]^{3-}$

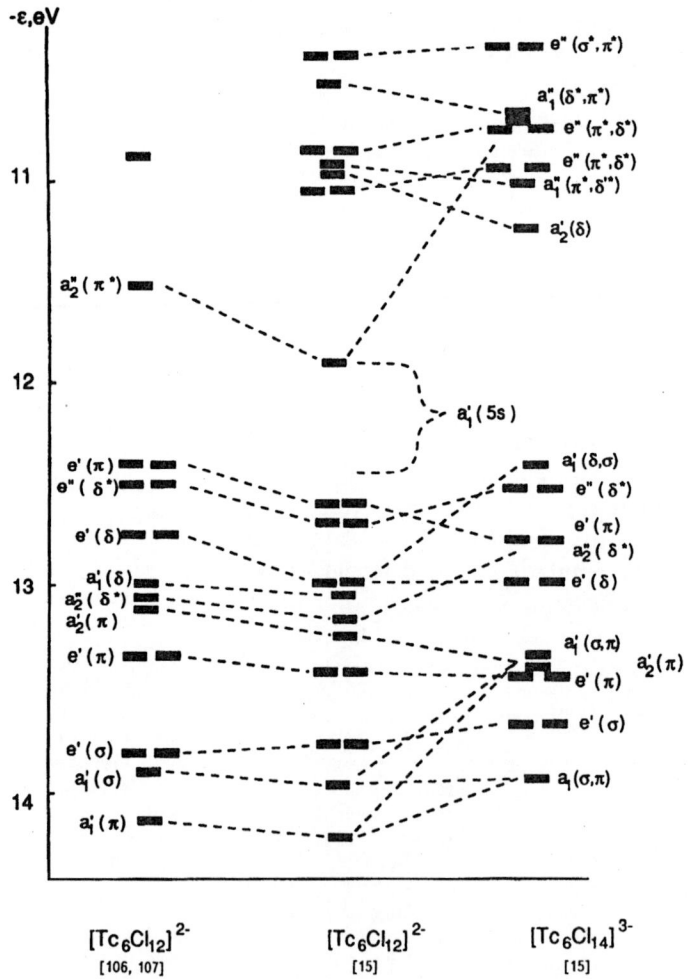

Fig. 11. Results of EHT calculations of technetium trigonal-prismatic clusters

and $[Tc_6Cl_{12}]^{2-}$ are 3.16(6) and 2.66(6), respectively. In other words, the calculations show that these bonds should be stronger in $[Tc_6Cl_{14}]^{3-}$ than in $[Tc_6Cl_{12}]^{2-}$, which is in agreement with experiment [15]. Moreover, if the relative energy location of MO's is supposed to be the same for $[Tc_6Cl_{14}]^{2-}$ and $[Tc_6Cl_{14}]^{3-}$, the M–M bond distances in $[Tc_6Br_{14}]^{2-}$ should be intermediate between those of $[Tc_6Cl_{14}]^{3-}$ and $[Tc_6Cl_{12}]^{2-}$. The experiment shows that this is the change in the M–M bond distance in trigonal-prismatic clusters [12, 15, 77].

At first glance, additional EHT calculations of the $[Tc_6Cl_{14}]^{3-}$ cluster [15, 113] seem to remove all discrepancies between theory and experiment. However, this is not so. Moreover, new discrepancies that cannot be explained in terms of the EHT arise. Thus, as was mentioned above, the EHT calculations fail to explain the weak paramagnetism of $[Tc_6Cl_{12}]^{2-}$ (the same is true of $[Tc_6Br_{14}]^{2-}$). In [113], it was shown that the weak paramagnetism was due to the nature of these clusters, and not to the presence of paramagnetic impurities. It was also noted that the ESR spectrum could not be satisfactorily interpreted in terms of EHT calculations. It follows from [142] that none of the possible variants of the $[Tc_6Cl_{14}]^{3-}$ electronic structure (taking into account the energy rearrangements of MO's in the HOMO–LUMO region, provided the formal multiplicities of the M–M bonds are close to those observed experimentally) explain the values: $g_{\parallel} \sim 2.00$, $g_{\perp} \sim 1.65$ [113]. Explanation of the experimental g-factors is only possible if the a'_1-MO's formed mainly by 5s(5p)-AO's of technetium atoms enter the HOMO–LUMO region shown by the dotted brackets in Fig. 11. This possibility was studied in [142], using a simplified model of the multiple M–M bonds in clusters for dimers of M_2^{n+} metals. It was shown that in the cases where technetium has minimal effective charges in its clusters, the probability that the outer 5s(5p)-AO's are involved in the formation of an additional $2\sigma'$-component of the M–M bond is the highest of all acido-clusters of d-transition elements (see section 5.1).

The above-reported data lead to the conclusion that the HOMO in $\{ [Tc_6(\mu\text{-Br})_6Br_6]Br_2] \}^{3-}$ is the a'_1-MO composed mainly of 5s(5p$_z$)- and 4d$_z^2$-AO's of technetium and having $2\sigma'$-bonding character with respect to short M–M bonds. In other words, the EHT calculations [106, 107, 113] overestimate the interaction of δ^*- and π^*-MO's of dimers in the horizontal directions and underestimate the participation of 5s- and 5p$_z$-AO's in the formation of M–M bonds [15, 142].

More detailed descriptions of the electronic structure of trigonal-prismatic technetium clusters, given in [15, 142] allow explanation of the main changes in their structures upon addition or removal of electrons from "metallic" MO's. However, based on these descriptions, it is impossible to establish a more exact relationship between the specific MO scheme in $[Tc_6X_{12}]^{n-}$ and the M–M bond distances in them, because this problem can only be solved when the whole energy is taken into account in multiconfigurational approximations of the MO method. In fact, in the section 5.1 it has already been shown that even in binuclear technetium clusters, it is necessary to take into account the CI's. This is supported by the experimental fact that in d^4–d^5 technetium complexes, the M–M distance is shorter than those in d^4–d^4 complexes with higher formal multiplicity of M–M bonds. Analogous phenomena were also observed for $[Tc_6X_{14}]^{3-}$ clusters in which multiple M–M bonds are shorter than those $[Tc_6X_{14}]^{2-}$ and $[Tc_6X_{12}]^{2-}$. Similar effects also take place in octanuclear technetium clusters [44, 53, 73, 74] (see the next section).

To conclude this section, we consider interpretation of the magnetic and other physicochemical properties of some of trigonal-prismatic technetium clusters. Analysis of the magnetic susceptibility of $[(CH_3)_4N]_2[Tc_6(\mu\text{-Cl})_6Cl_6]$ [15, 113] shows that it is described by the Curie–Weiss law with $\mu_{eff} = 1.24\ \mu_B$ and $\theta = -120$ K. On the other hand, the ESR spectra of $[Tc_6(\mu\text{-Cl})_6Cl_6]^{2-}$ show a signal with $g_{eff} \sim 1.97$ and low relative intensity (0.2 of the intensity of the $[Tc_6X_{14}]^{3-}$ signal). The compound in question is thus paramagnetic [113]. Since $[Tc_6(\mu\text{-Cl})_6Cl_6]^{2-}$ has an even number of "metallic" electrons, it has to be in a ground triplet state in order to display paramagnetism. The low-intensity ESR signals and low μ_{eff} compared to the theoretical values for a triplet state lead to the suggestion that in real crystals, only part of the $[Tc_6(\mu\text{-Cl})_6Cl_6]^{2-}$ clusters are paramagnetic. Assuming the relative intensity of the ESR signal (0.2) to be due to the fraction of the paramagnetic triplet clusters, we can determine the magnetic moment of the paramagnetic cluster ($\mu = 2.77\ \mu_B$) , using the formula $\mu^2 = \mu_{eff}^2/0.2$ [172]. This value agrees with the theoretical value of the magnetic moment for a triplet particle with $g_{eff} = 1.97$. Attempts to explain the temperature dependence of magnetic susceptibility and the ESR signal intensity by the Boltzmann population of an excited triplet state, intermolecular exchange interactions, or by the splitting of the triplet-state levels in the ligand crystal field do not give satisfactory results.

The simultaneous presence of singlet and triplet $[Tc_6(\mu\text{-}Cl)_6Cl_6]^{2-}$ clusters in crystals may be accounted for by the sensitivity of the cluster multiplet ground state to minor changes in its structure. In real crystals, the transition from a singlet to a triplet ground state may take place owing to the presence of defects (vacancies, substitution of one ligand by another, dislocations and the like), or a fast kinetic S–T exchange due to the delocalized MO's that are formed by mainly the 5s(5p)-AO's of the technetium atom [15].

The long-wavelength IR spectra of trigonal prismatic technetium clusters and a number of unusual physico-chemical properties of the clusters with ferricinium cations [108] support the latter assumption. The discovered properties of the clusters with ferricinium cations may be accounted for by the formation of the conductivity bands and, probably, hard-fermion bands in these compounds by the 5s(5p)-AO's of technetium atoms and 4s(4p)-AO's of the iron atoms. The formation of these bands may be supported by the following facts: the ESR spectra of these compounds with g_{eff} close to that of a free electron; temperature independent conductivity; and an unusual temperature dependence of the Mossbauer and X-ray photoelectron spectra [108].

5.3.2 Tetragonal-Prismatic Clusters

Trigonal-prismatic clusters are most similar to tetragonal-prismatic clusters in their molecular structure and their mechanism of formation [9, 12, 13, 15, 77]. Obviously, the electronic structures of these two types of clusters are, on the whole, similar. Calculations of the electronic structure of a hypothetical tetragonal-prismatic cluster with the geometry of the metal skeleton analogous to that of the bromide compound $\{[Tc_8(\mu\text{-}Br)_8Br_4]Br\}$ were carried out in [15, 107, 113]. Inaccurate literature data on the parameters of Br atoms did not allow EHT calculations of the electronic structure of the cluster with bromide ligands. The results of these calculations are given in Fig. 12. As in the case of trigonal-prismatic clusters, we can also observe: (1) MO's responsible for mainly M–M bonding in the direction of the short edges of a prism formed by σ_{M-M}- and π_{M-M}-MO of Tc_2^{n+} dimers and somewhat perturbed by the neighboring dimers; (2) MO's almost completely delocalized along the metal skeleton of the cluster and formed by mainly the δ_{M-M}-MO of Tc_2^{n+} dimers. Unfortunately, the energy sequence of the MO's cannot be guaranted, because the inaccuracy in the geometry of the hypothetical cluster ion may have a much more significant effect on the electronic structure. However, it is quite sure that MO's formed (as is the case of trigonal-prismatic and d^5–d^5 binuclear clusters) mainly by σ'_{M-M}-MO's of the Tc_2^{n+} dimers should be located somewhere in the range close to the

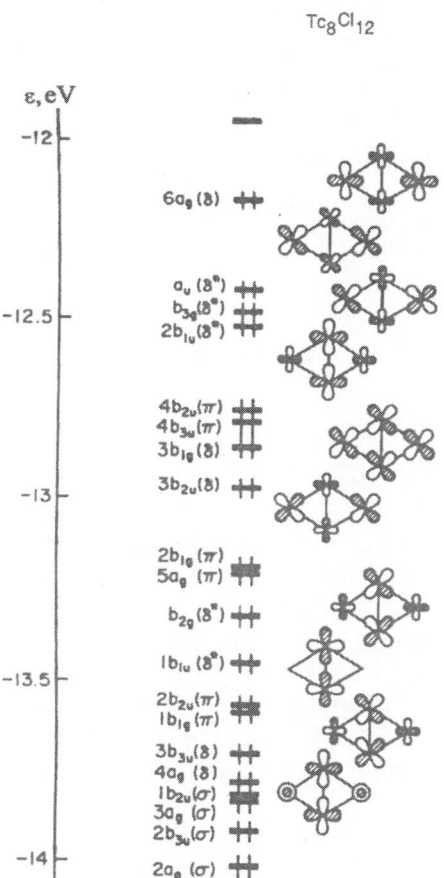

Fig. 12. Results of EHT calculations of hypothetical chloride technetium tetragonal-prismatic clusters [107]

HOMO–LUMO. This may be supported by low Z_{eff} of technetium atoms in tetragonal-prismatic clusters [103, 127] and formation of compounds with ferricinium cations with some unusual physico-chemical properties, as is the case of trigonal-prismatic clusters [108, 128].

The main structural difference between trigonal- and tetragonal-prismatic clusters [12, 77] is that not all technetium atoms of the latter are structurally equivalent: a pair of technetium atoms situated on the long diagonal of the rhombic base in tetragonal-prismatic clusters have the 2.0 + formal oxidation state, whereas the technetium atoms situated on the short diagonal of these rhombs have the formal oxidation state 1.0 + . However, X-ray photoelectron data [103] show that the electronic density is levelled on all technetium atoms. This indicates that a system of delocalized multicenter M–M bonds is formed in the $[Tc_8(\mu\text{-}Br)_8Br_4]^{n+}$ clusters (where $n = 0, 1$).

The electronic structure of the $[Tc_8(\mu\text{-}Br)_8Br_4]^{n+}$ clusters (where $n = 0, 1$) shows that their system of M–M bonds in them should be sensitive to the addition (or removal) of "excess" electrons, change of electronegativity of the ligands, and their crystal field intensity, because these changes should first influence the weak δ-components of M–M bonds and lead to their weakening (or strengthening) as a result of the strengthening (or weakening) of the system of multicenter M–M bonds formed by the same types of technetium AO's ($4d_{xy}$ and $4d_{x^2-y^2}$) as the δ-components of the quadruple M–M bonds. Eventually, these changes should lead to either decomposition of the cluster, followed by formation of binuclear $[Tc_2X_8]^{n-}$ complexes or to a complete rearrangement of the cluster and formation of a system of essentially equivalent M–M bonds [53].

5.3.3 Octahedral Clusters

The electronic structure of the octahedral clusters $[M_6(\mu_3\text{-}X)_8]^{4+}$ (where M = Mo, W; X = Cl, Br, I) is described [27, 173, 174] by the formation of 12 bonding "metallic" multicenter MO's composed of the nd_{xz}-, nd_{yz}-, nd_{z^2}-, and $nd_{x^2-y^2}$-AO's. In the case of $[Tc_6(\mu_3\text{-}Br)_5Br_6]^{2-}$, there are 9 "excess" (compared to $[M_6(\mu\text{-}X)_8]^{4+}$) "metallic" electrons [31]. Fig. 13 shows the results of the EHT calculations or hypothetical chloride octahedral technetium clusters with distorted and nondistorted ligands surrounding and structurally analogous to real bromide clusters. The EHT calculations show that the "excess" electrons occupy the MO's formed by mainly technetium $4d_{xy}$-AO's and sp^3-hybrid AO's of chlorine bridging atoms and they simultaneously have a weak M–M bonding and Tc–$(\mu_3\text{-}Cl)$ antibonding character. As a result, part of the bonds of the technetium atoms with the bridging chlorine atoms are ruptured. This is accompanied by a simultaneous increase in the strength of the bonding components of Tc–Tc bonds and part of the remaining $Tc(\mu_3\text{-}Cl)$ bonds situated on the opposite side of the Tc_6^{9+} octahedron relative to the ruptured $Tc(\mu_3\text{-}Cl)$ bonds [11, 12]. In other words, the limiting case of the Jahn–Teller effect significantly decreases the number of unpaired electrons on degenerated MO's and leads to a total decrease in the electronic energy, compared to the hypothetical nondistorted cluster with the same formal oxidation state of technetium atoms.

To conclude this section, we consider one more probable way for the stabilization of octahedral clusters, which, as was shown in the previous section, is characteristic of binuclear, trigonal-prismatic and, obviously, all types of technetium acido-clusters [10]. According to [12, 77], all technetium clusters with an odd number of "metallic" electrons have shorter multiple M–M bonds than those in analogous structures, but with an even number of "metallic" electrons. In our opinion [10, 15], this effect of an odd number of "metallic" electrons is essentially analogous to the effect of the increase in M–M bond

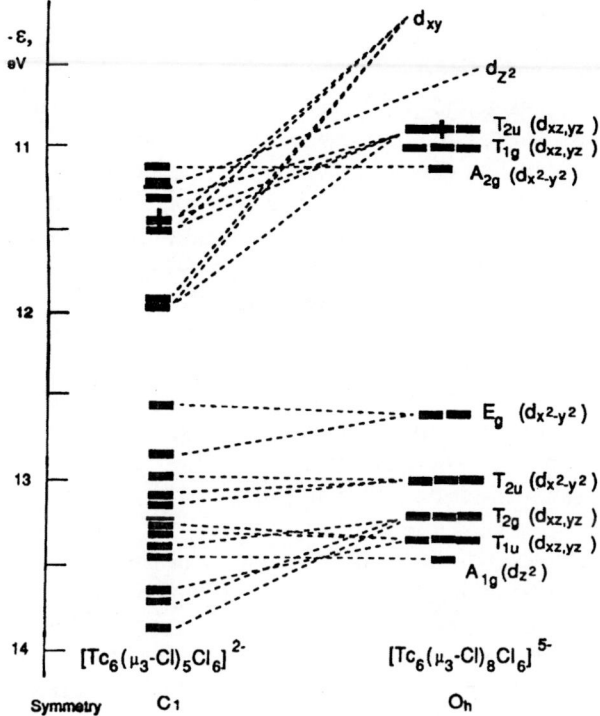

Fig 13. Results of EHT calculations of hypothetical chloride technetium octahedral clusters with disturbed and undisturbed ligands surrounding [15]

strength in binuclear d^4–d^5 clusters (compared to d^4–d^4 clusters) and is accounted for by a large contribution of the CI's in the case of an even number of "metallic" electrons and by small contribution of the CI's in the case of an odd number. Because all the high-lying MO's have M–M antibonding character, the addition of excited states due to CI's to the ground state should lead to an increase in the M–M bond distances. In this case, if the lowlying excited states also have antibonding M–X, M–(μ_3-X) or M–M (delocalized along the metal skeleton) character, the addition of these states to the ground state will decrease the strength of these bonds (e.g., for octanuclear tetragonal-prismatic clusters) or otherwise, increase their strength (e.g., for hexanuclear trigonal-prismatic or octahedral clusters). In all polynuclear clusters, the outer 5s(5p)-AO's of technetium atoms are most likely involved in the formation of additional components of fully symmetrical MO's delocalized along the metal skeleton (because of the low Z_{eff} of technetium atoms, see Sect. 5.1). This stabilizes the cluster system and increases the probability of their formation owing to symmetry prohibition during the reaction of cycloaddition of multiple M–M bonds [15].

6 Conclusion

This review has focussed on experimental and theoretical proofs of the participation of the outer 5s(5p)-AO's of technetium atoms in the formation of the bonding "metallic" MO's of its clusters with a formal oxidation state not higher than $2.0 +$. This effect leads to an increase in the M–M bond strength in technetium clusters in which there is formally an "excess" (in addition to the occupied shell of the bonding MO mainly composed of 4d-type AO's) amount of "metallic" electrons in them. On the other hand, it results in a stronger shielding of the positive charges of the nuclei (with a decrease in Z_{eff}) of technetium atoms in its clusters, thus decreasing the Coulomb repulsion upon formation of M–M bonds, and increasing the contribution of the CI's to the ground electronic state of the system; i.e., it also contributes to the cluster-forming properties of technetium. This interpretation is in good agreement with the "anomalous" properties of technetium clusters earlier described in [10].

Acknowledgements. I am grateful to the Russian Foundation of Fundamental Researches for a research grant (code: 93-03-18683) and the International Science Foundation for the research grants SAM000, MDF300 and MDF000. I also wish to thank the USA National Science Foundation and Nobel Laureate Prof. R. Hoffmann (USA) for their financial support of part of the investigations. An IBM PC 486 "Leo" computer used for carrying out the calculations and writing the paper was purchased from the finances of the NSF grant.

8 References

1. Cotton FA, Walton RA (1982) Multiple Bonds between Metal Atoms. Wiley, New York
2. Spitsyn VI, Kuzina AF (1981) Tekhnetsiy Nauka, Moscow
3. Deutsch E (1989) J Nuclear Medicine 33: 279
4. Deutsch E, Libson K, Vanderheyden J-L (1990) in: Nicolini M, Bandoli G, Mazzi U (eds) Technetium and rhenium in chemistry and nuclear medicine, vol 3. Raven Press, New York, p 13
5. Johnson BFG (ed) (1980) Transition Metal Clusters. Wiley, New York
6. Gubin SP (1984) Vestnik Akad Nauk SSSR 1: 55
7. Gubin SP (1987) Zhurn Mendeleev Vses Khim Obsh 32: 3
8. Gubin SP (1987) Khimiya klasterov. Nauka, Moscow, p 263
9. Kryutchkov SV (1983) Phisico-khimicheskie svoystva soedineniy technetsiya. Doctor Thesis, Institute of Physical Chemistry, USSR Academy of Sciences, Moscow
10. Kryutchkov SV (1985) Koord Khimiya 11: 631
11. Spitsyn VI, Kuzina AF, Kryutchkov SV (1985) Uspekhi Khimii 54: 637
12. Kryutchkov SV, Grigoriev MS, German KE (1990) in: Nicolini M, Bandoli G, Mazzi U (eds) Technetium and rhenium in chemistry and nuclear medicine, vol 3. Raven Press, New York, p 253
13. Kryutchkov SV, Kuzina AF, German KE (1990) in: Nicolini M, Bandoli G, Mazzi U (eds) Technetium and rhenium in chemistry and nuclear medicine, vol 3. Raven Press, New York, p 275

14. Kryutchkov SV (1990) in: Abstracts of posters of XXIII Intern conf coord chem 13–18 August 1990, Gera, DDR, vol 1 P 5–7
15. Kryutchkov SV (1991) Khimiya klasternikh soedineniy technetsiya. Second Doctoral Thesis, Institute of Physical Chemistry, Russian Academy of Sciences, Moscow
16. Tronev VG, Bondin SM (1952) Dokl Akad Nauk SSSR 86: 87
17. Kotel'nikova AS, Tronev VG (1958) Zhurn Neorgan Khimii 3: 1008
18. Babeshkina GK, Tronev VG (1962) Zhurn Neorgan Khimii 7: 215
19. Kotel'nikova AS, Vinogradova GA (1963) Dokl Akad Nauk SSSR 152: 621
20. Koz'min PA, Kuznetsov VG (1963) in: IV Vsesoyuznoe soveshanie po kristallokhimii, 19–23 December 1961, Abstracts of Papers, Shtiintsa, Kishinev, p 74
21. Kuznetsov VG, Koz'min PA (1963) Zhurn Strukt Khimii 4: 55
22. Glinkina MI, Kuzina AF, Spitsyn VI (1973) Zhurn Neorgan Khimii 18: 403
23. Bertrand JA, Cotton FA, Dollase WA (1963) J Am Chem Soc 85: 1349
24. Eakins JD, Huphreys DG, Mellish CE (1963) J Chem Soc 6012
25. Robinson WT, Fergusson JE, Penfold BR (1963) Proc Chem Soc 116
26. Cotton FA, Curtis NF, Harris CB, Johnson BFG, Lippard SJ, Mague JT, Robinson WR, Wood JS (1964) Science 145(3638): 1305
27. Cotton FA, Haas TE (1964) Inorg Chem 3: 10
28. Cotton FA, Curtis NF, Johnson BFG, Robinson WR (1965) Inorg Chem 4: 326
29. Cotton FA (1969) Accounts Chem Res 2: 240
30. Cotton FA, Pedersen E (1975) Inorg Chem 14: 383
31. Cotton FA (1975) Chem Soc Rev 4: 27
32. Schwochau K, Hedwig K, Schenk HJ (1977) Inorg Nucl Chem Letters 13: 77
33. Cotton FA (1978) Accounts Chem Res 11: 225
34. Preetz W, Peters G (1980) Z Naturforsch 35b: 797
35. Cotton FA, Daniels L, Davison A, Orvig C (1981) Inorg Chem 20: 3051
36. Kotel'nikova AS, Misailova TV, Babichevskaya IZ, Lebedev VG (1978) Zhurn Neorgan Khimii 23: 2402
37. Koz'min PA, Surazhskaya MD, Larina TB, Kotel'nikova AS, Ahmedov EF (1979) Zhurn Neorgan Khimii 24: 3383
38. Shtemenko AV, Kotel'nikova AS, Koz'min PA, Surazhskaya MD, Larina TB (1980) Zhurn Neorgan Khimii 25: 2300
39. Ahmetov EL, Kotel'nikova AS, Efstaf'eva ON (1981) Koord Khimiya 7: 568
40. Koz'min PA, Surazhskaya MD, Larina TB, Shtemenko AV, Golovaneva NF (1981) Koord Khimiya 7: 792
41. Koz'min PA, Surazhskaya MD, Larina TB, Shtemenko AV, Kotel'nikova AS (1981) Koord Khimiya 7: 1271
42. Spitsyn VI, Kuzina AF, Oblova AA, Belyaeva LI (1977) Dokl Akad Nauk SSSR 237: 1126
43. Spitsyn VI, Kuzina AF, Oblova AA, Kryutchkov SV, Belyaeva LI (1977) Dokl Akad Nauk SSSR 237: 1412
44. Spitzin VI, Kryutchkov SV, Grigoriev MS, Kuzina AF (1988) Z anorg allg Chem 563: 136
45. Cotton FA, Gage LD (1977) Nouv J Chim 1: 441
46. Cotton FA, Fanwick PE, Gage LD (1980) J Am Chem Soc 102: 1570
47. Zaytseva LL, Kotel'nikova AS, Rezvov AA (1980) Zhurn Neorgan Khimii 25: 2624
48. Spitsyn VI, Bairl B, Kryutchkov SV, Kuzina AF, Varen M (1981) Dokl Akad Nauk SSSR 256: 608
49. Kryutchkov SV, Kuzina AF, Spitsyn VI (1982) Dokl Akad Nauk SSSR 266: 127
50. Kryutchkov SV, Grigoriev MS, Yanovskiy AI, Struchkov YuT, Spitsyn VI (1987) Dokl Akad Nauk SSSRI 297: 867
51. Spitsyn VI, Kuzina AF, Oblova AA, Kryutchkov SV, German KE, Belyaeva LI (1982) in: Abstracts of Papers XXII International Conference on Coordination Chemistry, 1982, Budapest, p 811
52. Spitsyn VI, Kuzina AF, Kryutchkov SV (1980) Zhurn Neorgan Khimii 25: 741
53. Kryutchkov SV, Kuzina AF, Spitzin VI (1988) Z anorg allg Chem 563: 153
54. Bürgi HB, Anderegg G, Blauenstein P (1981) Inorg Chem 20: 3829
55. Anderegg G, Muller E, Zollinger K, Burgi HB (1983) Helv Chim Acta 66: 1593
56. Colmanet SF, Mackay MF (1987) J Chem Soc, Chem Commun 705
57. Linder KE, Dewan JC, Davison A (1989) Inorg Chem 28: 3820
58. Baturin NA, German KE, Grigoriev MS, Kryutchkov SV (1991) Koord Khimiya 17: 1375

59. Kryutchkov SV, Simonov AE (1990) Koord Khimiya 16: 339
60. Cotton FA, Shive LW (1975) Inorg Chem 14: 2032
61. Cotton FA, Bratton WK (1965) J Am Chem Soc 87: 921
62. Bratton WK, Cotton FA (1970) Inorg Chem 9: 780
63. Cotton FA, Davison A, Day VW, Fredich MF, Orvig C, Swanson R (1982) Inorg Chem 21: 1211
64. Grigoriev MS, Kryutchkov SV, Struchkov YuT, Yanovskiy AI (1990) Koord Khimiya 16: 90
65. Koz'min PA, Larina TB, Surazhskaya MD (1981) Koord Khimiya 7: 1719
66. Koz'min PA, Larina TB, Surazhskaya MD (1982) Koord Khimiya 8: 851
67. Koz'min PA, Larina TB, Surazhskaya MD (1983) Koord Khimiya 9: 1114
68. Kryutchkov SV, Grigoriev MS, Kuzina AF, Gulev BF, Spitsyn VI (1986) Dokl Akad Nauk SSSR 288: 389
69. Cotton FA, Daniels LM, Falvello LR, Grigoriev MS, Kryutchkov SV (1991) Inorg Chim Acta 189: 53
70. Koz'min PA, Larina TB, Surazhskaya MD (1983) Dokl Akad Nauk SSSR 271: 1157
71. Koz'min PA, Surazhskaya MD, Larina TB (1985) Koord Khimiya 11: 1559
72. German KE, Kryutchkov SV, Kuzina AF, Spitsyn VI (1986) Dokl Akad Nauk SSSR 288: 381
73. Kryutchkov SV, Grigoriev MS, Kuzina AF, Gulev BF, Spitsyn VI (1986) Dokl Akad Nauk SSSR 288: 893
74. Koz'min PA, Larina TB, Surazhskaya MD (1987) Koord Khimiya 13: 415
75. Kryutchkov SV, Grigoriev MS, Yanovskiy AI, Struchkov YuT., Spitsyn VI (1988) Dokl Akad Nauk SSSR 301: 618
76. Kryutchkov SV, Grigoriev MS, Kuzina AF, Gulev BF, Spitsyn VI (1986) Dokl Akad Nauk SSSR 290: 865
77. Grigoriev MS, Kryutchkov SV (1993) Radiochim Acta 63: 187
78. Hileman JC, Huggins DK, Kaesz HD (1961) J Am Chem Soc 63: 2953
79. Cotton FA, Pedersen E (1975) Inorg Chem 14: 383
80. Kryutchkov SV, Kuzina AF, Spitsyn VI (1983) Zhurn Neorgan Khimii 28: 1984
81. Koz'min PA, Novitskaya GN (1972) Zhurn Neorgan Khimii 17: 3138
82. Koz'min PA, Novitskaya GN (1975) Koord Khimiya 1: 248
83. Yukhnevich GV (1965) in: Trudy komissii po spectroscopii AN SSSR, vol 1. VINITY, Moscow, p 235
84. Basile A, La Bonville P, Ferraro JR, Williams J (1974) J Chem Phys 60: 1981
85. Izmaylov NA (1961) Elektrokhimiya rastvorov, Khimiya, Moscow
86. Lundren JO, Williams JM (1973) J Chem Phys 58: 788
87. Kryutchkov SV, Simonov AE (1987) Izvestiya AN SSSR, ser. Khim 2140
88. Cotton FA, Gage LD (1979) Inorg Chem 18: 1716
89. Koz'min PA, Surazhskaya MD (1980) Koord Khimiya 6: 653
90. Koz'min PA (1981) Koord Khimiya 7: 659
91. Koz'min PA, Surazhskaya MD, Larina TB, Kotel'nikova AS, Misailova TV (1980) Koord Khimiya 6: 1256
92. German KE, Grigoriev MS, Kuzina AF, Spitsyn VI (1987) Zhurn Neorgan Khimii 32: 1089
93. Spitsyn VI, Kuzina AF, German KE, Grigoriev MS (1987) Dokl Akad Nauk SSSR 293: 101
94. Meyer G, Hoppe R (1976) Z anorg allgew Chem 420: 40
95. German KE, Grigoriev MS, Kuzina AF, Gulev BF, Spitsyn VI (1986) Dokl Akad Nauk SSSR 287: 650
96. Cotton FA, Frenz BA, Shive W (1975) Inorg Chem 14: 649
97. Calvo C, Jayadevan NC, Lock CJL, Restivo R (1970) Canad J Chem 48: 219
98. Kannellakopulos B, Nuber B, Raptis K, Ziegler ML (1989) Angew Chem 101: 1055
99. Kannellakopulos B, Nuber B, Raptis K, Ziegler ML (1989) Angew Chem Int Ed Engl 28: 1055
100. Magneli A, Andersson G (1955) Acta Chem Scand 9: 1378
101. Koz'min PA, Larina TB, Surazhskaya MD (1982) Dokl Akad Nauk SSSR 265: 1420
102. Gerasimov VN, Kryutchkov SV, Kuzina AF, Kulakov VM, Pirozhkov SV, Spitsyn VI (1982) Dokl Akad Nauk SSSR 286: 148
103. Gerasimov VN, Kryutchkov SV, German KE, Kulakov VM, Kuzina AF (1990) in: Nicolini M, Bandoli G, Mazzi U (eds) Technetium and rhenium in chemistry and nuclear medicine, vol 3. Raven Press, New York, p 231
104. Kryutchkov SV, Grigoriev MS, Kuzina AF, Spitsyn VI (1987) Zhurn Neorgan Khimii 32: 2944
105. Kryutchkov SV, Grigoriev MS, Kuzina AF, Spitsyn VI (1987) Zhurn Neorgan Khimii 32: 2953

106. Wheeler R, Hoffmann R (1986) Angew Chem Int Ed Engl 25: 822
107. Wheeler R, Hoffmann R (1986) J Am Chem Soc 108: 6605
108. Antipov VN, Kryutchkov SV, Gerasimov VN, Grigoriev MS, Kazin PE, Kharitonov VV, Maksimov VG, Moisa VS, Sergeev VV, Yurik TK (1994) Radiochim Acta 64: 191
109. Koz'min PA, Kotel'nikova AS, Larina TB, Mekhtiev MM, Surazhskaya MD, Bagirov ShA, Osmanov NS (1987) Dokl Akad Nauk SSSR 295: 647
110. Koz'min PA, Kotel'nikova AS, Surazhskaya MD, Osmanov RS, Larina TB, Abbassova TA, Mekhtiev MM (1989) Koord Khimiya 15: 1216
111. Koz'min PA, Osmanov RS, Larina TB, Kotel'nikova AS, Surazhskaya MD, Abbassova TA (1989) Dokl Akad Nauk SSSR 306: 378
112. Osmanov RS, Kotel'nikova AS, Surazhskaya MD, Larina TB, Abbassova TA, Koz'min PA (1989) Koord Khimiya 15: 1340
113. Kryutchkov SV, Rakytin YuV, Kazin PE, Zhirov AI, Konstantinov NYu (1990) Koord Khimiya 16: 1230
114. Koltunov VS, Gomonova TV, Shapovalov MP, Abramova IG (1988) Radiokhimiya 30: 751
115. Huber EW, Heinemen WR, Deutsch E (1987) Inorg Chem 26: 3718
116. Chisholm MH, Cotton FA (1978) Accounts Chem Res 11: 356
117. Chisholm MH, Kirpatrick CE, Huffman JC (1981) Inorg Chem 20: 871
118. Kryutchkov SV, Kuzina AF, Spitsyn VI (1986) Dokl Akad Nauk SSSR 287: 1400
119. McGinnis RN, Ryan TR, McCarley RE (1978) J Am Chem Soc 100: 7900
120. Aufdembrink BA, McCarley RE (1986) J Am Chem Soc 108: 2474
121. Oblova AA, Kuzina AF, Belyaeva LI, Spitsyn VI (1982) Zhurn Neorgan Khimii 27: 2814
122. Simonov AE (1986) in: I Moskovskaya konferentsiya molodich uchenich po radiokhimii, May 1986, Abstracts of Papers, Inst Phys Chem Russian Acad. Sci., Moscow, p 36
123. Spitsyn VI, Kuzina AF, Kryutchkov SV, Simonov AE (1987) Zhurn Neorgan Khimii 32: 2180
124. Kryutchkov SV, German KE, Simonov AE (1991) Koord Khimiya 17: 480
125. Vinogradov IV, Zaytseva LL, Konarev MN, Shepel'kov SV (1980) in: I Vsesoyuznoye soveshanie "Poluchenie i videlenie radioaktivnikh izotopov" (July 1980, Tashkent). Press FAN, Tashkent, p 33
126. Vinogradov IV, Konarev MN, Zaytseva LL, Shepel'kov SV (1976) Zhurn Neorgan Khimii 21: 131
127. Makarov LL, Kryutchkov SV, Zaytsev YuM, Sablina NO, German KE, Kuzina AF (1990) in: Nicolini M, Bandoli G, Mazzi U (eds) Technetium and rhenium in chemistry and nuclear medicine, vol 3. Raven Press, New York, p 265
128. Antipov VN, Kryutchkov SV, Gerasimov VN, Grigoriev MS, Kazin PE, Kharitonov VV, Maksimov VG, Moisa VS, Sergeev VV, Yurik TK (1993) in: Abstracts of Reports of Topical Symposium on the Behavior and Utilisation of Technetium'93, 18–20 March 1993, Tohoku University, Sendai, Japan, report No 1-P-24, p 56
129. Grigoriev MS, Kryutchkov SV (1993) in: Abstracts of Reports of Topical Symposium on the Behavior and Utilisation of Technetium'93, 18–20 March 1993, Tohoku University, Sendai, Japan, report No 3-I-2, p 23
130. Huber H, Kundig EP, Moskovits M, Ozin GA (1975) J Am Chem Soc 97: 2097
131. Kundig EP, Moskovits M, Ozin GA (1975) Nature 254(5500): 503
132. Busby R, Klotzbucher W, Ozin GA (1976) J Am Chem Soc 98: 4013
133. Ozin GA (1976) Appl Spectrosc 30: 573
134. Ford TA, Huber H, Klotzbucher W, Kundig EP, Moskovits M, Ozin GA (1977) J Chem Phys 66: 524
135. Klotzbucher W, Ozin GA (1977) Inorg Chem 16: 984
136. Klyagina AP, Gutsev GL, Fursova VD, Levin AA (1984) Zhurn Neorgan Khimii 29: 2765
137. Klyagina AP, Levin AA (1984) Koord Khimiya 10: 579
138. Weltner W, Van Zee RJ (1984) Ann Rev Phys Chem 35: 291
139. Kostikova GP, Korol'kov DV (1985) Uspekhi Khimii 54: 591
140. Morse MD (1986) Chem Rev 86: 1049
141. Klyagina AP, Fursova VD, Levin AA, Gutsev GL (1987) Dokl Akad Nauk SSSR 292: 122
142. Kryutchkov SV, Mironov VS, Antipov BG, Plekhanov YuV (1991) Metallorgan Khimiya 14: 26
143. Kostikova GP, Kostikova YuP, Troyanov SI, Korol'kov DV (1978) Inorg Chem 17: 2279
144. Bursten BE, Cotton FA (1980) Faraday Discuss R Chem Soc Faraday Symp (No 14) 180
145. Bursten BE, Cotton FA, Hall MB (1980) J Am Chem Soc 102: 6348

146. Voronovitch NS, Zacheslavskaya RH, Korol'kov DV (1977) Vestnik LGU, ser Phys khim (No 4) 69
147. Rakytin YuV, Kryutchkov SV, Alexandrov AI, Kuzina AF, Nemtsev NV, Ershov BG, Spitsyn VI (1983) Dokl Akad Nauk SSSR 269: 1123
148. Rakytin YuV, Nefedov VI (1985) Zhurn Neorgan Khimii 29: 510
149. Cotton FA, Kalbacher B (1977) Inorg Chem 16: 2386
150. Nefedov VI, Kozmin PA (1982) Inorg Chim Acta Letters 64: L177
151. Cotton FA, Fanwick PE, Gage LD, Kalbacher B, Martin DS (1977) J Am Chem Soc 99: 5642
152. Noodleman L, Norman JG (1979) J Chem Phys 70: 4903
153. Benard M (1978) J Am Chem Soc 100: 2354
154. Mello PC, Edwards WD, Zarner MC (1982) J Am Chem Soc 104: 1440
155. Etkins P (1977) Kvanty. Spravochnik kontseptsiy. Mir, Moscow
156. Norman JG, Kolari HJ, Gray HB, Trogler WC (1977) Inorg Chem 16: 987
157. Burstein BE, Cotton FA, Fanwick PE, Stanley GG, Walton RA (1983) J Am Chem Soc 105: 2606
158. German KE, Mironov VS (1986) in: I Moskovskaya konferentsiya molodich uchenich po radiokhimii, May 1986, Abstracts of Papers, Inst Phys Chem, Russian Acad Sci, Moscow, p 34
159. Cotton FA, Frenz BA, Ebner JR, Walton RA (1976) Inorg Chem 15: 1630
160. Cotton FA, Dunbar KR, Falvello LR, Tomas M, Walton RA (1983) J Am Chem Soc 105: 4950
161. Cotton FA, Stanley GG, Walton RA (1978) Inorg Chem 17: 2099
162. Cotton FA, Extine MW (1978) J Am Chem Soc 100: 3788
163. Cotton FA, Ilsley WH, Kaim W (1980) J Am Chem Soc 102: 1918
164. Cotton FA, Thompson JL (1980) Inorg Chim Acta Letters 44: L247
165. Cotton FA, Thompson JL (1980) J Am Chem Soc 102: 6437
166. Fanwick PE, King MK, Tetrick SM, Walton RA (1985) J Am Chem Soc 107: 5009
167. Cotton FA (1983) Chem Soc Rev (London) 12: 35
168. Missner H, Korol'kov DV (1972) Zhurn Strukt Khimii 13: 689
169. Shustorovich EM, Korol'kov DV (1972) Zhurn Strukt Khimii 13: 682
170. Shaik S, Hoffmann R, Fisel CR, Summerville RH (1980) J Am Chem Soc 102: 4555
171. Spitsyn VI, Kazin PE, Subbotin MYu, Aslanov LA, Zelentsov VV, Zhirov AI, Felin MG (1986) Dokl Akad Nauk SSSR 287: 134
172. Kalinnikov VT, Rakytin YuV (1980) Vvedenie v magnetokhimiyu. Metod staticheskoy magnitnoy vospriimchvosty. Nauka, Moscow
173. Bursten BE, Cotton FA, Stanley GG (1980) Israel J Chem 19: 132
174. Manning MC, Trogler WC (1981) Coord Chem Rev 38: 89

Substitution Reactions of Technetium Compounds

Takashi Omori

Radiochemistry Research Laboratory, Shizuoka University, 836 Ohya, Shizuoka 422, Japan

Topics in Current Chemistry, Vol. 176
© Springer-Verlag Berlin Heidelberg 1996

1 Introduction

From the viewpoint of coordination chemistry, a substitution reaction can be defined as a process whereby a ligand in a complex is replaced by another ligand from outside the coordination sphere [1]. Substitution reactions by metal complexes have been classified by Saito [2] according to Taube's definition [3] of inertness. Saito classified metal ions into three groups as follows:

Group A

Group A consists of alkali metals, alkaline earth metals, most of the lanthanoids, and Cr(II), Mn(II), Co(II), Cu(II), Zn(II), Cd(II), Pb(II), Ti(III), Fe(III), and Ti(IV). Thus, d^0, low valency d^{10}, and high-spin complexes are included in this group.

Group B

Group B consists of a few of the lanthanoids, and includes V(II), Fe(II), Ni(II), Ru(II), Os(II), Ru(III), Al(III), Ga(III), In(III), Tl(III), Mo(IV), Si(IV), Ge(IV), and Sn(IV). Several d^3, d^6 and d^8 complexes, high oxidation state d^{10} complexes, and a few high-spin complexes belong to this group. Substitution reactions of group B are greatly influenced by the ligands and solvents.

Group C

Group C consists of low-spin complexes, several d^3 complexes, and Cr(III), Mo(III), Co(III), Rh(III), Ir(III), Re(IV), Ir(IV), and Pt(IV). This group exhibits inertness to substitution reactions.

According to this classification, low spin Tc(III)[d^4] and Tc(IV)[d^3] complexes are considered to be inert. The electronic structure of the Tc(V) complexes possessing a Tc=O core renders these kinetically inert [4]. However, the rate of ligand substitution on Tc(V) depends remarkably on the ligands. In addition, the rate is strongly affected by the structure of the Tc(V) complex. Many Tc(V) complexes containing the Tc=O core display a square-pyramidal structure in the absence of a group *trans* to the Tc=O group. Since this vacant position is labile, hydrolysis or solvolysis with alcohols in the absence of strong acids occurs at this position, to form *trans*-TcO(OR)$^{2+}$ or *trans*-TcO$_2^+$ [4]. For example, tetrachlorooxotechnetate(V) and KCN yield [TcO(OMe)(CN)$_4$]$^{2-}$ in methanol [5]. This is considered to be a trapped intermediate in the hydrolysis of TcO(CN)$_5^{2-}$ to TcO$_2$(CN)$_4^{3-}$.

Technetium is usually supplied in the form of heptavalent pertechnetate. Consequently, the syntheses of technetium complexes is necessarily accompanied by the reduction of pertechnetate. When concentrated hydrochloric acid is employed as a reductant, tetrachlorooxotechnetate(V) complexes can easily be obtained. A further reduction procedure is required to obtain hexachlorotechnetate(IV). Using these complexes, a number of technetium complexes have been synthesized by ligand substitution. The importance of preparative substitution reactions also increases in the light of the design and preparation of radiopharmaceuticals labelled with ^{99m}Tc and ^{188}Re.

The technetium(III) complexes are synthesized by the direct reduction of pertechnetate with an appropriate reductant in the presence of the desired ligand. However, when sodium dithionite is used as a reductant, the oxidation state of the synthesized complex varies from III to V, depending significantly on the nature of the coexisting ligand.

Although these complicated features of the redox reactions of technetium are important topics, the present article is limited to the susbtitution kinetics of technetium complexes. Some redox kinetics of technetium complexes have been discussed by Koltunov and Gomonova [6]. The redox potentials for analogous technetium and rhenium complexes, based on their differences, have been compiled by Deutsch et al. [7].

2 Isotopic Exchange Reactions

The most fundamental approach to assessing lability of complexes is by determination of the rate of isotopic exchange reactions. In the technetium-complex systems, no study of the exchange reaction on the central metal ion has been reported, but several reports have been published on isotopic exchange by ligand substitution.

The isotopic exchange reaction by ligand substitution can be expressed in general as

$$ML_n + {}^*L \rightleftharpoons ML_{n-1}{}^*L + L, \tag{1}$$

where M is the central metal ion, L is the ligand and *L is the ligand labelled with a tracer. The overall exchange rate R is described by the McKay equation [8]:

$$Rt = -\frac{ab}{a+b}\ln(1-F) \tag{2}$$

where t is time, a is the concentration of metal complex, b is the concentration of uncomplexed ligand, and F is the degree of exchange. Thus, R can be obtained by plotting $\ln(1-F)$ against time. Furthermore, net reaction processes can be evaluated by the dependence of R on a and b. Thus, the general expression of R is usually given by

$$R = \sum k_i [H^+]^p a^q b^r. \tag{3}$$

2.1 Isotopic Exchange in Tc(IV) and Re(IV) Complexes

Early systematic studies on the isotopic exchange reaction by ligand substitution were carried out by Schwochau [9, 10] in the following systems:

$$[MCl_5{}^{36}Cl]^{2-} + Cl^- \rightleftharpoons [MCl_6]^{2-} + {}^{36}Cl^-$$

$$[MBr_5{}^{82}Br]^{2-} + Br^- \rightleftharpoons [MBr_6]^{2-} + {}^{82}Br^-$$

(4)

where M designates Tc or Re.

Hexachlororhenate(IV) is very stable and resisted to both oxidation and reduction in an acidic solution. However, rhenium(IV) complexes containing Re–O bonds were readily oxidized [11]. At room temperature no exchange of radiochloride ion with hexachlororhenate(IV) was observed [12, 13].

The rates of ligand exchange of 0.01 M hexahalo-technetium(IV) and hexahalo-rhenium(IV) complexes were measured in the 8 M solutions of their corresponding acid at 60 °C. The overall exchange rates R were determined as follows:

$$R = 8.1 \times 10^{-7} \ M \ s^{-1} \ \text{for} \ TcCl_6^{2-},$$

$$= 3.58 \times 10^{-8} \ M \ s^{-1} \ \text{for} \ ReCl_6^{2-},$$

$$= 1.40 \times 10^{-4} \ M \ s^{-1} \ \text{for} \ TcBr_6^{2-},$$

$$= 2.27 \times 10^{-6} \ M \ s^{-1} \ \text{for} \ ReBr_6^{2-}.$$

The results show that the technetium complexes are more labile than the rhenium complexes. The overall reaction rate depended on the concentrations of both hexachlororhenate and hydrogen ion, but was independent of the chloride concentration.

2.2 Isotopic Exchange in Tc(V) and Re(V) Complexes

The complexes of Tc(V) and Re(V) containing an M=O core in the coordination sphere exhibit a square pyramidal structure, while those containing an O=M=O core are octahedral.

Oxygen-exchange reactions of square-pyramidal metal(V) complexes have been reported [14]. In monooxotechnetium(V) and monooxorhenium(V) complexes with N,N'-bis(mercaptoacetyl)butane-1,4-diamine(DBDS), the exchange of oxygen between the complexes and water in DMSO medium:

$$MO(DBDS)^- + H_2{}^{17}O \rightarrow M^{17}O(DBDS)^- + H_2O$$

has been studied by means of ^{17}O NMR. In addition to the fact that both the [TcO(DBDS)]$^-$ and [ReO(DBDS)]$^-$ complexes were very stable toward ligand substitution, no exchange of oxygen was observed. However, on addition of sodium methoxide as a catalyst, slow exchange of oxygen was discerned. The reaction was independent of $[H_2O]$; the exchange rate R and rate constant k at 298 K could be expressed as follows:

$$R = k[MO(DBDS)^-][CH_3O^-],$$

$$k = 7.6 \times 10^{-3} \text{ M}^{-1}\text{s}^{-1} \text{ for } [ReO(DBDS)]^-,$$

$$= 7.1 \times 10^{-4} \text{ M}^{-1}\text{s}^{-1} \text{ for } [TcO(DBDS)]^-.$$

Taking into consideration the fact that the ratio of k_{Tc}/k_{Re} was less than 0.1 and the kinetic parameter $\Delta S^{\ddagger}$ for $[ReO(DBDS)]^-$ was positive, it is clear that the exchange proceed in the associative mode. This was confirmed by the shorter bond length for Tc=O (1.657Å) than for Re=O (1.681Å), exhibiting a larger structural *trans* effect.

In marked contrast, $\Delta S^{\ddagger}$ of the exchange reaction of oxygen in a six-coordinated complex, *trans*-$[ReO_2(py)_4]^+$ is negative, ($-32 \text{ J K}^{-1} \text{ mol}^{-1}$), suggesting that the reaction proceeds by a dissociative mechanism. This is also supported by the kinetic analysis of the exchange reaction of pyridine on *trans*-$[MO_2(py)_4]^+$ (M = Tc or Re), which was monitored by ^{1}H-NMR [15]:

$$\textit{trans-}[MO_2(py)_4]^+ + {}^*py \rightleftharpoons \textit{trans-}[MO_2{}^*py(py)_3]^+ + py$$

where *py represents deuterated pyridine. The pyridine exchange rate was found to be first-order with respect to the concentration of the complex, but it was independent of the pyridine concentration. The ratio of the rate constants for pyridine exchange, k_{Tc}/k_{Re}, was about 8,000 at 298 K. A similar value for the ratio, k_{Tc}/k_{Re}, was reported in the unimolecular racemization of Tc(V) and Re(V) penicillamine (pen) complexes [16]. [MO(D-pen) (L-pen)]$^-$ (M = Tc or Re) racemized by exchange of carboxylates at the site *trans* to the oxo ligand. The rate constants were determined to be $k_{Tc} = 940 \text{ s}^{-1}$ and $k_{Re} = 16.5 \text{ s}^{-1}$ at 298 K, i.e., their ratio was about 60.

Oxygen exchange was also found in the case of Tc(VII) and Re(VII) complexes. For the permetallate, $[M^{VII}O_4]^-$, the rate law of oxygen exchange can be expressed as

$$R = k[MO_4^-][H^+]^2.$$

This means that there is a possibility of forming a hexa-coordinate intermediate. The finding that the exchange rate constant for Re(VII) is higher than that for Tc(VII) suggests that ReO_4^- can expand its coordination number from 4 to 6 more readily [7, 17].

2.3 Ligand Exchange Reactions of Tc(III) Complexes

An interesting investigation of a ligand exchange reaction for Tc(III) complex has been reported in the system of tris(acetylacetonato)technetium(III)

(Tc(acac)$_3$) [18]. Tc(acac)$_3$ is known to have the low-spin d^4 configuration [19]. Metal ions with d^4 configuration are classified as substitution-inert [2]. The substitution-inert character of Tc(acac)$_3$ has already been pointed out in the course of its nuclear synthesis by the ^{97}Ru(acac)$_3$(γ, n) reaction [20].

The exchange reaction of Tc(acac)$_3$ in acetylacetone solution is described as follows:

$$^{99}Tc(*acac)_3 + Hacac \rightleftharpoons\ ^{99}Tc(acac)_3 + H*acac \tag{5}$$

where H* acac denotes Hacac [2-^{14}C]. Solutions of the ^{14}C-labeled complex in acetylacetone were sealed in glass tubes. At appropriate intervals, free acetyl-acetone was separated from Tc(acac)$_3$ by vacuum distillation. Thus, radioactivity of ^{14}C was effectively measured with a liquid scintillation counter without interference from ^{99}Tc radioactivity. The observed rate was proportional to the concentration of Tc(acac)$_3$, but it was independent of [H$_2$O]. Therefore, the rate R was expressed as

$$R = k_1[Tc(acac)_3]$$

where k_1 is the observed first-order rate constant. The mechanism of the exchange reaction is summarized below.

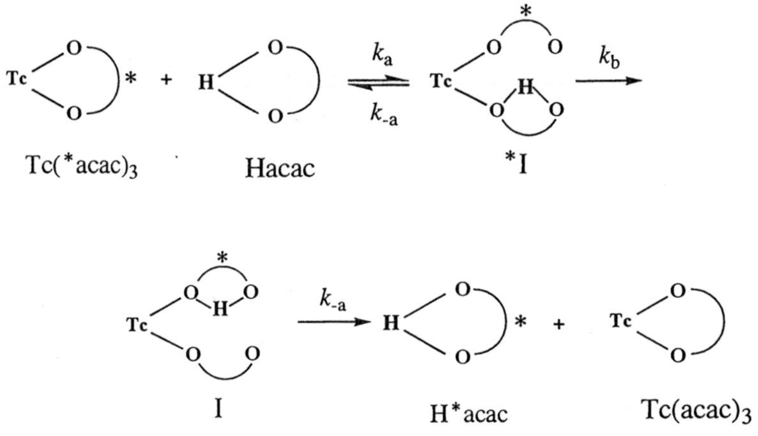

Scheme 1.

The exchange proceeds in three steps. The first step is substitution of the free ligand at one end of the chelate to give the intermediate *I(k_a). The second is intramolecular proton transfer between the unidentate ligand in *I(k_b). The third is the reverse of the first (k_{-a}). Consequently, application of the steady-state approximation to the intermediate *I, whose concentration is reasonably assumed to be very low, provides

$$k_1 = k_a \frac{k_b}{k_{-a} + k_b}. \tag{6}$$

Taking into consideration the deuterium isotope effect $(k_1(H)/k_1(D) = 2.3)$, they concluded that $k_a \ll k_{-a} \ll k_b$ and that the rate-determining step was the first substitution. Ligand substitution was thought to proceed by the I_a mechanism, on the basis of the negative $\Delta S^{\ddagger}$ and the independence of the rate on the concentration of acetylacetone. This feature is compatible with the results with tris(acetylacetonato)metal(III) previously obtained [21]. Furthermore, in the second transition series the k_1 value decreases in the order

$$\text{Mo(III)} \gg \text{Tc(III)} > \text{Ru(III)} > \text{Rh(III)}.$$
$$d^3 \qquad d^4, \text{LS} \quad d^5, \text{LS} \quad d^6, \text{LS}$$

Kido and Saito [21] have treated these values quantitatively by comparison with those of substitution reactions of the corresponding aqua complexes in water. They found the following linear relation between $\log k_1(\text{acac}^-)$ and $\log k_1(H_2O)$:

$$\log k_1(\text{acac}^-) = \log k_1(H_2O) - 4.5 \tag{7}$$

Furthermore, they introduced the *metal ion lability constant* σ as a measure of the effect of the metal ion on lability. The lability constants for various tervalent metal ions obtained by Eq. (8) are summarized in Table 1.

$$\sigma = \tfrac{1}{2}[\log k_1(\text{acac}^-)/s^{-1} + 4.5 + \log k_1(H_2O)/s^{-1}] \tag{8}$$

In general, the rate constant k_1 can be decomposed into two terms, σ and κ.

$$\log k_1/s^{-1} = \sigma + \kappa$$

If $k = 0$ for $k_1(H_2O)$ or $\kappa = -4.5$ for $k_1(\text{acac}^-)$, the k_1 can be evaluated. This relation is also valid for the exchange reaction of $[M(DMF)]^{3+}$ in dimethylformamide (DMF) with $\kappa = -1$. The metal-ion lability constant σ may be used as a measure of the ease of M–O bond loosening associated with bond making between an incoming nucleophile and the metal ion, to form an activated complex via the I_a mechanism. The reaction factor κ is, however, affected by the environment and contributes to k_1, regardless of σ. From Eq. (7), it is possible to

Table 1. Metal Ion Lability Constant σ for Tervalent Metal Ions*

M(III)[a]	σ	M(III)[b]	σ
Sc	6.5	Al	0.5
Mn(HS)	5	Mo	−1.5
In	5	Tc(LS)	−5[c]
Ti	5[d]	Cr	−5.5
Ga	2	Ru(LS)	−5.5
Fe(HS)	1.5	Co(LS)	−6
V	1	Rh(LS)	−7

[a] HS, high spin. [b] LS, low spin. [c] Estimated only from $k_1(\text{acac}^-)$. [d] Estimated only from $k_1(H_2O)$.
*Taken from Ref. 21.

estimate unknown k_1 values. For example, $\log k_1$ for the water-exchange reaction of $[Tc(H_2O)_6]^{3+}$ is expected to be ~ -5 at 25 °C.

3 Solvolytic Reactions of Technetium Complexes

3.1 Aquation of Hexahalotechnetate(IV)

A spectrophotometric investigation of the aquation of Tc(IV) complexes was carried out for the hexahalotechnetate(IV) systems [22]. On standing at 80 °C, the absorbance of hexachlorotechnetate(IV) in 4 M hydrochloric acid (molar absorption coefficient, $\varepsilon_{340nm} = 11{,}040\ M^{-1}\ cm^{-1}$) decreased to a constant value. Taking into consideration that the equilibrium spectrum could clearly be distinguished from the spectrum of pure $Tc(H_2O)Cl_5^-$, the equilibrium constant for the following equilibrium reaction between $TcCl_6^{2-}$ and $Tc(H_2O)Cl_5^-$ could be determined.

$$TcX_6^{2-} + H_2O \rightleftharpoons Tc(H_2O)X_5^- + X^- \quad (X = Cl, Br) \qquad (9)$$

When the concentration of chloride ion was below 3 M, further aquation reactions from $Tc(H_2O)Cl_5^-$ to $Tc(H_2O)_2Cl_4$, etc. were observed. Similarly, aquation of hexabromotechnetate(IV) was studied (molar absorption coefficient, $\varepsilon_{445\ nm} = 5720\ M^{-1}\ cm^{-1}$). The equilibrium constants K for Eq. (9) at different temperatures are summarized in Table 2. Analysis of the aquation rate gave the following equation:

$$k_{obsd} = k_f[H^+] + k_b[X^-]$$

where k_{obsd} is the observed rate constant, and k_f and k_b represent the overall rate constants for aquation and anation, respectively.

3.2 Solvolysis of Technetium Complexes

3.2.1 Solvolysis of Tris(β-Diketonato)Metal(III)

The substitution-inert character of the metal(III) ion in the second transition series has already been discussed in 2.3. However, interesting behavior has been reported by Kasahara et al. [23], who found that a β-diketone coordinated to the central Ru(III) could easily be replaced by an acetonitrile with the aid of a strong acid. When the reaction was conducted in acetonitrile, its stoichiometry was confirmed by means of spectrophotometric titration as follows:

$$[Ru^{III}L_3] + 2CH_3CN + H^+ \rightarrow [Ru^{III}L_2(CH_3CN)_2]^+ + HL$$

Table 2. Equilibrium constant K between TcX_6^{2-} and $Tc(H_2O)X_5^-$ in 4 M HX (X = Cl, Br)[a]

Chloro-complex system				
Temp./°C	75.0	80.5	84.5	90.0
K	1.25	1.29	1.33	1.39
Bromo-complex system				
Temp./°C	40.0	45.0	50.0	54.5
K	0.756	0.901	1.02	1.14

[a]Taken from Ref. 22.

where L means β-diketonate ion. They also found that bis(acetonitrile)bis(β-diketonato)ruthenium(III) complexes were convenient intermediates for the synthesis of mixed-ligand β-diketonato ruthenium(III) complexes of the $[Ru^{III}L_2L']$ type.

These procedures were further extended to tris(acetylacetonato)technetium(III) [24]. In an acetonitrile solution of Tc(acac)₃, the absorbances at the characteristic absorption maxima at 348, 375, 505 and 535 nm decreased with time, while an increase in the absorbance at 272 nm corresponded to an increase of free acetylacetone liberated during the substitution reaction. The final absorption spectra of the reaction mixture exhibited absorption maxima at 271, 325 and 387 nm. The first order rate constant k for decomposition was found to be $k = (8.86 \pm 0.08) \times 10^{-4}\,s^{-1}$ at $[H^+] = 2.0$ M at 30 °C.

3.2.2 Solvolysis of *cis*-Chlorobis(8-Quinolinolato)Oxotechnetium(V)

Although a number of Tc(V) complexes possessing the $Tc=O^{3+}$ core have a square pyramidal structure, an unusual example of a distorted octahedron was found in *cis*-chlorobis(2-methyl-8-quinolinolato)oxotechnetium(V) [25]. One of the oxygen atoms of a 2-methyl-8-quinolinol ligand is located *trans* to the Tc =O bond, while a chloride ion occupies a *cis* position. The fact that the chloride ion in the complex is susceptible to solvolysis was demonstrated by HPLC. In a methanol solution of *cis*-chlorobis(8-quinolinolato)oxotechnetium(V) [TcOCl(ox)₂], the chromatograms exhibited a decrease in the main peak with time and a corresponding increase in a second peak, which was believed to be a product of methanolysis. The reaction proceeds by a first-order process with a half-life of 1.9 h. The bromo complex [TcOBr(ox)₂] decomposes much faster than the chloro complex. This was expected, since the bromo ligand is more readily susceptible to substitution than the chloro ligand. Taking into account that the decomposition product of the bromo complex detected on the chromatogram was the same as that of the chloro complex, the methanolysis product of these complexes was considered to be [TcO(OCH₃)(ox)₂]. In DMF solution of [TcOCl(ox)₂], no detectable change in the chromatogram was observed over a 24 h period.

3.3 Base Hydrolysis of Technetium β-Diketone Complexes

Tc(III), Tc(IV) and Tc(V) β-diketonate complexes are stable in acid solution. In fact, when a chloroform solution of $TcCl_2(acac)_2$ was shaken with 1 M hydrochloric acid solution, no detectable change in the distribution ratio of the complex – defined as the ratio of the concentration of technetium in the organic phase to that in the aqueous phase – was observed over a 24 h period [26]. However, when the technetium complexes were backextracted into aqueous alkaline solution, decomposition occurred [26–29]. In all the cases studied, spectrophotometric investigation revealed that pertechnetate was formed quantitatively as a final product.

The kinetic behavior of the base hydrolysis of $TcCl_2(acac)_2$ is described as an example [26]. A plot of the logarithm of the concentration of the technetium complex in the organic phase against time gives a straight line. Thus, the reaction rate of the base hydrolysis of $TcCl_2(acac)_2$ is expressed as

$$-\frac{d[TcCl_2(acac)_2]}{dt} = k_{app}[TcCl_2(acac)_2].$$ (10)

The apparent rate constant k_{app} depends on the concentration of hydroxide ion as is shown in Fig. 1. The absorption maxima of $TcCl_2(acac)_2$ in chloroform appear at 281, 314(sh), 340(sh), 382 and 420 nm. On the other hand, the spectrum of the aqueous phase exhibits absorption maxima at 292, 350 and 540 nm. The absorbances at 350 and 540 nm increase with time, but decrease after reaching maxima. This suggests that the chemical species which is formed by the backextraction of $TcCl_2(acac)_2$ decomposes with time. In order to clarify the behavior of chloride ion liberated from the complex, an electrochemical method was introduced for the homogeneous system. In acetonitrile, no detectable change in the spectrum of $TcCl_2(acac)_2$ was observed. On the addition of an aqueous solution of hydroxide, however, the brown solution immediately turned redviolet, and exhibited absorption maxima at 292, 350 and 540 nm. The red-violet

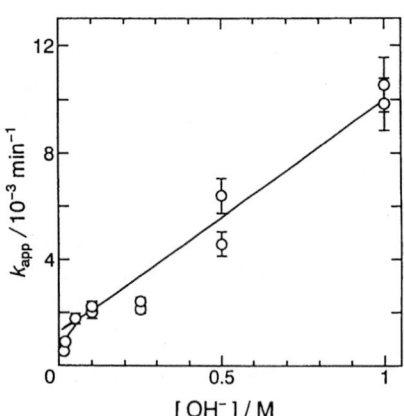

Fig. 1. Dependence of the apparent decomposition rate constant of $TcCl_2(acac)_2$ on the concentration of hydroxide at 25 °C[a]. The solid line was calculated by Eq. (13)

[a]Taken from Ref. 26

solution was gradually decolorized, until the spectrum of pertechnetate, characterized by maxima at 244 and 287 nm finally appeared. Furthermore, exactly twice as much free chloride ion as technetium was found in the earliest stage and its concentration remained constant during the kinetic run. These features indicated that the red-violet species is a bis(acetylacetonato)technetium(IV) complex splitting off the chloride ion. The presence of the bis(acetylacetonato)-technetium(IV) complex was also confirmed by the fact that the same absorption spectrum was obtained for the species extracted into the aqueous phase by base hydrolysis of $TcBr_2(acac)_2$. The absorbance at 350 and 540 nm decreased linearly with time as shown in Fig. 2. The remarkable finding that the rate was independent of the concentration of hydroxide ion revealed that decomposition of $Tc(acac)_2^{2+}$ obeys good first order kinetics. On the other hand, the absorbance at 292 nm increased at first, reached a maximum, and then gradually decreased, as shown in Fig. 3. This complicated behavior could be analyzed by considering

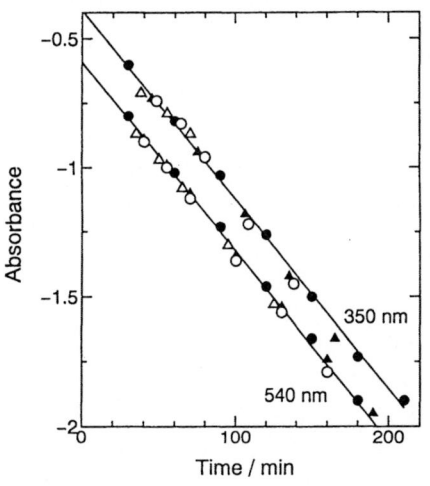

Fig. 2. Time dependence of the absorbance of the bisacetylacetonate complex in acetonitrile solution at 25 °C[a].

$\triangle$: $[OH^-] = 0.1$ M, $\bigcirc$: $[OH^-] = 0.25$ M, $\blacktriangle$: $[OH^-] = 0.5$ M, $\bullet$: $[OH^-] = 1.0$ M. Upper line: 350 nm, lower line: 540 nm

[a]Taken from Ref. 26

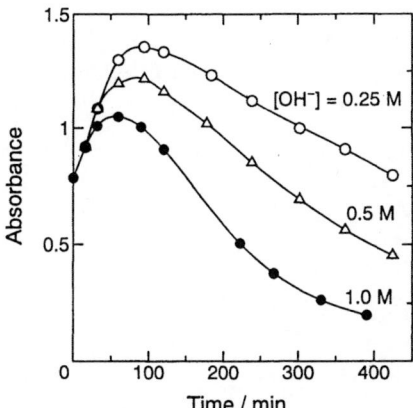

Fig. 3. Time dependence of the absorbance at 292 nm in acetonitrile solution at 25 °C[a].

$\bigcirc$: $[OH^-] = 0.25$ M, $\triangle$: $[OH^-] = 0.5$ M, $\bullet$: $[OH^-] = 1.0$ M

[a]Taken from Ref. 26

both the accumulation of free acetylacetonate liberated from the complex and its decomposition in an alkaline solution as follows:

$$\frac{d[acac^-]}{dt} = k_M[Tc(acac)_2^{2+}] - k_d[OH^-][acac^-]. \tag{11}$$

The best-fit values for k_M and k_d obtained by a non-linear least squares method were $3.8 \times 10^{-4}\,s^{-1}$ and $1.2 \times 10^{-4}\,M^{-1}s^{-1}$ respectively.

Thus, an overall model for the base hydrolysis processes could be constructed. At higher concentrations of hydroxide ion, the bisacetylacetonate complex, which exhibits absorption maxima at 350 and 540 nm, is formed by virtue of the attack of hydroxide ion as follows:

$$TcX_2(acac)_2 \xrightarrow{k_1,OH^-} Tc(acac)_2^{2+} \tag{12}$$

On the other hand, in the low $[OH^-]$ region the base hydrolysis processes could be described as follows:

$$TcX_2(acac)_2 \underset{k_3,Cl^-}{\overset{k_2}{\rightleftharpoons}} TcX(acac)_2(H_2O)^+ \xrightarrow{k_4,OH^-} Tc(acac)_2^{2+}$$

When the steady-state approximation is applied to the concentration of $TcX(acac)_2(H_2O)^+$, the decomposition rate of $TcX_2(acac)_2$ can be expressed as:

$$-\ln\left(\frac{[TcX_2(acac)_2]_t}{[TcX_2(acac)_2]_0}\right)_{org} = \left(k_1[OH^-] + \frac{k_2k_4[OH^-]}{k_3[Cl^-] + k_4[OH^-]}\right)t. \tag{13}$$

Studies of the base-hydrolysis mechanism for hydrolysis of technetium complexes have further been expanded to an octahedral tris(acetylacetonato)technetium(III) [30]. Although a large number of studies dealing with base hydrolysis of octahedral metal(III) complexes have been published [31], the mechanism of the tris(acetylacetonato)metal complex is still unclear. The second-order base hydrolysis of the cationic complex tris(acetylacetonato)silicon(IV) takes place by nucleophilic attack of hydroxide ion at carbonyl groups, followed by acetylacetone liberation, and finally silicon dioxide production [32]. The kinetic runs were followed spectrophotometrically by the disappearance of the absorbance at 505 nm for $Tc(acac)_3$. The rate law has the following equation:

$$Rate = (k_1[OH^-] + k_2[OH^-]^2)[Tc(acac)_3] \tag{14}$$

where k_1 and k_2 at 25 °C are $(1.67 \pm 0.11) \times 10^{-5}\,M^{-1}s^{-1}$ and $(1.71 \pm 0.14) \times 10^{-5}\,M^{-2}s^{-1}$, respectively.

3.4 Protonation of Tc(V) and Tc(VI) Complexes

Protonation reactions of the type of complex, $[MO_2(CN)_4]^{n-}$ (M = Mo(IV), W(IV), and Re(V)), give the hydroxo, oxo, and aqua-oxo complexes. Tetracyanodioxotechnetate(V) was prepared by the following steps [33].

$$[TcOCl_4]^- \xrightarrow{\text{py}} [TcO_2(py)_4]^+ \xrightarrow{\text{KCN}} [TcO_2(CN)_4]^{3-}$$

In acidic aqueous solution, protonation reactions of tetracyanodioxotechnetate(V) give a complicated equilibration, leading to formation of $[TcO(OH)(CN)_4]^-$ and $[TcO(H_2O)(CN)_4]^-$. At pH values less than 1, these monomer species are fairly stable, while at pH 2–5, $[Tc_2O_3(CN)_8]^{4-}$ is formed rapidly. This complicated feature is seen in a plot of k_{obsd} against pH (Fig. 4). When thiocyanate ion is added to this system at pH 1, it replaces a water molecule or hydroxy group in the coordination site.

$$[TcO(H_2O)(CN)_4]^- + NCS^- \underset{k_{-1}}{\overset{k_1}{\rightleftharpoons}} [TcO(NCS)(CN)_4]^{2-}$$

$$K_{a1} \ -H^+ \quad +H^+ \qquad k_2 \ -OH^- \quad k_{-2} \ +OH^-$$

$$[TcO(OH)(CN)_4]^{2-} + NCS^-$$

$$\downarrow k_3, \text{slow}$$

$$[Tc_2O_3(CN)_8]^{4-}$$

The observed first-order rate constant is derived as follows:

$$k_{obsd} = \frac{(k_1 + k_2 K_{a1}/[H^+])[NCS^-]}{1 + K_{a1}/[H^+]} + k_{-1} + k_{-2}[OH^-] \tag{15}$$

where at $15\,^\circ\text{C}$ $k_1 = 9.21\ \text{M}^{-1}\text{s}^{-1}$, $k_{-1} = 0.13\ \text{s}^{-1}$, $k_2 < 10^{-5}\ \text{M}^{-1}\text{s}^{-1}$, $pK_{a1} = 2.90$. When $[H^+] \gg K_{a1}$, Eq. (15) may be simplified as follows:

$$k_{obsd} = k_1[NCS^-] + k_{-1}. \tag{16}$$

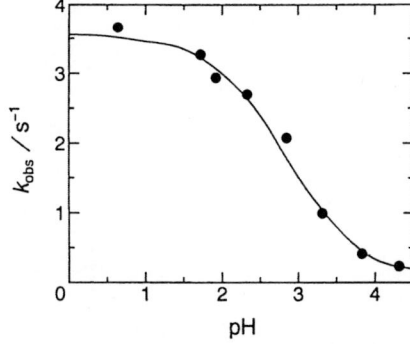

Fig. 4. Plot of k_{obsd} vs. pH at $15.0\,^\circ\text{C}$, $\mu = 1.0$ M (KNO$_3$), [NCS$^-$] = 0.4 M, and [Tc(V)] $= 5 \times 10^{-5}$ M[a]

[a]Taken from Ref. 33

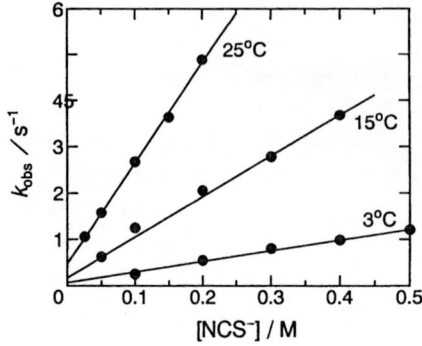

Fig. 5. Plot of k_{obsd} vs. [NCS$^-$] at [HNO$_3$] = 0.25 M, μ = 1.0 M (KNO$_3$), λ = 410 nm, and [Tc(V)] = 5×10^{-5} M. Temperature is accurate to $\pm$ 0.1 °C[a]

[a]Taken from Ref. 33

The validity of Eq. (16) can be seen in Fig. 5, that shows a linear relation between k_{obsd} and [NCS$^-$] at [H$^+$] = 0.25 M at various temperatures.

Some recent interest in the technetium chemistry has been focused on complexes possessing a Tc≡N^{3+} core. Tetrachloronitridotechnetate(VI) complexes can easily be synthesized by the reaction of pertechnetate with sodium azide in concentrated hydrochloric acid [34]. Although its square-pyramidal structure resembles that of tetrachlorooxotechnetate(V) complexes, stable character of the nitrido complexes in aqueous solution shows a remarkable contrast to the oxo complexes. However, when a strong acid and a coordinating ligand are absent, the interconversion of di(μ-oxo)nitridotechnetium(VI) complexes to the monomeric form occurs in the following complicated manner [35]

$$[\{TcN(OH)(H_2O)\}_2(\mu\text{-O})_2]$$
$$H^+ \downarrow \uparrow$$

$$[\{TcN(H_2O)_3\}_2(\mu\text{-O})_2]^{2+}$$
$$L^- \downarrow \uparrow$$

$$[\{TcNL_2\}_2(\mu\text{-O})_2]^{2-}$$
$$H^+ \downarrow \uparrow$$

$$[\{TcNL_2(H_2O)\}_2(\mu\text{-O})]^{2-}$$
$$L^- \downarrow \uparrow$$

$$[\{TcNL_3\}_2(\mu\text{-O})]^{4-}$$
$$H^+ \downarrow \uparrow$$
$$[TcNL_4]^-$$

The dark-brown and chloride-free precipitate, $[\{TcN(OH)(H_2O)\}_2(\mu\text{-O})_2]$, can be formed by hydrolysis of Cs$_2$[TcNCl$_5$] in a large amount of water [36]. In a weakly coordinating acid solution, a different situation appears. After 27 hours in 7.5 M CF_3SO_3H aqueous solution, an intense visible absorption appeared at 474 nm, which is a characteristic absorption of μ-oxo dimers. This orange complex was considered to be $[\{TcN(H_2O)_4\}_2(\mu\text{-}O)]^{4+}$.

4 Ligand Substitution Reactions of Technetium Complexes

4.1 Ligand Substitution Reactions of Hexakis(Thiourea)Technetium(III)

The reduction of pertechnetate with concentrated hydrochloric acid finally yields the tetravalent state, and no further reduction to the tervalent state takes place. Therefore, the tervalent technetium complex has usually been synthesized by the reduction of pertechnetate with an appropriate reductant in the presence of the desired ligand. Recently, the synthesis of tervalent technetium complexes with a new starting complex, hexakis(thiourea)technetium(III) chloride or chloropentakis(thiourea)technetium(III) chloride, has been developed. Thus, tris(β-diketonato)technetium(III) complexes (β-diketone: acetylacetone, benzoyl-acetone, and 2-thenoyltrifluoroacetone) were synthesized by the ligand substitution reaction on refluxing $[TcCl(tu)_5]Cl_2$ with the desired β-diketone in methanol [28].

Crystals of $[Tc(tu)_6]Cl_3$ or $[TcCl(tu)_5]Cl_2$ are often employed for the synthesis of technetium(III) complexes. However, since the direct reduction of pertechnetate with excess thiourea in a hydrochloric acid solution yields $[Tc(tu)_6]^{3+}$ in high yield [37], direct use of the aqueous solution of the thiourea complex would be preferable for the synthesis of the technetium(III) complex without isolation of the crystals of the thiourea complex. In fact, technetium could be extracted from the aqueous solution of the Tc-thiourea complex with acetylacetone-benzene solution in two steps [38]. More than 95% extraction of technetium was attained using the following procedure [39]: First a pertechnetate solution was added to a 0.5 M thiourea solution in 1 M hydrochloric acid. The solution turned red-orange as the Tc(III)-thiourea complex formed. Next, a benzene solution containing a suitable concentration of acetylacetone was added. After the mixture was shaken for a sufficient time (preliminary extraction), the pH of the aqueous phase was adjusted to 4.3 and the aqueous solution was shaken with a freshly prepared acetylacetonebenzene solution (main extraction). The extraction behavior of the technetium complex is shown in Fig. 6. The chemical species extracted into the organic phase seemed to differ from tris(acetylacetonato)technetium(III). Kinetic analysis of the two step extraction mechanism showed that the formation of 4,6-dimethylpyrimidine-

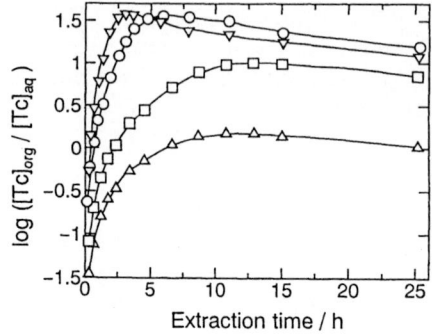

Fig. 6. Dependence of the distribution ratio on time at various concentrations of acetylacetone in a preliminary extraction (preliminary extraction: pH 0/ main extraction: pH 4.3, [Hacac] = 0.5 M)[a]

[a]Taken from Ref. 39

2(1H)-thione (Hdmpt) occurs in the preliminary extraction as

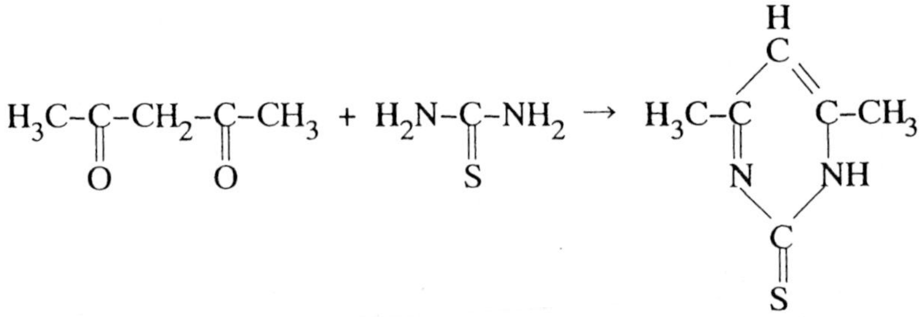

$$H_3C-\underset{O}{\underset{\|}{C}}-CH_2-\underset{O}{\underset{\|}{C}}-CH_3 + H_2N-\underset{S}{\underset{\|}{C}}-NH_2 \rightarrow$$

and the complex which has been formed by the substitution of $Tc(tu)_6^{3+}$ by $dmpt^-$ is then extracted into the organic phase. This feature was further confirmed by the extraction of $Tc(dmpt)_3$ from $[Tc(tu)_6]^{3+}$ with Hdmpt. The overall extraction mechanism is summarized in Scheme 2.

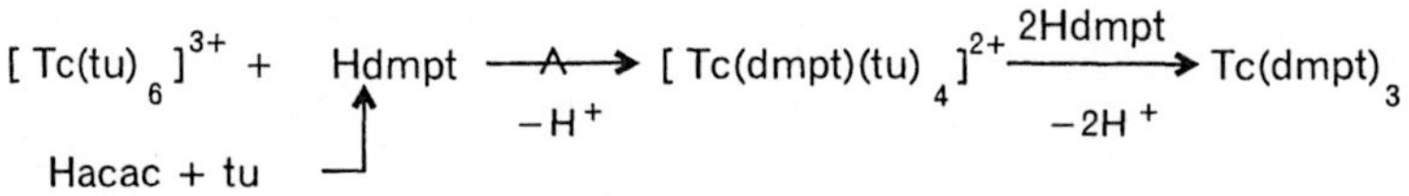

$$[Tc(tu)_6]^{3+} + Hdmpt \xrightarrow[-H^+]{A} [Tc(dmpt)(tu)_4]^{2+} \xrightarrow[-2H^+]{2Hdmpt} Tc(dmpt)_3$$

Hacac + tu

Scheme 2.

Another attempted synthesis of Tc(III)-EDTA and Tc(III)-HEDTA complexes (EDTA: ethylenediaminetetraacetic acid; HEDTA: *N*-(2-hydroxymethyl)ethylenediamine-*N*,*N*′,*N*′-triacetic acid) was carried out using $[Tc(tu)_6]^{3+}$ as the starting complex [40]. Technetium-EDTA complexes have been synthesized by the direct reduction of pertechnetate with a suitable reductant in the presence of excess EDTA [41–43]. On addition of EDTA to the $Tc(tu)_6^{3+}$ solution, the intensity of the absorption spectrum decreased with time and the solution color changed from reddish orange to light brown. An electrophoretic analysis for the Tc(III)-EDTA complex showed that more than 70%

of the reaction products exist as neutral species. It is worthy of note that in the HEDTA complex system, $Tc(tu)_6^{3+}$ was reformed when hydrochloric acid was added to a solution of [Tc-HEDTA] containing an excess of thiourea. However, the [Tc-HEDTA] solution contained only 21.5% neutral species and a fairly large part was found in the cationic fraction (43.5%). The remaining fraction contained an anionic form. The cationic species, [Tc-HEDTA]$^+$, could be separated as a tetraphenylborate salt. The presence of thiourea in the [Tc(III)-HEDTA]$^+$ tetraphenylborate was suggested by IR spectroscopy. It was also confirmed by elemental analysis.

The rate law for the substitution reaction was proven to be expressed in the presence of a large excess of EDTA as

$$-\frac{d[Tc(tu)_6^{3+}]}{dt} = k_{obs}[Tc(tu)_6^{3+}] . \tag{17}$$

Since the substitution reaction is generally expressed as

$$[Tc(tu)_6^{3+}] + H_n edta^{(4-n)-} \xrightarrow{k_n} [Tc(edta)]^- + nH^+ , \tag{18}$$

the observed rate constant k_{obs} is derived as

$$k_{obsd} = \left\{ k_0 + \sum_{n=1}^{4} k_n[H^+]^n/K_n \right\}[edta^{4-}] \tag{19}$$

where K_n is the dissociation constant of free EDTA defined by

$$K_n = [H^+]^n[edta^{4-}]/[H_n edta^{(4-n)-}]. \tag{20}$$

On the basis of Eq. (19), the plot of $k_{obs}/[edta^{4-}]$ as a function of the hydrogen ion concentration on a logarithmic scale gave a straight line of slope -2.0. The slope shows the most probable value for n. Thus, at pH 2 to 4 the substitution reaction proceeds as

$$[Tc(tu)_6]^{3+} + H_2 edta^{2-} \xrightarrow{k_2} ([Tc(tu)_2 H_2 edta]^+) \rightarrow [Tc(tu)(Hedta)]^0 . \tag{21}$$

The formation rate constant in Eq. (21) was $k_2 = (3.2 \pm 0.3) \times 10^{-1} \, M^{-1} s^{-1}$ at 25 °C.

In the formation reaction of Tc(III)-HEDTA complex, a similar plot gives a slope of -1.1. Thus, the rate constant, k_1, for the reaction

$$[Tc(tu)_6]^{3+} + Hhedta^{2-} \xrightarrow{k_1} Tc(III)\text{-}HEDTA \tag{22}$$

was determined to be $k_1 = (2.1 \pm 0.2) \times 10^{-1} \, M^{-1} s^{-1}$ at 25 °C.

4.2 Ligand Substitution Reactions of Tc(V) Complexes in Nonaqueous Solvents

In addition to complexes of the type, $trans$-$[TcO_2(pyr)_2]^+$ (pyr: pyridine or imidazole), various species, such as $trans$-$[TcO(RO)X_2(pyr)_2]$ (R: CH_3 or CH_3CH_2; X: Cl or Br) were detected in alcohol. Further complicated mixed-valence species, $[X_2(pyr)_3Tc-O-Tc(pyr)_2X_3]$ and $[X(pyr)_4Tc-O-Tc(pyr)X_4]$, appeared on long standing or heating in pyridine [44, 45]. Rather peculiar features were found in the substitution reaction of $trans$-$[TcO_2(py)_4]^+$ with 4-aminopyridine (apy) in mixtures of methanol and toluene in the presence of excess pyridine ($[py] \sim 0.14$ M) [46]. Its rate was expressed as

$$\frac{d[TcO_2(apy)_4]}{dt} = k[MeOH][TcO_2(py)_4^+].$$ (23)

Accordingly, the observed rate constant increased linearly with an increase in the mole fraction of methanol in toluene and was essentially independent of the concentrations of incoming and leaving ligands. In contrast to this solvent-mediated mechanism, a dissociative reaction was reported for the substitution of thiol on $trans$-$[TcO_2(py)_4]^+$ in methanol containing a higher concentration of pyridine (2.5 M) [47]. It has also been reported that the rate was strongly affected by the nature and concentration of both the incoming and leaving ligands, when the substitution of 4-aminopyridine on $trans$-$[TcO_2(pyr)_4]^+$ was carried out in DMF (pyr: pyridine(py), 4-picoline(pic) or 3,5-lutine(lu)). The observed rate was derived as

$$\frac{1}{\text{Rate}} = \left(\frac{1}{k_2 K} \frac{[pyr]}{[apy]} + \frac{1}{k_1} \right) \frac{1}{[TcO_2(pyr)_4^+]}$$ (24)

where k_1 is the rate constant for loss of the leaving group, k_2 is the rate constant for the entering ligand and K is the association constant for the leaving ligand.

Interesting examples of the ligand substitution reaction of Tc(V) containing a $Tc=O^{3+}$ core can be seen in the tridentate Schiff-base complexes, which show a distorted square pyramidal configuration with a monodentate ligand such as the halogen atom [48]. A phenolate group in a bidentate Schiff-base or 8-quinolinol can easily replace the coordinated monodentate ligand in alcohol solution, and a neutral imine group coordinates to complete the hexa-coordination. The reaction products, which precipitate immediately from the solutions, can be recrystallized from dichloromethane and acetonitrile.

4.3 Ligand Substitution Reactions in the Synthesis of ^{99m}Tc-Labeled Complexes

Preparative methods of ^{99m}Tc-labeled radiopharmaceuticals are discussed in detail by Volkert and Jurrison in this book [49]. Accordingly, limited examples of kinetics for the preparation of ^{99m}Tc-labeled compounds by ligand substitution are discussed in this section.

Labeling by the ligand exchange method has the advantage of minimizing the non-specific labeling associated with the direct reduction-complex formation method. However, it is difficult to find kinetic studies which refer to the ligand-substitution mechanism. In fact, in a large number of studies, the final yields of ^{99m}Tc-labeled complexes prepared by ligand-substitution are compared with those by the direct reduction of pertechnetate with Sn(II) in the presence of ligand. The hydroxycarbonate complexes of technetium(V) are usually used as starting compounds for ligand substitution. Early studies revealed that the reactivity of Tc(V)-hydroxycarbonates decreases in the order: α-hydroxyisobutyrate > tartrate > malate > gluconate > citrate [50]. Starting from Tc(V)-α-hydroxyisobutyrate, desired complexes with N-donor ligands, such as 1,10-phenanthroline, 8-quinolinol etc. could be prepared in 10–20 min at room temperature. However, when Tc(V)-gluconate was employed, a longer incubation time for preparation of such complexes was required; as long as 60 min even at 50 °C.

In the ligand substitution of technetium complexes, the interchange mechanism seems to play an important role. According to this mechanism, the substitution of a free ligand for one end of a coordinated ligand should take place as the first step. Therefore, it is reasonable that labelling of multidentate amine-phenol ligand was achieved more successfully with ^{99m}Tc-citrate than with ^{99m}Tc-DTPA (DTPA: diethylenetriamine-N,N,N',N'',N''-pentaacetic acid) [51]. When 1,3-bis(2-hydroxybenzylamino)-2,2-dimethylpropane is used as a tetradentate amine-phenol ligand, the maximum yield was 90% with ^{99m}Tc-citrate, whereas it was only 10% with ^{99m}Tc-DTPA. In addition, a surprisingly large yield (more than 40%) was obtained by the substitution of pentadentate 1,7-bis(2-hydroxybenzyl)-4-benzyldiethylenetriamine with ^{99m}Tc-DTPA. In comparison with non-specific labeling, a disadvantage of the ligand-substitution procedure is considered to be the additional time required for the preparation of the desired complex from the preformed complex. Consequently, elucidation of the mechanism in the ligand substitution reaction is desirable for effective preparation of ^{99m}Tc-labeled radiopharmaceuticals.

5 Summary

In inert systems such as technetium and rhenium, ligand substitution reactions – including solvolysis – proceed under virtually irreversible conditions. Thus, the nature of the reaction center, the nature of the leaving group, and the nature and position of the other ligands in the complex affect the rates and activation parameters in a complicated manner. Most substitution reactions take place via interchange mechanisms. This is not too surprising when the solvent is water – or water-like – and where, in order to compete with the solvent, other nucleophiles must be in position before the substitution step is initiated [1].

Although systematic studies on the substitution reaction have not yet been carried out, the fundamental importance of understanding the kinetics and mechanism is now recognized. In particular, comparison of the kinetics of the substitution reactions of technetium complexes with those of rhenium complexes is essential for the determination and evaluation of the fundamental parameters that control the substitution reactions [52]. In general, a great number of the ligand substitution reactions of technetium(V) and rhenium(V) complexes proceed via interchange mechanisms, according to the Lanford-Gray nomenclature [53]. Furthermore, the rates for technetium(V) complexes are more than three orders larger than those for analogous rhenium(V) complexes, irrespective of the experimental conditions. Suitable and effective reaction conditions for the preparation of rhenium(V) radiopharmaceuticals should be identified.

In closing, recovery of technetium from waste solution should be touched upon. Studies of the base hydrolysis of technetium β-diketone complexes revealed that all of the complexes studied decompose in an alkaline solution even at room temperature, until technetium is finally oxidized to pertechnetate. These phenomena are very important for the management of technetium in waste solutions. Since most metal ions precipitate in alkaline solution, only technetium and some amphoteric metal ions can be present in the filtrate [29]. A further favorable property of pertechnetate is its high distribution coefficient to anion exchangers. Consequently, it is possible to concentrate and separate technetium with anion exchangers from a large volume of waste solution; this is especially effective using an alkaline solution [54].

Acknowledgements. Preparation of this review was supported by the Ministry of Education, Culture and Science of Japan through Grant-in-Aid No. 05453041 for Scientific Research.

References

1. Tobe ML (1987) In: Wilkinson G (ed) Comprehensive Coordination Chemistry, Vol. 1, Pergamon Press, Oxford, p 281
2. Saito K (1976) Chemistry of Octahedral Complexes (in Japanese), Tokyo Univ. Press, p 178.
3. Taube H (1952) Chem Rev 50: 69.
4. Davison A (1983) In: Deutsch E, Nicolini M, Wagner Jr HN (eds) Technetium in Chemistry and Nuclear Medicine, Cortina International, Verona, p 3.
5. Trops HS, Jones AG, Davison A (1980) Inorg Chem 19: 1993.
6. Koltunov VS, Gomonova TV (1990) In: Nicolini M, Bandoli G, Mazzi U (eds) Technetium and Rhenium in Chemistry and Nuclear Medicine, Cortina International, Verona, p 35.
7. Deutsch E, Libson K, Vanderheyden J-L (1990) In: Nicoloni M, Bandoli G, Mazzi U (eds) Technetium and Rhenium in Chemistry and Nuclear Medicine, Cortina International, Verona, p 13.

8. McKay HAC (1938) Nature 142: 997.
9. Schwochau K (1965) Z Naturforsch 20a: 1286
10. Schwochau K (1986) In: Nicolini M, Bandoli G, Mazzi U (eds) Technetium in Chemistry and Nuclear Medicine, Cortina International, Verona, p. 13.
11. Maun EK, Davison N (1950) J Amer Chem Soc 72: 2254
12. Jezowska-Trzebiatowska B, Nawojska J, Wajda S (1957) Bull Acad Polon Sci Classe III, 5: 1081
13. Casey JA, Murmann RK (1967) Inorg Chem 6: 1053
14. Chen B, Heeg MJ, Deutsch E (1992) Inorg Chem 31: 4683.
15. Helm L, Deutsch K, Deutsch EA, Merbach AE (1992) Helv Chim Acta 75: 210.
16. Johnson DL, Fritzberg AR, Hawkins BL, Kasina S, Eshima D (1984) Inorg Chem 23: 4204.
17. Gamsjager H, Murmann RK (1983) In: Sykes AG (ed) Advances in Inorganic and Bioinorganic Mechanisms, Academic Press, N.Y., p 318
18. Kido H, Hatakeyama Y (1988) Inorg Chem 27: 3623.
19. Abrams MJ, Davison A, Jones AG, Costello CE (1983) Inorg Chim Acta 77: L235.
20. Yoshihara K, Omori T, Kido H (1981) J Inorg Nucl Chem 43: 639.
21. Kido H, Saito K (1988) J Amer Chem Soc 110: 3187.
22. Kawashima M, Koyama M, Fujinaga T (1976) J Inorg Nucl Chem 38: 819
23. Kasahara Y, Hoshino Y, Shimizu K, Satô GP (1990) Chem Lett 381.
24. Mutalib A, Omori T, Yoshihara K (1992) J Radioanal Nucl Chem Lett 165: 19.
25. Wilcox BE, Heeg MJ, Deutsch E (1984) Inorg Chem 23: 2962.
26. Omori T, Yamada Y, Yoshihara K (1987) Inorg Chim Acta 130: 99
27. Omori T, Yamada Y, Iino S, Yoshihara K (1987) J Radioanal Nucl Chem Lett 119: 223.
28. Hashimoto K, Yamada Y, Omori T, Yoshihara K (1989) J Radioanal Nucl Chem Lett 135: 187.
29. Yoshihara K, Yamada-Maruo Y, Omori T, Kato-Azuma M (1990) In: Nicolini M, Bandoli G, Mazzi U (eds) Technetium and Rhenium in Chemistry and Nuclear Medicine, Cortina International, Verona, p 125.
30. Mutalib A, Omori T, Yoshihara K (1993) J Radioanal Nucl Chem, Articles, 170: 67.
31. Tobe ML (1970) Acc Chem Rev 3: 377; (1983) Adv Inorg Bioinorg Mech 2: 1.
32. Pearson RG, Edgington DN, Basolo F (1955) J Amer Chem Soc 77: 527.
33. Roodt A, Leipoldt JG, Deutsch EA, Sullivan JC (1992) Inorg Chem 31: 1080.
34. Baldas J, Bonnyman J (1985) Int J Appl Radiat Isotopes 36: 133.
35. Baldas J, Boas JF, Colmanet SF, Ivanov A, Williams GA (1993) Radiochim Acta 63: 111.
36. Baldas J, Boas JF, Bonnyman J, Colmanet SF, Williams GA (1991) Inorg Chim Acta 179: 151.
37. Abrams MJ, Davison A, Brodack JE, Jones AG, Faggiani R (1982) J Labelled Compd Radiopharm 14: 1596; Abrams MJ, Davison A, Faggiani A, Jones AG, Lock CJL (1984) Inorg Chem 23: 3284.
38. Hashimoto K, Sekine T, Omori T, Yoshihara K (1986) J Radioanal Nucl Chem Lett 130: 99.
39. Hashimoto K, Omori T, Yoshihara K (1990) Radiochim Acta 49: 69.
40. Hashimoto M, Wada H, Omori T, Yoshihara K (1993) Radiochim Acta 63: 173.
41. Steigman J, Meinkein G, Richard P (1975) J Appl Radiat Isotopes 26: 601.
42. Burgi B, Anderegg G, Bauenstein P (1981) Inorg Chem 20: 3829; Anderegg G, Gasch W, Zollinger K (1983) In: Deutsch E, Nicolini M, Wagner Jr HN (eds) Technetium in Chemistry and Nuclear Medicine, Cortina International, Verona, p 15.
43. Seifert S, Noll B, Münze (1983) Int J Appl Radiat Isotopes 34: 581.
44. Clarke MJ, Lu J (1990) In: Nicolini M, Bandoli G, Mazzi U (eds) Technetium and Rhenium in Chemistry and Nuclear Medicine, Cortina International, Verona, p 23.
45. Fackler P, Kastner ME, Clarke MJ (1984) Inorg Chem 23: 3968.
46. Lu J, Clarke MJ (1989) Inorg Chem 28: 3215.
47. de Varies N, Jones AG, Davison A (1989) Inorg Chem 28: 3728.
48. Bandoli G, Mazzi U, Moresco A, Nicolini M, Refosco F, Tisato F (1986) In: Nicolini M, Bandoli G, Mazzi U (eds) Technetium in Chemistry and Nuclear Medicine, Cortina International, Verona, p 73.
49. Volkert WA, Jurrison S, This volume, p. 000.
50. Seifert S, Muenze R, Johannsen B (1982) Radiochem Radioanal Lett 54: 153.
51. Kothari K, Pillai MRA (1993) Appl Radiat Isot 44: 911.
52. Libson, Helm L, Roodt A, Cutler C, Merbach AE, Sullivan JC, Deutsch E (1990) In: Nicolini M, Bandoli G, Mazzi U (eds) Technetium and Rhenium in Chemistry and Nuclear Medicine, Cortina International, Verona, p 31.
53. Langford CH, Gray HB (1965) Ligand Substitution Process, Benjamin, New York.
54. Kawasaki M, Omori T, Hasegawa K (1993) Radiochim Acta 63: 53.

Rhenium Complexes Labeled with ^{186,188}Re for Nuclear Medicine

Kazuyuki Hashimoto[1] and Kenji Yoshihara[2]

[1] Department of Radioisotopes, Japan Atomic Energy Research Institute, Tokai-mura, Ibaraki-ken 319-11, Japan
[2] Department of Chemistry, Tohoku University, Sendai 980-77, Japan

Table of Contents

Complexes labeled with ^{186}Re and ^{188}Re have been developed for the radiotherapy treatment of diseases because of the desirable nuclear properties of these radioisotopes and because rhenium possesses chemical properties similar to those of technetium, a well-established diagnostic agent. Production of ^{186}Re has been carried out by two methods: one using the ^{185}Re$(n,\gamma)^{186}$Re reaction in a nuclear reactor and the other using the ^{186}W$(p,n)^{186}$Re reaction. From the viewpoint of specific

Topics in Current Chemistry, Vol. 176
© Springer-Verlag Berlin Heidelberg 1996

activity, the former is useful when a high flux nuclear reactor is available, but the latter is practicable when a cyclotron is at hand. ^{188}Re has been produced from the double neutron capture reaction of ^{186}W. A generator of ^{188}W/^{188}Re can be successfully prepared when a high flux nuclear reactor is available. The preparation of rhenium complexes (HEDP and MDP) for treating painful bone cancer has been investigated in order to determine the optimum conditions in chemical procedures. Other complexes (DTPA and DMSA) of rhenium have been synthesized for the radiolabeling antibodies. These are reviewed in this paper as well as further studies on antibody labeling using bifunctional ligands.

1 Introduction

The utilization of radioisotopes in the field of nuclear medicine has been promoted for various purposes. Among them, diagnostic applications have had much success during the past two decades. Technetium-99m, thallium-210 and iodine-123, for example, have been used as radioisotopes for imaging studies.

In recent years, attention has been focused on radiopharmaceuticals for therapy. These compounds must be so designed that they accumulate selectively at malignant sites on the target organs and deposit high doses of radiation there, while keeping the radiation doses to surrounding normal, healthy cells to a minimum. The potential for utilizing monoclonal antibodies (MoAb) as vehicles of the radioisotopes for the selective destruction of tumors has been investigated. In order to optimize the therapeutic effect, selection of the radioisotopes is an important factor. The physical properties of the radioisotope itself, (mainly half-life, and the type and energy of the radiation emitted) should be taken into account in the selection [1–4].

It is important to match the half-life of a radioisotope labeling organs and monoclonal antibodies to the effective period defined by the in vivo pharmacokinetics of the labeled compound. If the half-life is too short, the nuclide has decayed out before the labeled monoclonal antibody accumulates in tumors (the tumor/background ratio should reach a maximum). It is commonly accepted that a period between one half and three days is required to reach the maximum tumor uptake. Conversely, if the half-life is too long, the nuclide will decay out after the labeled monoclonal antibody is shed from the tumors. This means that undesirable radiation doses are given to other tissues. For specimens with equal radioactivity concentrations, the absorbed dose rate of a radioisotope with a longer half-life is lower than that with a shorter half-life. Generally, the long-lived radioisotope has a large mass of radioisotopes, ligands or antibodies and this causes undesirable absorbed doses. Therefore, labeling with long-lived radioisotopes is not desirable.

The range in tissues and linear energy transfer (LET) depend on the type of radiation emitted and its energy. The potent lethality of Auger and low-energy conversion electrons is demonstrated by intranuclear localization of the radioisotope due to their short ranges (about one cell nucleus in diameter). Alpha particles have ranges of several cell diameters (40–90 μm) and are effective in

Table 1. Physical properties of therapeutic radionuclides

Radionuclide	Half-line	Maximum β-ray energy (Average)	γ-Ray energy (Abundance)	X_{90}^a
^{32}P	14.3 d	1.71 MeV (695 keV)	–	4.6 mm
^{67}Cu	2.58 d	0.58 MeV (142 keV)	185 keV (48.6%)	0.5 mm
^{89}Sr	50.56 d	1.49 MeV (583 keV)	909 keV (0.0095%)	4.2 mm
^{90}Y	2.67 d	2.28 MeV (934 keV)	–	6.0 mm
^{105}Rh	1.47 d	0.57 MeV (152 keV)	319 keV (19.0%), 306 keV (5.1%)	0.5 mm
^{111}Ag	7.45 d	1.04 MeV (354 keV)	342 keV (6.7%)	2.2 mm
^{131}I	8.04 d	0.81 MeV (182 keV)	364 keV (81.2%), 637 keV (7.3%)	0.7 mm
^{153}Sm	1.95 d	0.81 MeV (225 keV)	103 keV (28.3%)	1.3 mm
^{166}Ho	1.12 d	1.86 MeV (711 keV)	81 keV (6.2%)	4.7 mm
^{186}Re	3.78 d	1.08 MeV (323 keV)	137 keV (8.6%)	2.0 mm
^{188}Re	17.0 h	2.12 MeV (765 keV)	155 keV (14.9%)	5.5 mm

^{a} X_{90}: distance within which 90% of the radiation dose is deposited.

killing cells because of high LET. Theoretically, α particles are suitable for therapy, but unstable daughter nuclides restrict the use of many α emitters. A nonuniform distribution of monoclonal antibody on the tumor cell is demonstrated in many cases. These results also show the disadvantage of short-ranged α and Auger electron emitters for therapy. The longer range of β particles provides uniform irradiation to a tumor despite a nonuniform distribution of labeled compound within the tumor. Beta particles have penetration ranges in tissue in the order of a few millimeters to a few centimeters and are useful for the irradiation of small- to medium-size tumors. Pure gamma-emitting radioisotopes are not usually used for therapeutic purposes because of their long range and low LET, but the presence of gamma rays in the range of 100 to 300 keV makes external imaging possible. From these considerations, the β emitters which are most suitable for therapeutic purposes in the field of nuclear medicine are listed in Table 1. Among them, radioactive rhenium isotopes are especially useful because their energetic beta particles have a large penetrability (cf. X_{90} values) and imageable gamma rays are in low abundance. Rhenium and technetium belong to Group 7 (7A) of the periodic table and have similar chemical properties. Therefore, ^{186}Re and ^{188}Re can be expected to be as widely used as ^{99m}Tc.

In this review, we describe the production of ^{186}Re, ^{188}Re and rhenium complexes labeled with radioactive rhenium for nuclear medicine.

2 Production of Rhenium Radioisotopes

2.1 Production of ^{186}Re

Rhenium-186 is usually produced by the ^{185}Re(n,γ) reaction using a nuclear reactor. The cross section of this reaction is $1.12 \pm 0.02 \times 10^{-26}$m^2 ($112 \pm 2$

barns) [5]. Therefore, its specific activity is limited. The specific activity of a radioisotope produced by simple thermal neutron capture reaction depends on the neutron flux in the nuclear reactor. For example, the specific activity of ^{186}Re produced in the University of Missouri Research Reactor (neutron flux: 4.5×10^{18} m^{-2} s^{-1}) is 2–3 Ci (74–111 GBq)/mg ^{185}Re [6]. However, ^{186}Re is produced with lower specific activity because such high flux reactors are not readily available worldwide. Labeling with monoclonal antibodies requires higher specific activity of radioisotopes so that a higher therapeutic effect using higher specific activity is achieved. Jia and Ehrhardt [7] have investigated an inorganic-based Szilard-Chalmers method for the enhancement of the specific activity of ^{186}Re. An enrichment factor of ^{186}Re was obtained as 5–10 in an experiment using a natural rhenium target. Using a reactor (neutron flux: 1.5×10^{17} m^{-2} s^{-1}) at the Paul Scherrer Institute in Switzerland, Schubiger et al. [8] tried Szilard-Chalmers enrichment of ^{186}Re. They irradiated $C_5H_5ReO_3$ and obtained an enrichment factor of 400–800 which was an encouraging result for producing perrhenate labeled with ^{186}Re (the product of 2 MBq with specific activity of 0.8 MBq/μg).

With regard to the specific activity, ^{186}Re with very high specific activity is produced by the ^{186}W(p,n)^{186}Re reaction using a cyclotron. The chemical separation of ^{186}Re from tungsten gives no-carrier added ^{186}Re. Shigeta et al. [9] investigated the production method of no-carrier added ^{186}Re using an AVF cyclotron. The excitation function of the ^{186}W(p,n) ^{186}Re reaction was studied up to 20 MeV by a stacked-foil technique in order to estimate the produced radioactivity of ^{186}Re. The peak of the excitation function was found at 10 MeV (59 mb), which agreed with literature data [10]. A 99.79% enriched [^{186}W] WO$_3$ powder was used as a target for the production. The target (disk-shaped pellet) was irradiated with 13.6 MeV protons. The chemical separation of ^{186}Re was accomplished by an anion exchange method. By-products ^{183}Ta and ^{187}W as well as stable tungsten were eluted with solutions of 0.5 M NaOH/0.5 M NaCl and 1.5 M HCl. Finally, ^{186}Re was eluted from the column with 4 M HNO$_3$ and was recovered in high yield (approximately 90%). The product solution was obtained in the amount of 6.5 MBq (180 μCi) at EOB and its radiochemical purity was more than 99%.

2.2 Production of ^{188}Re

Tungsten-188 decays to ^{188}Re with a half-life of 69 days. Therefore, the carrier-free ^{188}Re is obtained from a ^{188}W/^{188}Re generator system. The parent ^{188}W is produced by double neutron capture reactions in enriched ^{186}W (tungsten trioxide).

$$^{186}W(n,\gamma) \; ^{187}W(n,\gamma) \; ^{188}W \xrightarrow{\beta^-} \; ^{188}Re \xrightarrow{\beta^-} \; ^{188}O_S \text{ (stable)}$$

The cross sections of the thermal neutron capture reaction of ^{186}W and ^{187}W are

$3.79 \pm 0.06 \times 10^{-27}$ and $6.4 \pm 1.0 \times 10^{-27} \, m^2$ (37.9 ± 0.6 and 64 ± 10 barns), respectively [5]. However, the calculated yield using these cross sections is not consistent with the experimental results. So the cross section values should be reevaluated [11, 12]. Tungsten-188 is only available in low specific activity and it is difficult to adsorb a sufficient amount of [188]W onto the compact-size column. Anion exchange resin, alumina and hydrous zirconium oxide have been used [13–17] as an adsorbent to separate rhenium from tungsten in the generator. Among them, the alumina generator system gives a [188]Re solution with a high efficiency (75–85%) using saline. Scaling up the chromatographic generators with these adsorbents using low specific activity of [188]W would require larger columns and result in larger elution volumes. In order to overcome this problem, Ehrhardt et al. [18] investigated the production of gel generators in which the irradiated target material ([186]W/[188]W) was processed to form the packing material (zirconyl tungstate) of the column in itself. The irradiated target [188]WO$_3$ was dissolved in a strong basic solution. The zirconyl nitrate (or zirconyl chloride) in an acid solution was added to the target solution and the precipitate of zirconyl tungstate was formed. The [188]Re solution was eluted with normal saline. This gel generator exhibited excellent elution profiles and low breakthrough of [188]W. In another approach, the alumina/anion-exchange tandem generator system [19] was investigated. The [188]Re was eluted from the alumina column with an ammonium nitrate solution and then was adsorbed onto the anion-exchange column before final elution with nitric acid. This tandem generator system would be useful even for low specific activity [188]W. Therefore, the [188]Re specific volume (mCi/ml) obtained from the tandem generator is independent of the size of the alumina column and specific activity of [188]W. In addition, highly purified and concentrated solutions of carrier-free [188]Re are obtained using an anion-exchange column.

Knapp Jr. et al. [11, 12] investigated the production of curie-scale [188]W/[188]Re generators for clinical application using [186]W-enriched tungsten oxide and metal as target materials. Relatively long irradiation periods are required for the large-scale production of [188]W even in high flux reactors. The traditional technique for processing alumina-based [188]W/[188]Re generators involves a dissolution process of reactor-irradiated [186]W-oxide targets in 0.1 M NaOH. After long irradiation (> 21 days) of large amounts of [186]WO$_3$ in the ORNL High Flux Isotope Reactor (HFIR) at a steady-state thermal neutron flux of about $2 \times 10^{19} \, m^{-2} \, s^{-1}$, however, the irradiated target solution contained a NaOH-insoluble [188]W-labeled black solid (approx. 30–50% of total activity). The insoluble [188]W black material significantly reduced the [188]W production yields and the specific activity. Although the black material was insoluble in ordinary acidic and basic solutions, it could be dissolved in a 0.1 M NaOH solution containing 5% sodium hypochlorite (NaOCl). The complete dissolution of this insoluble material resulted in a significant increase in the production yield and specific activity of [188]W. Alternatively, irradiated [186]W-metal targets were readily dissolved in a sodium hydroxide solution by the addition of < 30% hydrogen peroxide. An alumina generator (alumina: 7 g) loaded with 1038 mCi of ^{188}W was prepared from hypochlorite-processed ^{186}W-oxide or peroxide-processed ^{186}W-metal targets and it provided over 700 mCi (> 70% yield) of ^{188}Re (30–35 mCi/ml) almost without ^{188}W breakthrough.

3 Simple Complexes

3.1 Re Complexes for Treating Painful Bone Cancers – Re-HEDP, MDP

The need for and ethics of cancer therapy with radiopharmaceuticals are now being carefully discussed. The success in diagnosis of cancer using ^{99m}Tc has provoked interest in the possibility of cancer therapy with radiorhenium. Even if complete recovery from the disease has not yet been achieved, positive changes in the quality of life (in about 80% of all patients) were reported by rhenium treatment for bone metastases.

Diphosphonic acid derivatives (Fig. 1) labeled with ^{99m}Tc have been widely used as imaging agents for bone disease. Studies of diphosphonates labeled with ^{186}Re, produced by the ^{185}Re(n,γ) reaction, and their biodistribution behavior indicate that ^{186}Re-labeled diphosphonate compounds are equally as good as bone seeker agents as ^{99m}Tc-diphosphonates are. In particular, ^{186}Re-HEDP (HEDP: (1-hydroxyethylidene)diphosphonic acid) has been investigated clinically [20–31] because of its specifically high labeling yields in a neutral pH range (pH 5–8). The ^{186}Re-HEDP demonstrated effective palliation of bone pain due to metastases of primary carcinomas. Eisenhut [32] showed that the labeling of ^{186}Re-MDP (MDP: methylene diphosphonic acid) required low pH (1.4–1.6), and no labeling was observed in the neutral pH range. These results meant that the optimum pH ranges for the syntheses of Re-HEDP and Re-MDP were different even though both HEDP and MDP are diphosphonic acid derivatives. The synthesis of ^{188}Re-MDP labeled compounds using carrier-free ^{188}Re from

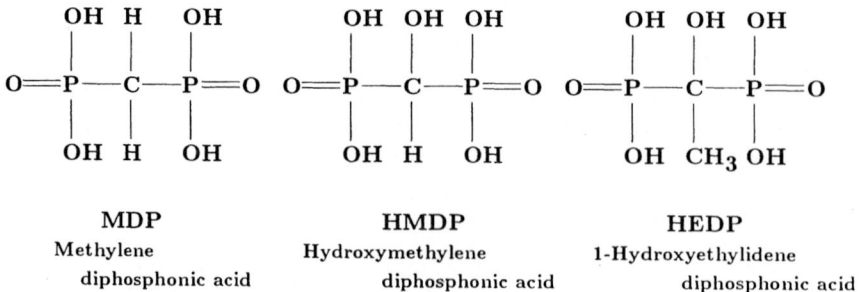

Fig. 1. Chemical formulas of diphosphonates

the [188]W/[188]Re generator was recently investigated [33] in order to compare the results with and without its carrier. The dependence of the labeling yield upon pH is shown in Fig. 2. The maximum labeling yield of 85–89% was obtained in the pH range from 0.6 to 0.8. Below pH 0.5 and above pH 1 the yield decreased. The influence on the labeling yield after adding the carrier was observed. The yield with the carrier was 92–95%. The influences on the yield of a pH change from 0.6 (the optimum) to 3–5 were investigated without and with the carrier. The stability of [188]Re-MDP against pH is shown in Fig. 3. The yield decreased with increasing pH and fell to about 50% around pH 4 in the case of carrier-free [188]Re. However, the stability of Re-MDP for carrier-added [188]Re

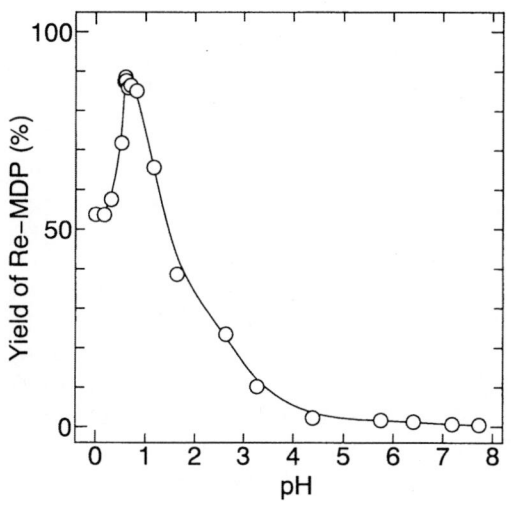

Fig. 2. Influence of pH on yields of [188]Re-MDP

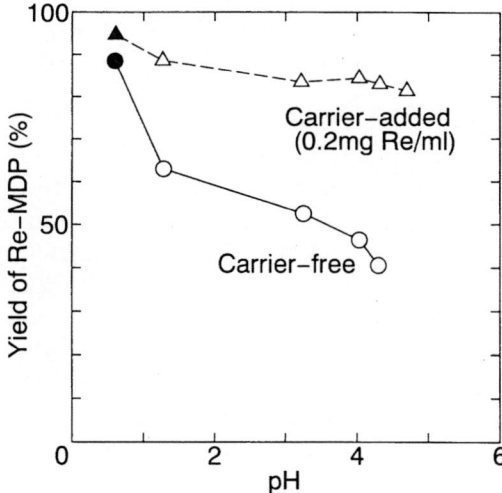

Fig. 3. Influence of pH on stability of [188]Re-MDP

281

was higher than that for no-carrier-added ^{188}Re. The yield with carrier was more than 80% even at pH 4.

The structures of Re-diphosphonate complexes (Re-HEDP, Re-MDP etc.) have not been confirmed yet. The chromatogram of HPLC analysis for Re-diphosphonate complexes indicated many peaks [22, 23, 32] and the Re-labeled diphosphonate compound was thought to include oligomeric and polymeric complexes. The results for Tc-diphosphonate complexes [34–40] showed that the number of these peaks and the ratio depended on reducing agents, carrier (carrier-free or carrier-added), pH and ligand concentration. Detailed EXAFS studies [41] on solid samples and frozen aqueous solutions of Re-HEDP indicated the presence of a strong metal-metal interaction due to Re-Re bonding. The short Re-Re distance (2.4 Å) suggested an Re–Re triple bond and a valence state of IV (or lower). In this structure, Re was surrounded by six oxygen atoms (four Re–O: 2.0 Å, tow Re–O: 2.1 Å) from HEDP ligands and water ligands. These oxygen atoms which had longer Re–O distances would be bridged over two Re centers. These results indicated that Re-HEDP involved a unique radiopharmaceutical structure.

3.2 Re-DTPA, DMSA Complexes

Majali [42] studied the preparation of ^{186}Re-DTPA complexes (DTPA: diethylenetetraaminepentaacetic acid) using low specific activity of ^{186}Re for antibody labeling. Rhenium labeling required a high temperature ($\sim 100\,^{\circ}$C) for reduction of Re and subsequent complexation, unlike technetium, which can be easily reduced and subsequently complexed with DTPA. A larger amount of stannous chloride and an acidic pH were also required for the reduction of the perrhenate. The optimum pH for the preparation of ^{186}Re-DTPA was about 3. Labeling yields were low at neutral and alkaline pH values. Furthermore, the labeling yield depended on the methods of producing ^{186}Re, which included extraction with methyl ethyl ketone (MEK) or not, even under the same labeling conditions. The extraction of ^{186}Re with MEK before complexation provided better labeling yields.

The ^{99m}Tc-DMSA (DMSA: meso-1,2-dimercaptosuccinic acid) has been widely used for diagnoses of renal diseases. This radiopharmaceutical also has been shown to localize in medullary thyroid carcinoma [43, 44] etc. Therefore, DMSA labeled with ^{186}Re or ^{188}Re is possibly suitable for tumor therapy. Bisunadan et al. [45] investigated labeling with ^{186}Re using a commercial DMSA kit. Heating at 100 °C for 30 min was required for the preparation of ^{186}Re-DMSA, whereas the technetium complexes were formed instantaneously at room temperature. This complex was identified as $[^{186}$Re] [Re(V) O(DMSA)$_2$]$^-$ by TLC, NMR, optical and IR spectroscopy, and elemental analysis. The [Re(V) O(DMSA)$_2$]$^-$ ion exists in three isomers, which differ according to the relative orientation of the carboxyl groups, as shown in Fig. 4. Singh et al. [46, 47] separated these isomers by high-performance liquid chromatography

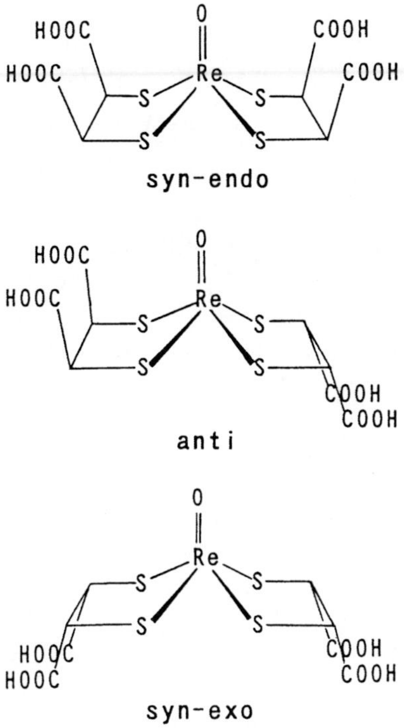

Fig. 4. Isomers of $[ReO(DMSA)_2]^-$

[Hamilton PRP-1 column (reverse-phase column), a gradient solvent system comprising 0.1% trifluoroacetic acid in water and 0.1% trifluoroacetic acid in acetonitrile] and identified three peaks as the *anti-*, *syn-endo-* and *syn-exo-* isomers in order of elution. For the carrier-added preparations the ratios of *anti-*, *syn-endo-* and *syn-exo-*isomers were approximately 45:45:10. For the carrier-free ^{188}Re preparations under optimum conditions, the *syn-endo*-isomer was found to be more dominant. On different occasions the yields of the three isomers were 75–87% (*syn-endo*), 11–25% (*anti-*) and 0–1% (*syn-exo*), respectively. The ratio of the three isomers also depended on pH and the ratio of DMSA:Sn:Re. The crystal structure of *syn-endo*-$[NEt_4][ReO(DMSA)_2] \cdot 1\frac{1}{2}$ H_2O was determined [46]. The complex has an approximate square-pyramidal configuration (slightly distorted towards trigonal bipyramidal). The Re-O distance was 1.699(8) Å, where the digit in the parenthesis denotes the experimental error, and the average Re-S distance was 2.310 Å (2.301(4), 2.322(4), 2.286(3) and 2.329(4) A). Lisic et al. [48] reported another route for synthesizing the Re-DMSA complex using carrier-free ^{188}Re (Fig. 5), The $[^{188}Re]$ $[ReOCl_3(PPh_3)_2]$ was prepared with a yield of more than 95% by the reaction of perrhenate ($^{188}ReO_4^-$) with triphenyl phosphine (PPh_3) and HCl with subsequent extraction into CH_2Cl_2 or $CHCl_3$. The reaction of $[^{188}Re]$ $[ReOCl_3(PPh_3)_2]$ with DMSA at room temperature yielded $[^{188}Re]$ $[ReO(DMSA)_2]^-$ in a high yield. This compound was identical to the Re-DMSA complex prepared using stannous chloride as a reducing agent [45].

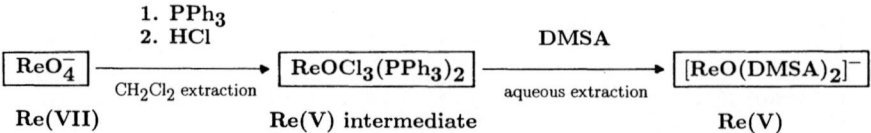

Fig. 5. Formation reaction of $[^{188}ReO(DMSA)_2]^-$ via a reactive intermediate

Katti et al. [49] reported the labeling of organometallic phosphinimine with ^{188}Re. Reactions of the organometallic phosphinimine $Ph_3P = N-SiMe_3$ in CH_3CN with aqueous $^{188}ReO_4^-$ at room temperature yielded an ion pair $Ph_3P = NH_2^{+\,188}ReO_4^-$ (a) in a 98% yield. Heating the toluene extract of the ion pair complex (a) at 80 °C for 30 min produced the neutral Re(VII) species: $Ph_3P = N-^{188}ReO_3$ (b) in a yield $> 95\%$. The solid state structure of $Ph_3P = NH_2^+ ReO_4^-$ showed the presence of hydrogen bonding between the iminato hydrogen atoms and the oxygen atoms of the ReO_4^- anion. The 4 Re–O bond distances were 1.710(8), 1.690(7), 1.699(8) and 1.727(7) Å. The P–N bond distance (1.636(7) Å) found in the ion pair complex was within the range observed for a non-metallated free phosphinimine ion pair of the type: $R_2R'P = NH_2^+ Cl$ reported recently [50]. The ion pair complex (a) and the neutral complex (b) were soluble in organic solvents and were quite stable in dry toluene, no decomposition was observed for up to 18 h. These complexes would have prospects for introducing a new class of radiopharmaceuticals.

4 Antibody Labeling Using Bifunctional Ligands

Bifunctional ligands have been used for the labeling of monoclonal antibodies with metallic radioisotopes. The reaction of DTPA with acetic anhydride, as was described earlier, yields bicyclic anhydride of DTPA (cDTPAA) with a bifunctional property. Quadri and Wessels [51] studied the labeling of human serum albumin, anti-human serum albumin antibody and monoclonal antibody with ^{186}Re using $SnCl_2 \cdot 2H_2O$, $Na_2S_2O_4$ and H_3PO_2 (50%) as reducing agents. These experiments showed that stannous chloride ($SnCl_2 \cdot 2H_2O$) and sodium dithionite ($Na_2S_2O_4$) reduction methods provided overall labeling yields between 5 and 18% with an associated immunoreactivity of 12–40%. The hypophosphorous acid (H_3PO_2) reduction method, however, yielded no usable radiolabeled product.

More recently the labeling of antibodies has been investigated using a diamide dimercaptide (N_2S_2) ligand and a triamide mercaptide (N_3S) ligand as bifunctional ligands (Fig. 6). Stannous chloride was used as a reducing agent in these cases. The attachment between the monoclonal antibody (MoAb) and the Re complex was carried out by conjugation of active ester groups of the Re-complex ($Re-N_2S_2$ or $Re-N_3S$) with the lysine amino groups ($MoAb-NH_2$) of

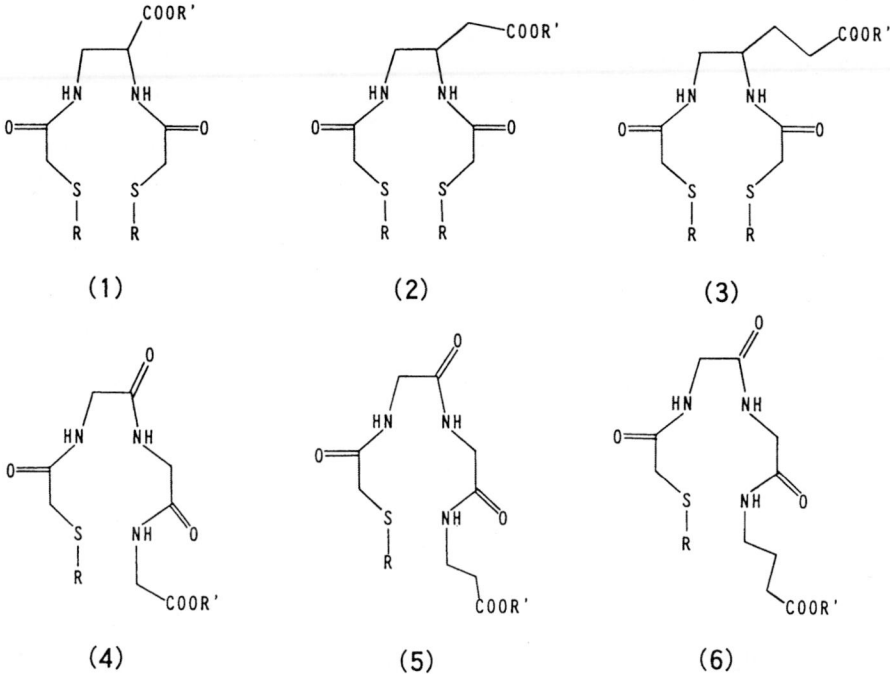

Fig. 6. Chemical formulas of the diamide dimercaptide and triamide mercaptide.
(1) C_3-N_2S_2, (2) C_4-N_2S_2, (3) C_5-N_2S_2, (4) MAG_3, (5) MAG_2-ala, (6) MAG_2-GABA, R = thiobenzonate, R' = H or tetrafluorophenyl (TFP)

the proteins, following the scheme shown in Fig. 7. Vanderheyden et al. [52] investigated the preparation of [186]Re, [188]Re-map (map: 2,3-bis(mercaptoacetoamido)propanoic acid [C_4-N_2S_2]) and the biodistribution of rhenium-labeled antibodies. In the biodistribution studies, tumor-to-blood ratios greater than 10 were obtained. Antibody labeling with [186]Re using N_3S ligands (MAG_2-GABA and MAG_3) [53–57] indicated that stable labeled antibodies were obtained in a high yield and localized in tumors. Visser et al. [55] investigated the stability of [186]Re-MAG_3-E48 IgG in 0.9% NaCl with or without antioxidant (ascorbic acid or gentisic acid). The [186]Re-MAG_3-MoAb conjugate appeared to be fairly stable at a low radioactivity concentration (0.12 mCi/ml); after 2, 24 and 120 h at room temperature, the percentage of protein-bound [186]Re was 95.3%, 91.0% and 82.8%, respectively. But at higher radioactivity concentrations (2.5 mCi/ml), at 24 and 120 h, the percentage of protein-bound [186]Re was 65.3% and 10.7%, respectively. The use of antioxidants was shown to suppress the radiolytic decomposition. Addition of ascorbic acid or gentisic acid at low pH ensured the stability of [186]Re-MAG_3-MoAb. The optimal antioxidative capacities of ascorbic acid and gentisic acid were found at pH 3.5 (92.5% protein bound [186]Re at 120 h) and pH 2.3 (92.4% at 120 h), respectively. Studies [58–61] on the structures of Re-N_2S_2 and

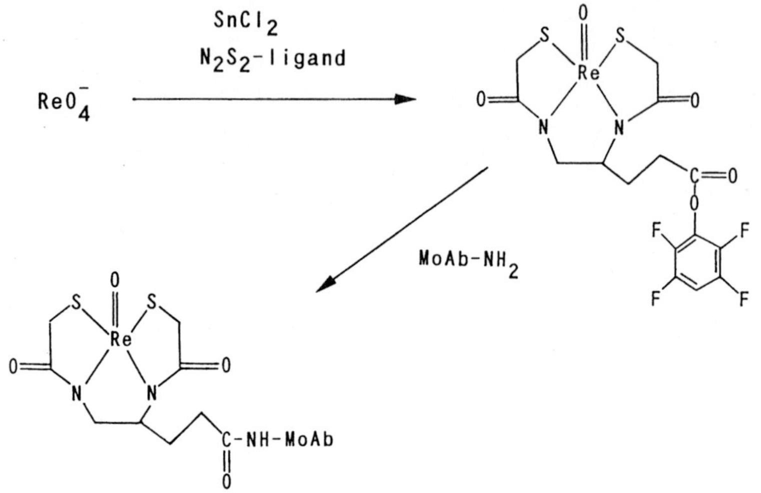

Fig. 7. General scheme for radiolabeling antibodies with rhenium using N_2S_2-ligand (N_3S-ligand)

Re-N_3S complexes revealed the presence of a Re(V) = O core in both complexes. The selected bond lengths and other parameters are listed in Table 2. In all cases, an Re atom is coordinated to five atoms (N, N, S, S, O for Re-N_2S_2 or N, S, S, S, O for Re-NS_3) in an approximately sqaure-pyramidal geometry. The rhenium atoms (N, N, S, S for Re-N_2S_2 or N, S, S, S for Re-NS_3). The Re-N_2S_2 and Re-N_3S complexes could exist in two isomers due to *syn-* and *anti*-orientations of the carboxyl group of the ligand relative to the rhenium-oxo group. In the cases of [ReO(map)]$^-$ and [ReO(mapt)]$^-$ (mapt: 4,5-bis(mercaptoacetamido) pentanoic acid [C_5-N_2S_2]), *syn-* and *anti*-isomers were separated by HPLC (reverse-phase C_{18} column, solvent: 1–5% CH_3CN/0.01 M phosphate (pH = 7)), respectively [58, 62]. The *anti*-isomer was eluted first on the C_{18} reverse-phase column followed by the *syn*-isomer for rhenium complexes, while the *syn*-isomer was eluted first for technetium complexes.

Najafi et al. [63, 64] recently synthesized a new chelating agent N_2S_4 (N, N, N′,N′-tetrakis(2-mercaptoethyl) ethylenediamine), as shown in Fig. 8, and investigated the antibody labeling with ^{186}Re using this N_2S_4 ligand by a stannous chloride reduction method. The antibody radiolabeling using a N_2S_4 ligand was simple (pH = 7) and resulted in a radiolabeling yield greater than 95%. The labeled antibody required no further purification. The in vitro stability study of ^{186}Re-labeled N_2S_4 coupled with MoAb was performed by mixing the labeled antibody with a solution of human serum albumin (HSA) at 37 °C. More than 90% of total radioactivity was still bound to the antibody after 5 days of incubation in HSA, and about 30% was bound with the antibody after 15 days. Immunoreactivity studies were performed by live cell assay. The percent binding of radioactivity to the cells was found to be higher than 85%.

Table 2. Structures of Re-N$_2$S$_2$ and Re-N$_3$S complexes

Complex	Bond distance (Å)			Distance (Å) of Re from the basal plane	Re = O stretch (cm^{-1})	Reference
	Re–O	Re–S	Re–N			
N$_2$S$_2$						
[ReO (map)]$^-$	1.662 (5)	2.282 (2), 2.284 (3)	1.959 (7), 2.006 (7)	0.747 ± 0.005[d]	970	Rao et al.[58]
[ReO (tedadt[a])]$^-$	1.709 (2)	2.283 (1), 2.288 (1)	1.921 (3), 2.148 (3)		920	Jackson et al.[59]
N$_3$S						
[ReO(MAG$_3$)]$^-$	1.68 (1)	2.285 (7)	1.98 (2)–2.04 (2)	0.72	975	Rao et al.[60]
[ReO(MAG$_2$-oABAH[b])]$^-$	1.684 (4)	2.295 (1)	1.976 (4)–2.019 (6)	0.73	966	Hansen et al.[61]
[ReO(MAG$_2$-pABAH[c])]$^-$	1.670 (5)	2.283 (2)	1.984 (5)–2.024 (6)	0.74	970	Hansen et al.[61]

[a] tedadt: 1,1,8,8-tetraethyl-3,6-diazaoctane-1,8-dithiolato,
[b] MAG$_2$-oABAH: mercaptoacetylglycylglycyl-o-aminobenzoato.
[c] MAG$_2$-pABAH: mercaptoacetylglycylglycyl-p-aminobenzoato.
[d] The average value for [TcO (map)]$^-$ and [ReO(map)]$^-$.

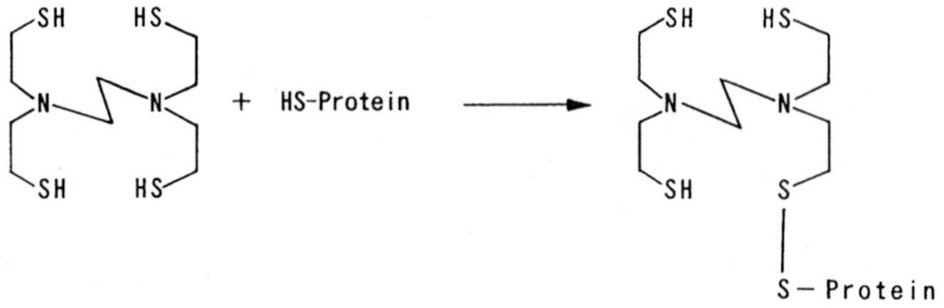

Fig. 8. Scheme for coupling of a new N_2S_4-ligand to antibody

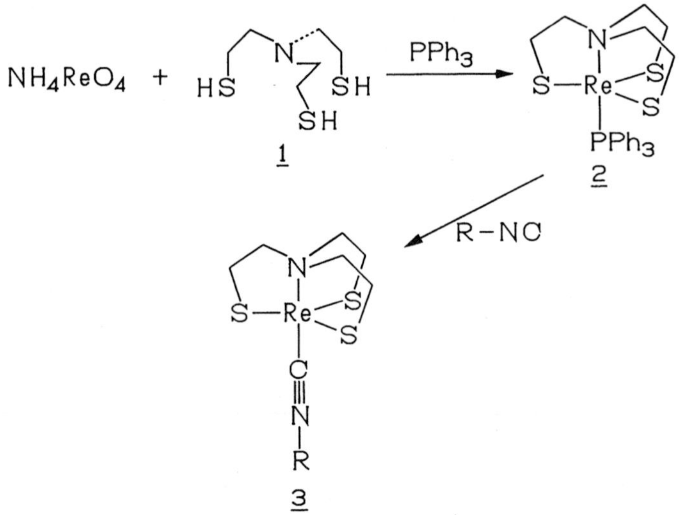

Fig. 9. Scheme for syntheses of rhenium complexes with tetradentate NS_3 tripod ligands (R-NC: $R = CH_2COOCH_3$)

The coupling of N_2S_4 to the antibody was believed to be via a disulfide bond (-S-S-). The therapeutic trials were performed using nude mice treated with 0.5 mCi [186]Re-labeled MoAbs. Three out of seven mice (42%) had complete regression of their tumors following a single treatment with [186]Re-labeled MoAb (Lym-1), and two additional mice were free to tumor following a second treatment. Thus, for five out of seven (71%) complete tumor responses were observed within 2 weeks after a second treatment.

In the bifunctional radiopharmaceuticals, Tc and Re complexes that are suitable for coupling to antibodies are almost square-pyramidal oxo complexes of tetradentate ligands, as mentioned earlier. Spies et al. [65] recently studied the syntheses and structures of rhenium complexes with a tetradentate NS_3 tripod ligand 2, 2, ', 2''-nitrilotris(ethanethiol) (1 in Fig. 9) as an interesting alternative to the oxo complex. Such ligands could lead to weakly polar, trigonal-bipyramidal complexes in which the metal atom is more strongly shielded than in the square-pyramidal oxo complexes. Schemes for the syntheses of rhenium complexes are shown in Fig. 9. The Re-NS$_3$ tripod ligand complex $\underline{2}$ was obtained by reducing the perrhenate using triphenylphosphane (PPh$_3$) in the presence of NS$_3$ tripod ligand $\underline{1}$ in an ethanol/water solution made acidic with hydrochloric acid. The rate of reaction was found to be highly dependent on the type of phosphane. Complex $\underline{3}$ was obtained by the direct exchange of the phosphane ligand in $\underline{2}$ for one equivalent of isocyanide (R-NC, R = CH$_2$COOCH$_3$) in CH$_2$Cl$_2$. Complex $\underline{3}$ was also obtained by the ligand exchange reaction of rhenium(III) complexes (*trans*-[ReOCl$_3$(PPh$_3$)$_2$] or [ReCl$_3$(PPh$_3$)$_2$(MeCN)]). The X-ray structure analyses on crystals of $\underline{2}$ and $\underline{3}$ showed that they had pentacoordinate trigonal-bipyramidal geometries. The tetradentate ligand $\underline{1}$ in complexes $\underline{2}$ and $\underline{3}$ occupied three equatorial and one axial position, whereas the second axial position by a triphenylphosphane ligand or isocynanide ligand. These complexes were new types of neutral, trigonal-bipyramidal and mixed-ligand complexes of rhenium(III). Complex $\underline{3}$ in particular could be coupled to biomolecules by the formation of an amide via an ester function.

5 Concluding Remarks

The radiopharmaceutical chemistry of rhenium for cancer therapy is a new and developing field. The rapid increase in the numbers of papers concerned with ^{186,188}Re-complexes for nuclear medicine unquestionably shows the importance of the chemistry of rhenium complexes, as mentioned by Deutsch in his lecture at the Fourth International Symposium on Technetium in Chemistry and Nuclear Medicine held in Bressanone, Italy in 1994.

However, problems are encountered in production of rhenium radionuclides and work is being done to increase the yields of the radionuclides to meet urgent demands for their use in therapy. Moreover, rhenium is not as reactive as technetium. This situation makes rhenium chemistry somewhat specific – optimum conditions in the preparation of rhenium complexes or in antibody labeling using bifunctional ligands must be identified.

Despite such problems many people are now interested in the radiopharmaceuticals of rhenium. We hope the coming decade will be characterized by more successful applications of this therapy.

Acknowledgements. The authors thank the Director of the Department of Radioisotopes of the Japan Atomic Energy Research Institute, Mr. H. Yamabayashi for his interests in this work. They also thank Dr. Carol Kikuchi for her reading the manuscript.

6 References

1. Srivastava SC, Mausner LF, Mease RC (1991) BNL-46589, CONF-9109287-1
2. Vera-Ruiz H (1993) IAEA Bulletin 35: 24
3. Lewington VJ (1993) Eur J Nucl Med 20: 66
4. Mausner LF, Srivastava SC (1993) Med Phys 20: 503
5. Mughabghab SF (1984) Neutron Cross Sections Vol. I Part B. Academic Press, New York
6. Goldenberg DM, Griffiths GL (1992) J Nucl Med 33: 1110
7. Jia W, Ehrhardt GJ (1994) J Nucl Biol Med 38: 447
8. Schubiger PA, Cahn RM, Alberto RA, Bläuenstein P (1993) J Labeled Compd Radiopharm XXXII: 215
9. Shigeta N, Matsuoka H, Osa A, Koizumi M, Izumo M, Kobayashi K, Hashimoto K, Sekine T (1993) JAERI TIARA Annual Report 1993: 163
10. Thomas RJ Jr., Bartolini W (1968) Nucl Phys A106: 323
11. Knapp FF Jr., Mirzadeh S, Beets AL, Sharkey R, Griffiths G, Juweid M, Goldenberg DM (1994) J Nucl Biol Med 38: 448
12. Knapp FF Jr., Callahan AP, Beets AL, Mirzadeh S, Hsieh B-T (1994) Appl Radiat Isot 45: 1123
13. Blachot J, Herment J, Moussa A (1969) Int J Appl Rad Isotopes 20: 467
14. Botros N, El-Garhy M, Abdulla S, Aly HF (1986) Isotopenpraxis 22: 368
15. Malyshev KV, Smirnov VV (1975) Radiokimya 17: 249
16. Callahan AP, Rice DE, Knapp FF Jr. (1989) NucCompact 20: 3
17. Kodina G, Tulskaya T, Gureev E, Brodskaya G, Gapurova O, Drosdovsky B (1990) in: Nicolini M, Bandoli G, Mazzi U (eds) Technetium and Rhenium in Chemistry and Nuclear Medicine 3. Cortina International, Verona, Italy, p 635
18. Ehrhardt GJ, Ketring AR, Turpin TA, Razavi M-S, Vanderheyden J-L, Su F-M, Fritzberg AR (1990) in: Nicolini M, Bandoli G, Mazzi U (eds) Technetium and Rhenium in Chemistry and Nuclear Medicine 3. Cortina International, Verona, Italy, p 631
19. Kamioki H, Mirzadeh S, Lambrecht RM, Knapp R Jr., Dadachova K (1994) Radiochim Acta 65: 39
20. Mathieu L, Chevalier P, Galy G, Berger M (1979) Int J Appl Radiat Isot 30: 725
21. Weininger J, Ketring AR, Deutsch E, Maxon HR, Goeckeler WR (1983) J Nucl Med 24: P125
22. Deutsch E, Libson K, Vanderheyden J-L, Ketring AR, Maxon HR (1986) Nucl Med Biol 13: 465
23. Ketring AR (1987) Nucl Med Biol 14: 223
24. Maxon HR, Deutsch EA, Thomas SR, Libson K, Lukes SJ, Williams CC, Ali S (1988) Radiology 166: 501
25. Maxon III HR, Schroder LE, Thomas SR, Hertzberg VS, Deutsch EA, Scher HI, Samaratunga RC, Libson KF, Williams CC, Moulton JS, Schneider HJ (1990) Radiology 176: 155
26. Zonnenberg BA, De Klerk JMH, Van Rijk PP, Quirijnen JMSP, Van Het Schip AD, Van Dijk A, Ten Kroode NFJ (1991) J Nucl Med 32: 1082
27. Maxon III HR, Schroder LE, Hertzberg VS, Thomas SR, Englaro EE, Sammaratunga R, Smith H, Moulton JS, Williams CC, Ehrhardt GJ, Schneider HJ (1991) J Nucl Med 32: 1877
28. De Klerk JMH, Van Dijk A, Van Het Schip AD, Zonnenberg BA, Van Rijk PP (1992) J Nucl Med 33: 646
29. Srivastava SC, Meinken GE, Mausner LF, Cutler C, Atkins HL, Deutsch E (1994) J Nucl Biol Med 38: 465
30. Shukla SK, Limouris GS, Cipriani C, Argiró G, Atzei G, Boccardi F, Boemi S (1994) J Nucl Biol Med 38: 484
31. Van Pijk PP, De Klerk JMH, Zonnenberg BA, Van Het Schip AD, Van Dijk A, Quirijnen JMSP (1994) J Nucl Biol Med 38: 487
32. Eisenhut M (1982) Int J Appl Radiat Isot 33: 99
33. Hashimoto K, Bagiawati S, Izumo M, Kobayashi K (1996) Appl Radiat Isot in press
34. Pinkerton TC, Ferguson DL, Deutsch E, Heineman WR, Libson K (1982) Int J Appl Radiat Isot 33: 907
35. Tanabe S, Zodda JP, Deutsch E, Heineman WR (1983) Int J Appl Radiat Isot 34: 1577
36. Tanabe S, Zodda JP, Libson K, Deutsch E, Heineman WR (1983) Int J Appl Radiat Isot 34: 1585
37. De Groot GJ, Das HA, De Ligny CL (1986) Appl Radiat Isot 37: 23

38. Mikelsons MV, Pinkerton TC (1986) Anal Chem 58: 1007
39. De Groot GJ, Das HA, De Ligny CL (1987) Appl Radiat Isot 38: 611
40. Huigen YM, Gelsema WJ, De Ligny CL (1987) Appl Radiat Isot 38: 615
41. Pipes D, Yuan J, Helmer B, Deutsch K, Elder LC, Deutsch E (1993) J Nuc Med 34: 38P
42. Majali MA (1993) J Radianal Nucl Chem 170: 471
43. Ohta H, Yamamoto K, Endo K, Mori T, Hamanaka D, Shimazu A, Ikekubo K, Makimoto K, Iieda Y, Konishi J, Morita R, Hata N, Horiuchi K, Yokoyama A, Torizuka K, Kuma K (1984) J Nucl Med 25: 323
44. Clarke SEM, Lazarus CR, Wraight P, Sampson C, Maisey MN (1988) J Nucl Med 29: 33
45. Bisunadan MM, Blower PJ, Clarke SEM, Singh J, Went MJ (1991) Appl Radiat Isot 42: 167
46. Singh J, Powell AK, Clarke SEM, Blower PJ (1991) J Chem Soc Chem Commun: 1115
47. Singh J, Reghebi K, Lazarus CR, Clarke SEM, Callahan AP, Knapp Jr FF, Blower PJ (1993) Nucl Med Commum 14: 197
48. Lisic EC, Mirzadeh S, Knapp Jr FF (1993) J Labeled Compd Radiopharm XXXIII: 65
49. Katti KV, Singh PR, Barnes CL, Katti KK, Kopicka K, Ketring AR, Volkert WA (1993) Z Naturforsch 48b: 1381
50. Katti KV, Pinkerton AA, Cavell RG (1991) Inorg Chem 30: 2631
51. Quadri SM, Wessels BW (1986) Nucl Med Biol 13: 447
52. Vanderheyden J-L, Rao TN, Kasina S, Wester D, Su F-M, Fritzberg AR (1990) in: Nicolini M, Bandoli G, Mazzi U (eds) Technetium and Rhenium in Chemistry and Nuclear Medicine 3. Cortina International, Verona, Italy p 623
53. Goldrosen MH, Biddle WC, Pancook J, Bakshi S, Vanderheyden J-L, Fritzberg AR, Morgan Jr AC, Foon KA (1990) Cancer Research 50: 7973
54. Breitz HB, Weiden PL, Vanderheyden J-L, Appelbaum JW, Bjorn MJ, Fer MF, Wolf SB, Ratliff BA, Seiler CA, Foisie DC, Fisher DR, Schroff RW, Fritzberg AR, Abrams PG (1992) J Nucl Med 33: 1099
55. Visser GWM, Gerretsen M, Herscheid JDM, Snow GB, Van Dongen G (1993) J Nucl Med 34: 1953
56. Guhlke S, Diekmann D, Zamora PO, Knapp FF Jr, Biersack HJ (1994) J Nucl Biol Med 38: 444
57. Su F-M, Lyen L, Breitz HB, Weiden PL, Fritzberg R (1994) J Nucl Biol Med 38: 485
58. Rao TN, Adhikesavalu D, Camerman A, Fritzberg AR (1990) J Am Chem Soc 112: 5798
59. Jackson TW, Kojima M, Lambrecht M (1993) Aust J Chem 46: 1093
60. Rao TN, Adhikesavalu D, Camerman A, Fritzberg AR (1991) Inorg Chim Acta 180: 63
61. Hansen L, Cini R, Taylor A Jr, Marzilli LG (1992) Inorg Chem 31: 2801
62. Rao TN, Brixner DI, Srinivasan A, Kasina S, Vanderheyden J-L, Wester DW, Fritzberg AR (1991) Appl Radiat Isot 42: 525
63. Najafi A, Alauddin MM, Siegel ME, Epstein AL (1991) Nucl Med Biol 18: 179
64. Najafi A, Alauddin MM, Sosa A, Ma GQ, Chen DCP, Epstein AL, Siegel ME (1992) Nucl Med Biol 19: 205
65. Spies H, Glaser M, Pietzsch H-J, Hahn FE, Kintzel O, Lügger T (1994) Angew Chem Int Ed Engl 33: 1354

Author Index Volumes 151-176

The volume numbers are printed in italics

Hashimoto, K., and Yoshihara, K.: Rhenium Complexes Labeled with [186/188]Re for Nuclear Medicine. *176*, 275-292 (1996).

Hadjiarapoglou, L., see Adam, W.: *164*, 45-62 (1993).

Hart, H., see Vinod, T. K.: *172*, 119-178 (1994).

Harbottle, G.: Neutron Acitvation Analysis in Archaecological Chemistry. *157*, 57-92 (1990).

Hatlevig, S.A., see Freeman, P.K.: *168*, 47-91 (1993).

Hauser, A., see Colombo, M. G.: *171*, 143-172 (1994).

Hayashida, O., see Murakami, Y.: *175*, 133-156 (1995).

He, W.C., and He, W.J.: Peak-Valley Path Method on Benzenoid and Coronoid Systems. *153*, 195-210 (1990).

He, W.J., see He, W.C.: *153*, 195-210 (1990).

Heaney, H.: Novel Organic Peroxygen Reagents for Use in Organic Synthesis. *164*, 1-19 (1993).

Heinze, J.: Electronically Conducting Polymers. *152*, 1-19 (1989).

Helliwell, J., see Moffat, J.K.: *151*, 61-74 (1989).

Hennig, H., see Billing, R.: *158*, 151-199 (1990).

Hesse, M., see Meng, Q.: *161*, 107-176 (1991).

Hiberty, P.C.: The Distortive Tendencies of Delocalized π Electronic Systems. Benzene, Cyclobutadiene and Related Heteroannulenes. *153*, 27-40 (1990).

Hladka, E., Koca, J., Kratochvil, M., Kvasnicka, V., Matyska, L., Pospichal, J., and Potucek, V.: The Synthon Model and the Program PEGAS for Computer Assisted Organic Synthesis. *166*, 121-197 (1993).

Ho, T.L.: Trough-Bond Modulation of Reaction Centers by Remote Substituents. *155*, 81-158 (1990).

Höft, E.: Enantioselective Epoxidation with Peroxidic Oxygen. *164*, 63-77 (1993).

Hoggard, P. E.: Sharp-Line Electronic Spectra and Metal-Ligand Geometry. *171*, 113-142 (1994).

Holmes, K.C.: Synchrotron Radiation as a source for X-Ray Diffraction-The Beginning.*151*, 1-7 (1989).

Hopf, H., see Kostikov, R.R.: *155,* 41-80 (1990).

Indelli, M.T., see Scandola, F.: *158*, 73-149 (1990).

Inokuma, S., Sakai, S., and Nishimura, J.: Synthesis and Inophoric Properties of Crownophanes. *172*, 87-118 (1994).

Itie, J.P., see Fontaine, A.: *151*, 179-203 (1989).

Ito, Y.: Chemical Reactions Induced and Probed by Positive Muons. *157*, 93-128 (1990).

Johannsen, B., and Spiess, H.: Technetium(V) Chemistry as Relevant to Nuclear Medicine. *176*, 77-122 (1996).

John, P., and Sachs, H.: Calculating the Numbers of Perfect Matchings and of Spanning Tress, Pauling's Bond Orders, the Characteristic Polynomial, and the Eigenvectors of a Benzenoid System. *153*, 145-180 (1990).

Jucha, A., see Fontaine, A.: *151*, 179-203 (1989).

Jurisson, S., see Volkert, W. A.: *176*, 77-122 (1996).

Kaim, W.: Thermal and Light Induced Electron Transfer Reactions of Main Group Metal Hydrides and Organometallics. *169*, 231-252 (1994).

Kavarnos, G.J.: Fundamental Concepts of Photoinduced Electron Transfer. *156*, 21-58 (1990).

297